Collected Papers

Volume V

Springer
New York
Berlin
Heidelberg
Barcelona
Hong Kong
London
Milan
Paris
Singapore
Tokyo

Serge Lang

Collected Papers

Volume V

with Jay Jorgenson

1993–1999

Springer

Serge Lang
Department of Mathematics
Yale University
10 Hillhouse Ave.
PO Box 208283
New Haven, CT 06520-8283
USA

Jay Jorgenson
Mathematics Department
City College of New York
138 St. and Convent Ave.
New York, NY 10031
USA

Mathematics Subject Classification (1991): 11Gxx, 14Kxx, 11Fxx, 11Jxx, 14H

Library of Congress Cataloging-in-Publication Data
Lang, Serge, 1927–
  Collected papers / Serge Lang.
      p.   cm.
  Includes bibliographical references.
  Contents: v. 1. 1952–1970 — v. 2. 1971–1977 — v. 3. 1978–1990 — v. 4. 1990–1996
  — v. 5. 1993–1999.
     ISBN 0-387-98802-5 (v. 1 : hardcover : alk. paper) — ISBN 0-387-98803-3
  (v. 2 : hardcover : alk. paper) — ISBN 0-387-98800-9 (v. 3 : hardcover : alk. paper)
  — ISBN 0-387-98804-1 (v. 4 : hardcover : alk. paper) — ISBN 0-387-95030-3
  (v. 5 : hardcover : alk. paper)
     1. Mathematics.   I. Title.
  QA7.L288   2001
  510—dc21   99-17359

Printed on acid-free paper.

© 2001 Springer-Verlag New York, Inc.
All rights reserved. This work may not be translated or copied in whole or in part without the written permission of the publisher (Springer-Verlag New York, Inc., 175 Fifth Avenue, New York, NY 10010, USA), except for brief excerpts in connection with reviews or scholarly analysis. Use in connection with any form of information storage and retrieval, electronic adaptation, computer software, or by similar or dissimilar methodology now known or hereafter developed is forbidden.
The use of general descriptive names, trade names, trademarks, etc., in this publication, even if the former are not especially identified, is not to be taken as a sign that such names, as understood by the Trade Marks and Merchandise Marks Act, may accordingly be used freely by anyone.

Production managed by Michael Koy; manufacturing supervised by Joe Quatela.
Photocomposed copy prepared from the author's original manuscripts.
Printed and bound by Sheridan Books, Inc., Ann Arbor, MI.
Printed in the United States of America.

9 8 7 6 5 4 3 2 1

ISBN 0-387-95030-3   SPIN 10764973

Springer-Verlag New York Berlin Heidelberg
A member of BertelsmannSpringer Science+Business Media GmbH

Contents

Bibliography (through 1999)   vii

[1993a] On Cramér's Theorem for General Euler Products with Functional Equation 1

[1993b] *Basic Analysis of Regularized Series and Products* 35

[1994a] Artin Formalism and Heat Kernels 167

[1994b] *Explicit Formulas for Regularized Products and Series* 203

[1996c] Extension of Analytic Number Theory and the Theory of Regularized Harmonic Series from Dirichlet Series to Bessel Series 339

[1999a] Hilbert-Asai Eisenstein Series, Regularized Products, and Heat Kernels 389

Permissions   **423**

Bibliography (through 1999)

(*Boldface items are books or Lecture Notes.*)

[1952a] On quasi algebraic closure, *Ann. of Math.* **55** No. 2 (1952) pp. 373–390.

[1952b] Hilbert's nullstellensatz in infinite dimensional space, *Proc. AMS* **3** No. 3 (1952) pp. 407–410.

[1952c] (with J. TATE) On Chevalley's proof of Luroth's theorem, *Proc. AMS* **3** No. 4 (1952) pp. 621–624.

[1953] The theory of real places, *Ann. of Math.* **57** No. 2 (1953) pp. 378–391.

[1954a] Some applications of the local uniformization theorem, *Am. J. Math.* **76** No. 2 (1954) pp. 362–374.

[1954b] (with A. WEIL) Number of points of varieties in finite fields, *Am. J. Math.* **76** No. 4 (1954) pp. 819–827.

[1955] Abelian varieties over finite fields, *Proc. NAS* **41** No. 3 (1955) pp. 174–176.

[1956a] Unramified class field theory over function fields in several variables, *Ann. of Math.* **64** No. 2 (1956) pp. 285–325.

[1956b] On the Lefschetz principle, *Ann. of Math.* **64** No. 2 (1956) pp. 326–327.

[1956c] L-series of a covering, *Proc. NAS* **42** No. 7 (1956) pp. 422–424.

[1956d] Sur les séries L d'une variété algébrique, *Bull. Soc. Math. France* **84** (1956) pp. 385–407.

[1956e] Algebraic groups over finite fields, *Am. J. Math.* **78** (1956) pp. 555–563.

[1957a] (with J.-P. SERRE) Sur les revêtements non ramifiés des variétés algébriques, *Am. J. Math.* **79**, No. 2 (1957) pp. 319–330.

[1957b] (with W.-L. CHOW) On the birational equivalence of curves under specialization, *Am. J. Math.* **79**, No. 3 (1957) pp. 649–652.

[1957c] Divisors and endomorphisms on an abelian variety, *Am. J. Math.* **79** No. 4 (1957) pp. 761–777.

[1957d] Families algébriques de Jacobiennes (d'après IGUSA), *Séminaire Bourbaki* No. 155, 1957/1958.

[1958a] Reciprocity and correspondences, *Am. J. Math.* **80** No. 2 (1957) pp. 431–440.

[1958b] (with J. TATE) Principal homogeneous spaces over abelian varieties, *Am. J. Math.* **80** No. 3 (1958) pp. 659–684.

[1958c] (with E. KOLCHIN), Algebraic groups and the Galois theory of differential fields, *Am. J. Math.* **80** No. 1 (1958) pp. 103–110.

[1958d] *Introduction to algebraic geometry*, Wiley-Interscience, 1958.

[1959a] (with A. NÉRON) Rational points of abelian varieties over function fields, *Am. J. Math.* **81** No. 1 (1959) pp. 95–118.

[1959b] Le théorème d'irreductibilité de Hilbert, *Séminaire Bourbaki* No. 201, 1959/1960.

[1959c] *Abelian varieties*, Wiley-Interscience, 1959; Springer-Verlag, 1983.

[1960a] (with E. KOLCHIN) Existence of invariant bases, *Proc. AMS* **11** No. 1 (1960) pp. 140–148.

[1960b] Integral points on curves, *Pub. IHES* No. 6 (1960) pp. 27–43.

[1960c] Some theorems and conjectures in diophantine equations, *Bull. AMS* **66** No. 4 (1960) pp. 240–249.

[1960d] On a theorem of Mahler, *Mathematika* **7** (1960) pp. 139–140.

[1960e] L'équivalence homotopique tangentielle (d'apres MAZUR), *Séminaire Bourbaki* No. 222, 1960/1961.

[1961] Review: Elements de géométrie algébrique (A. Grothendieck). *Bull. AMS* **67** No. 3 (1961) pp. 239–246.

[1962a] A transcendence measure for E-functions, *Mathematika* **9** (1962) pp. 157–161.

[1962b] Transcendental points on group varieties, *Topology* **1** (1962) pp. 313–318.

[1962c] Fonctions implicities et plongements Riemanniens, *Séminaire Bourbaki* 1961/1962, No. 237, May 1962.

[1962d] *Introduction to Differential Manifolds*, Addison Wesley, 1962.

[1962e] *Diophantine Geometry*, Wiley-Interscience, 1962.

[1963] *Transzendente Zahlen*, Bonn Math. Schr. No. 21 (1963).

[1964a] Diophantine approximations on toruses, *Am. J. Math.* **86** No. 3 (1964) pp. 521–533.

[1964b] Les formes bilinéaires de Néron et Tate, *Séminaire Bourbaki* 1963/64 Fasc. 3 Exposé 274, Paris 1964.

[1964c] *First Course in Calculus*, Addison Wesley 1964; Fifth edition by Springer-Verlag, 1986.

[1964d] *Algebraic and Abelian Functions*, W.A. Benjamin Lecture Notes, 1964. See also [1982c].

[1964e] *Algebraic Numbers*, Addison Wesley, 1964; superceded by [1970d].

[1965a] Report on diophantine approximations, *Bulletin Soc. Math. France* **93** (1965) pp. 177–192.

[1965b] Division points on curves, *Annali Mat. pura ed applicata*, Serie IV **70** (1965) pp. 229–234.

[1965c] Algebraic values of meromorphic functions, *Topology* **3** (1965) pp. 183–191.

[1965d] Asymptotic approximations to quadratic irrationalities I, *Am. J. Math.* **87** No. 2 (1965) pp. 481–487.

[1965e] Asymptotic approximations to quadratic irrationalities II, *Am. J. Math.* **87** No. 2 (1965) pp. 488–496.

[1965f] (with W. ADAMS) Some computations in diophantine approximations, *J. reine angew. Math.* Band **220** Heft 3/4 (1965) pp. 163–173.

[1965g] Corps de fonctions méromorphes sur une surface de Riemann (d'après ISS'SA), *Séminaire Bourbaki* No. 292, 1964/65.

[1965h] *Algebra*, Addison Wesley, 1965; second edition 1984; third edition 1993.

[1966a] Algebraic values of meromorphic functions II, *Topology* **5** (1960) pp. 363–370.

[1966b] Asymptotic diophantine approximations, *Proc. NAS* **55** No. 1 (1966) pp. 31–34.

[1966c] *Introduction to transcendental numbers*, Addison Wesley, 1966.

[1966d] *Introduction to Diophantine Approximations*, Addison Wesley 1966; see [1995d].

[1966e] *Rapport sur la cohomologie des groupes*. Benjamin, 1966.

[1967] *Algebraic structures*, Addison Wesley 1967.

[1968] *Analysis I,* Addison Wesley, 1968; (superceded by [1983c]).

[1969] *Analysis II*, Addison Wesley, 1969; (superceded by [1993c]).

[1970a] (with E. BOMBIERI) Analytic subgroups of group varieties, *Inventiones Math.* **11** (1970) pp. 1–14.

[1970b] Review: L.J. Mordell's *Diophantine Equations*, *Bull. AMS* **76** (1970) pp. 1230–1234.

[1970c] *Introduction to Linear Algebra*, Addison Wesley 1970; see also [1986b].

[1970d] *Algebraic Number Theory*, Addison Wesley 1970; see also [1994c].

[1971a] Transcendental numbers and diophantine approximations, *Bull. AMS* **77** No. 5 (1971) pp. 635–677.

[1971b] On the zeta function of number fields, *Invent. Math.* **12** (1971) pp. 337–345.

[1971c] The group of automorphisms of the modular function field, *Invent. Math.* **14** (1971) pp. 253–254.

[1971d] *Linear Algebra*, Addison Wesley 1971; see also [1987b].

[1971e] *Basic Mathematics*, Addison Wesley 1971; Springer-Verlag 1988.

[1972a] Isogenous generic elliptic curves, *Amer. J. Math.* **94** (1972) pp. 661–674.

[1972b] (with H. TROTTER) Continued fractions for some algebraic numbers, *J. reine angew. Math.* **255** (1972) pp. 112–134.

[1972c] *Differential manifolds*, Addison Wesley, 1972.

[1972d] *Introduction to Algebraic and Abelian Functions*, Benjamin-Addison Wesley, 1972; second edition see [1982c].

[1973a] Frobenius automorphisms of modular function fields, *Amer. J. Math.* **95** (1973) pp. 165–173.

[1973b] *Calculus of Several Variables*, Addison Wesley 1973; Third edition see [1987d].

[1973c] *Elliptic functions*, Addison Wesley 1973; second edition see [1987d].

[1974a] Higher dimensional diophantine problems, *Bull. AMS* **80** No. 5 (1974) pp. 779–787.

[1974b] (with H. TROTTER) Addendum to "Continued fractions of some algebraic numbers," *J. reine angew. Math.* **267** (1974) pp. 219–220.

[1975a] Diophantine approximations on abelian varieties with complex multiplication, *Advances Math.* **17** (1975) pp. 281–336.

[1975b] Division points of elliptic curves and abelian functions over number fields, *Amer. J. Math.* **97** No. 1 (1975) pp. 124–132.

[1975c] (with D. KUBERT) Units in the modular function field I, Diophantine Applications, *Math. Ann.* **218** (1975) pp. 67–96.

[1975d] (with D. KUBERT) Units in the modular function field II, A full set of units, *Math. Ann.* **218** (1975) pp. 175–189.

[1975e] (with D. KUBERT) Units in the modular function field III, Distribution relations, *Math. Ann.* **218** (1975) pp. 273–285.

[1975f] La conjecture de Catalan d'après Tijdeman, *Séminaire Bourbaki* 1975/76 No. 29.

[1975g/85] $SL_2(\mathbf{R})$, Addison Wesley, 1975; Springer-Verlag corrected second printing, 1985.

[1976a] (with J. COATES) Diophantine approximation on abelian varieties with complex multiplication, *Invent. Math.* **34** (1976) pp. 129–133.

[1976b] (with D. KUBERT) Distribution on toroidal groups, *Math. Z.* **148** (1976) pp. 33–51.

[1976c] (with D. KUBERT) Units in the modular function field, in *Modular Functions in One Variable* V, Springer Lecture Notes **601** (Bonn Conference) 1976, pp. 247–275.

[1976d] (with H. TROTTER) *Frobenius distributions in* GL_2*-extensions*, Springer Lecture Notes **504**, Springer-Verlag 1976.

[1976e] *Introduction to Modular Forms*, Springer-Verlag, 1976.

[1977a] (with D. KUBERT) Units in the modular function field IV, The Siegel functions are generators, *Math. Ann.* **227** (1997) pp. 223–242.

[1977b] (with H. TROTTER) Primitive points on elliptic curves, *Bull. AMS* **83** No. 2 (1977) pp. 289–292.

[1977c] *Complex Analysis*, Addison Wesley; second Edition Springer-Verlag 1985; fourth edition Springer-Verlag 1999.

[1978a] (with D. KUBERT) The p-primary component of the cuspidal divisor class group of the modular curve $X(p)$, *Math. Ann.* **234** (1978) pp. 25–44.

[1978b] (with D. KUBERT) Units in the modular function field V, Iwasawa theory in the modular tower, *Math. Ann.* **237** (1978) pp. 97–104.

[1978c] (with D. KUBERT) Stickelberger ideals, *Math. Ann.* **237** (1978) pp. 203–212.

[1978d] (with D. KUBERT) The index of Stickelberger ideals of order 2 and cuspidal class numbers, *Math. Ann.* **237** (1978) pp. 213–232.

[1978e] Relations de distributions et exemples classiques, *Séminaire Delange-Pisot-Poitou (Théorie des Nombres)*, 1978 No. 40 (6 pages).

[1978f] *Elliptic curves: Diophantine Analysis*, Springer-Verlag 1978.

[1978g] *Cyclotomic Fields* I, Springer-Verlag, 1978.

[1979a] (with D. KUBERT) Cartan-Bernoulli numbers as values of L-series, *Math. Ann.* **240** (1979) pp. 21–26.

[1979b] (with D. KUBERT) Independence of modular units on Tate curves, *Math. Ann.* **240** (1979) pp. 191–201.

[1979c] (with D. KUBERT) Modular units inside cyclotomic units, *Bull. Soc. Math. France* **107** (1979) pp. 161–178.

[1980] *Cyclotomic Fields* II, Springer-Verlag 1980.

[1981a] (with N. KATZ) Finiteness theorems in geometric classfield theory, *Enseignement mathématique* **27** (3–4) (1981) pp. 285–314.

[1981b] (with DAN KUBERT) *Modular Units*, Springer-Verlag 1981.

[1982a] Représentations localement algébriques dans les corps cyclotomiques, Séminaire de Théorie des Nombres 1982, Birkhauser, pp. 125–136.

[1982b] Units and class groups in number theory and algebraic geometry, *Bull. AMS* **6** No. 3 (1982) pp. 253–316.

[1982c] *Introduction to algebraic and abelian functions, Second Edition*, Springer-Verlag, 1982.

[1983a] Conjectured diophantine estimates on elliptic curves, in Volume I of *Arithmetic and Geometry*, dedicated to Shafarevich, M. Artin and J. Tate editors, Birkhauser (1983) pp. 155–171.

[1983b] *Fundamentals of Diophantine Geometry*, Springer-Verlag 1983.

[1983c] *Undergraduate Analysis*, Springer-Verlag, 1983.

[1983d] (with GENE MURROW) *Geometry: A High School Course*, Springer Verlag, 1983 Second edition 1988.

[1983e] *Complex Multiplication*, Springer-Verlag 1983.

[1984a] Vojta's conjecture, *Arbeitstagung Bonn 1984*, Springer Lecture Notes **1111** 1985, pp. 407–419.

[1984b] Variétés hyperboliques et analyse diophantienne, *Séminaire de théorie des nombres*, 1984/85, pp. 177–186.

[1985a] (with W. FULTON) *Riemann-Roch Algebra*, Springer-Verlag, 1985.

[1985b] *The Beauty of Doing Mathematics*, Springer-Verlag, 1985 originally published as articles in the *Revue du Palais de la Découverte*, Paris, 1982–1984, specifically:
Une activité vivante: faire des mathématiques, *Rev. P.D.* Vol. **11** No. **104** (1983) pp. 27–62
Que fait un mathématicien pure et pourquoi?, *Rev. P.D.* Vol. **10** No. **94** (1982) pp. 19–44
Faire des Maths: grands problèmes de géométrie et de l'espace, *Rev. P.D.* Vol. **12** No. **114** (1984) pp. 21–72.

[1985c] *Math! Encounters with High School Students*, Springer-Verlag, 1985 (French edition *Serge Lang, des Jeunes et des Maths*, Belin, 1984).

[1986a] Hyperbolic and diophantine analysis, *Bulletin AMS* **14** No. 2 (1986) pp. 159–205.

[1986b] *Introduction to Linear Algebra*, Second Edition, Springer-Verlag 1986.

[1987a] Diophantine problems in complex hyperbolic analysis, *Contemporary Mathematics AMS* **67** (1987) pp. 229–246.

[1987b] *Linear Algebra*, Third Edition, Springer-Verlag 1987.

[1987c] *Undergraduate Algebra*, Springer-Verlag 1987.

[1987d] *Elliptic functions*, Second Edition, Springer-Verlag 1987.

[1987e] *Introduction to complex hyperbolic spaces*, Springer-Verlag 1987.

[1988a] The error term in Nevanlinna theory, *Duke Math. J.* **56** No. 1 (1988) pp. 193–218.

[1988b] *Introduction to Arakelov Theory*, Springer-Verlag 1988.

[1990a] The error term in Nevanlinna Theory II, *Bull. AMS* **22** No. 1 (1990) pp. 115–125.

[1990b] Old and new conjectured diophantine inequalities. *Bull. AMS* **23** No. 1 (1990) pp. 37–75.

[1990c] *Lectures on Nevanlinna theory*, in *Topics in Nevanlinna Theory*, Springer Lecture Notes **1433** (1990) pp. 1–107.

[1990d] *Cyclotomic Fields I and II*, combined edition with an appendix by Karl Rubin, Springer-Verlag, 1990.

[1991] *Number Theory III*, Survey of Diophantine Geometry, Encyclopedia of Mathematical Sciences, Springer-Verlag 1991.

[1993a] (with J. JORGENSON) On Cramér's theorem for general Euler products with functional equation, *Math. Ann.* **297** (1993) pp. 383–416.

[1993b] (with J. JORGENSON) *Basic analysis of regularized series and products*, Springer Lecture Notes **1564** (1993).

[1993c] *Real and Functional Analysis*, Springer-Verlag, 1993.

[1993d] *Algebra, Third Edition*, Addison Wesley 1993.

[1994a] (with J. JORGENSON) Artin formalism and heat kernels, *J. reine angew. Math.* **447** (1994) pp. 165–200.

Bibliography

[1994b] (with J. JORGENSON) *Explicit Formulas for regularized products and series*, in Springer Lecture Notes **1593** pp. 1–134.

[1994c] *Algebraic Number Theory, Second Edition*, Springer-Verlag 1994.

[1995a] Mordell's review, Siegel's letter to Mordell, diophantine geometry, and 20th century mathematics, *Notices AMS* March 1995 pp. 339–350.

[1995b] Some history of the Shimura-Taniyama conjecture, *Notices AMS* November 1995 pp. 1301–1307.

[1995c] *Differential and Riemannian Manifolds*, Springer-Verlag 1995.

[1995d] *Introduction to Diophantine Approximations, new expanded edition*, Springer-Verlag 1995.

[1996a] La conjecture de Bateman-Horn, *Gazette des mathématiciens* January 1996 No. 67 pp. 82–84.

[1996b] Comments on Chow's works, *Notices AMS* **43** (1996) No. 10 pp. 1117–1124.

[1996c] (with J. JORGENSON) Extension of analytic number theory and the theory of regularized harmonic series from Dirichlet series to Bessel series, *Math. Ann.* **306** (1996) pp. 75–124.

[1996d] *Topics in Cohomology of groups*, Springer-Verlag, Springer Lecture Notes **1625**, 1996 (English translation and expansion of *Rapport sur la Cohomologie des Groupes*, Benjamin, 1966).

[1997] *Survey of Diophantine Geometry*, Springer-Verlag 1997 (same as *Number Theory III*, with corrections and additions).

[1998] The Kirschner article and HIV: Scientific and journalistic (ir)responsibilities.

[1999a] (with J. JORGENSON) Hilbert-Asai Eisenstein series, regularized products, and heat kernels, *Nagoya Math. J.* **153** (1999) pp. 155–188.

[1999b] Response to the Steele Prize, *Notices AMS* **46** No. 4, April 1999 p. 458.

[1999c] *Fundamentals of Differential Geometry*, Springer-Verlag 1999.

[1999d] *Complex analysis*, fourth edition, Springer-Verlag 1999.

[1999e] *Math Talks for Undergraduates*, Springer-Verlag 1999.

The Zurich Lectures

I was invited by Wustholz for a decade to give talks to students in Zurich. I express here my appreciation, also to Urs Stammbach for his translations, and for his efforts in producing and publishing the articles in *Elemente der Mathematik*.

Primzahlen, *Elem. Math.* **47** (1992) pp. 49–61
Die abc-vermutung, *Elem. Math.* **48** (1993) pp. 89–99
Approximationssätze der Analysis, *Elem. Math.* **49** (1994) pp. 92–103
Die Wärmeleitung auf dem Kreis und Thetafunktionen, *Elem. Math.* **51** (1996) pp. 17–27
Globaler Integration lokal integrierbarer Vektorfelder, *Elem. Math.* **52** (1997) pp. 1–11
Bruhat-Tits-Raüme, *Elem. Math.* **54** (1999) pp. 45–63

Articles on Scientific Responsibility

The DOD, Government and Universities, in the collection *The Social Responsibility of the Scientist*, edited by Martin Brown, Free Press, 1971, pp. 51–79
Circular A-21. A history of bureaucratic encroachment, J. Society of Research Administrators, 1984
Questions de responsabilité dans le journalisme scientifique, *Revue du Palais de la Découverte* Paris February 1991 pp. 17–46
Questions of scientific responsibility: The Baltimore case, *J. Ethics and Behavior* **3(1)** (1993) pp. 3–72
The Kirschner article and HIV: Scientific and journalistic (ir)responsibilities, Refused publication by the *Notices AMS*, dated 5 January 1998

Books on Scientific Responsibility

The Scheer Campaign, W.A. Benjamin, 1966
The File, Springer-Verlag, 1981
Challenges, Springer-Verlag, 1998

Math. Ann. 297, 383–416 (1993)

© Springer-Verlag 1993

On Cramér's theorem for general Euler products with functional equation

Jay Jorgenson and Serge Lang

Department of Mathematics, Yale University, Box 2155, Yale Station, New Haven, CT 06520, USA

Received October 30, 1992; in revised form April 22, 1993

Mathematics Subject Classification (1991): 11M35, 11M41, 11M99, 30B50, 30D15, 35P99*

Contents

0. Background for regularized products 385
1. The formal expression 388
2. A contour integral and the proof of Theorem 1.1 392
3. Regularized products 397
4. Meromorphy assumptions and their consequences 401
5. Asymptotic development near $z = 0$ 404
6. Cramér's theorem 408
7. Examples 410

Introduction

In the two papers [JoL92a] and [JoL92b] we laid out the foundations of the analytic theory of regularized products insofar as Euler products and functional equations are not relevant. Here we begin the development of the next part of the theory, which depends on this additional structure. Specifically, this next part deals with three types of theorems that come most classically from analytic number theory:

the Cramér theorem;
the explicit formula;
the counting of zeros in the critical strip up to a certain height.

The present paper develops the Cramér theorem, which was brought to our attention by Deninger and Soulé via [De 92] and has also been considered in [Kur 88]. The other theorems will be presented in subsequent papers.

In [Cr 19], Cramér showed that if $\{\varrho_k\}$ ranges over the non-trivial zeros of the Riemann zeta function with $\text{Im}(\varrho_k) > 0$, then the series

$$V(z) = \sum e^{\varrho_k z}$$

---

* Actually the MSC does not, but should, include an item for regularized products. J.J. and S.L.

converges for $\mathrm{Im}(z) > 0$ and has a singularity at the origin of the type $\log z/(1 - e^{-z})$, by which we mean that the function

$$F(z) = 2\pi i V(z) - \frac{\log z}{1 - e^{-z}}$$

has a meromorphic continuation to all $\mathbf{C}$, with simple poles at the points $\pm \pi i n$ where n ranges over the integers, and at the points $\pm \log p^m$ where p^m ranges over the prime powers. We shall prove the analogous theorem under very general conditions of Euler product and functional equation on a meromorphic function Z so that it applies not only to the classical zeta and L-functions of number fields, but also to those coming from modular forms and to Selberg-type zeta functions, specifically as in the case of hyperbolic Riemann surfaces or certain higher dimensional manifolds. Recall that in the geometric context of Riemann surfaces with metrics of constant negative curvature, the role of a prime p is played by e^l, where l is the length of a primitive geodesic, so a prime power p^m is replaced by e^{ml}.

For an application of the results of this paper, suppose we start with a sequence $\{\lambda_k\}$ satisfying the basic axioms of [JoL 92a] involving an asymptotic expansion at the origin. (In Sect. 0, we shall recall relevant notation and terminology from [JoL 92a].) Assume now in addition that the associated zeta function Z has an Euler product and functional equation, which will be defined in general below, and assume the gamma-like factors in the functional equation are formed out of exponentials of polynomials and regularized products. Then the new sequence $\{\lambda_k'\} = \{\varrho_k/i\}$ formed with the set of zeros $\{\varrho_k\}$ of Z in the upper half of the critical strip satisfies the same axioms. In particular, this implies that Z itself can be expressed in terms of exponentials of polynomials and regularized products. This result can be viewed as going one step up the ladder of regularized products. As an example, this result implies that the Riemann zeta function can be written as

$$\Gamma(s/2)\zeta_{\mathbf{Q}}(s) = e^{P(s)}[s(s-1)]^{-1} D_+(-s/i) D_-(s/i),$$

where $D_+(z)$ is the regularized product formed from the non-trivial zeros $\{\varrho_k\}$ of $\zeta_{\mathbf{Q}}$ with $\mathrm{Im}(\varrho_k) > 0$, $D_-(z)$ is the regularized product formed from the non-trivial zeros $\{\varrho_k\}$ of $\zeta_{\mathbf{Q}}$ with $\mathrm{Im}(\varrho_k) < 0$, and $P(s)$ is a polynomial of degree one.

We may then apply all the results of [JoL 92a] and [JoL 92b] to the zeta function formed with the sequence $\{\lambda_k'\}$, in particular all the formulas including the "Lerch formula". One then sees that the Lerch formula for the case $\zeta_{\mathbf{Q}}$ amounts to a formula discovered by Deninger [De 92], Theorem 3.3. We observe that our general theory yields not only the formula (valid in a certain range of s) but its analytic continuation, also valid for all regularized products. Our generalized Cramér's theorem is used here only to prove that the theta function formed with the sequence $\{\lambda_k'\}$ satisfies the basic axiom **AS 2** (see Sect. 0). In the theory of manifolds, when one is dealing with operators (Laplace, elliptic, pseudo-differential), axiom **AS 2** follows from the asymptotic expansion of the heat kernel associated with the operator.

Deninger arrived at his formula by a formal argument, suggesting that the zeta function should be the regularized determinant of some (as yet unknown) operator ([De 92], formula 3.2) on a suitable cohomology space (also unknown). Although we prove that the zeta function is a regularized product in our sense (the sequence of zeros satisfies some analytic axioms, of which the basic one is **AS 2**), we do not propose either an operator or a space.

0 Background for regularized products

Let us briefly recall necessary background material from the theory of regularized products and series, as established in [JoL 92a] and [JoL 92b], to which we refer for details and further results. We let $L = \{\lambda_k\}$ and $A = \{a_k\}$ be sequences of complex numbers which may be subject to the following conditions.

DIR 1. For every positive real number c, there is only a finite number of k such that $\text{Re}(\lambda_k) \leq c$.

We use the convention that $\lambda_0 = 0$ and $\lambda_k \neq 0$ for $k \geq 1$. Under condition **DIR 1** we delete from the complex plane $\mathbf{C}$ the horizontal half lines going from $-\infty$ to $-\lambda_k$ for each k, together, when necessary, the horizontal half line going from $-\infty$ to 0. We define the open set:

$\mathbf{U}_L =$ the complement of the above half lines in $\mathbf{C}$.

If all λ_k are real and positive, then we note that $\mathbf{U}_L$ is simply $\mathbf{C}$ minus the negative real axis $\mathbf{R}_{\leq 0}$.

DIR 2. (a) The Dirichlet series

$$\sum_k \frac{a_k}{\lambda_k^\sigma}$$

converges absolutely for some real σ, say σ_0.

(b) The Dirichlet series

$$\sum_k \frac{1}{\lambda_k^\sigma}$$

converges absolutely for some real σ, say σ_1.

DIR 3. There is a fixed $\varepsilon > 0$ such that for all k sufficiently large, we have

$$-\frac{\pi}{2} + \varepsilon \leq \arg(\lambda_k) \leq \frac{\pi}{2} - \varepsilon.$$

We will consider a **theta series** or **theta function**, which is defined by

$$\theta_{A,L}(t) = \theta(t) = a_0 + \sum_{k=1}^{\infty} a_k e^{-\lambda_k t}.$$

and, for each integer $N \geq 1$, we define the **asymptotic exponential polynomials** by

$$Q_N(t) = a_0 + \sum_{k=1}^{N-1} a_k e^{-\lambda_k t}.$$

We are also given a sequence of complex numbers $\{p\} = \{p_0, \ldots, p_j, \ldots\}$ with

$$\text{Re}(p_0) \leq \text{Re}(p_1) \leq \ldots \leq \text{Re}(p_j) \leq \ldots$$

increasing to infinity, and, to every p in this sequence, we associate a polynomial B_p of degree n_p and set

$$b_p(t) = B_p(\log t).$$

We then define the **asymptotic polynomials at 0** by

$$P_q(t) = \sum_{\text{Re}(p)<\text{Re}(q)} b_p(t) t^p.$$

We define
$$m(q) = \max \deg B_p \quad \text{for} \quad \operatorname{Re}(p) = \operatorname{Re}(q),$$
$$n(q) = \max \deg B_p \quad \text{for} \quad \operatorname{Re}(p) < \operatorname{Re}(q),$$
$$n(q') = \max \deg B_p \quad \text{for} \quad \operatorname{Re}(p) \leq \operatorname{Re}(q).$$

We shall use the term **special case** to describe the instance when $n(q) = 0$ for all q. The **principal part** of $\theta(t)$ is defined to be
$$P_0 \theta(t) = \sum_{\operatorname{Re}(p)<0} b_p(t) t^p.$$

Let $\mathbf{C}\langle T\rangle$ be the algebra of polynomials in T^p with arbitrary complex powers $p \in \mathbf{C}$. Then, with this notation, $P_q(t) \in \mathbf{C}[\log t]\langle t\rangle$.

A general function f on $(0, \infty) = \mathbf{R}_{>0}$ may be subject to **asymptotic conditions**:

AS 1. Given a positive number C and $t_0 > 0$, there exists N and $K > 0$ such that
$$|f(t) - Q_N(t)| \leq K e^{-Ct} \quad \text{for} \quad t \geq t_0.$$

AS 2. For every q, we have
$$f(t) - P_q(t) = O(t^{\operatorname{Re}(q)} |\log t|^{m(q)}) \quad \text{for} \quad t \to 0,$$
which shall be written as
$$f(t) \sim \sum_p b_p(t) t^p.$$

AS 3. Given $\delta > 0$, there exists an $\alpha > 0$ and a constant $C > 0$ such that for all N and $0 < t \leq \delta$ we have
$$|\theta(t) - Q_N(t)| \leq C/t^\alpha.$$

We shall assume throughout that the theta series converges absolutely for $t > 0$. From **DIR 1** it follows that the convergence of the theta series is uniform for $t \geq \delta > 0$ for every δ.

The **Mellin transform** of a measurable function f on $(0, \infty)$ is defined by
$$\mathbf{M}f(s) = \int_0^\infty f(t) t^s \frac{dt}{t}.$$

The **Laplace transform** is defined to be
$$\mathbf{L}f(z) = \int_0^\infty f(t) e^{-zt} \frac{dt}{t}.$$

The **Laplace-Mellin transform** combines both, with the definition
$$\mathbf{LM}f(s, z) = \int_0^\infty f(t) e^{-zt} t^s \frac{dt}{t}.$$

Theorem 0.1. *Let f satisfy* **AS 1**, **AS 2**, *and* **AS 3**. *Then* $\mathbf{LM}f$ *has a meromorphic continuation for $s \in \mathbf{C}$ and $z \in \mathbf{U}_L$. For each z, the functions $s \mapsto \mathbf{LM}f(s, z)$ has poles only at the points $-(p+n)$ with $b_p \neq 0$ in the asymptotic expansion of f at 0. A* *pole at $-(p+n)$ has order at most $n(p') + 1$. In the special case when the asymptotic expansion at 0 has no log terms, the poles are simple.*

We shall use a systematic notation for the coefficients of the Laurent expansion of $\mathbf{LM}f(s,z)$ near $s = s_0$. Namely we let $R_j(s_0; z)$ be the coefficient of $(s - s_0)^j$, so that
$$\mathbf{LM}f(s,z) = \sum R_j(s_0; z)(s - s_0)^j.$$
The constant term $R_0(s_0; z)$ is so important that we give it a special notation, namely,
$$\mathrm{CT}_{s=s_0}\mathbf{LM}f(s,z) = R_0(s_0; z).$$
In particular, we define the **regularized harmonic series** to be the meromorphic function defined by
$$\mathrm{CL}_{s=1}\mathbf{LM}f(s,z) = R_0(1; z).$$

Theorem 0.2. *Let f satisfy* **AS 1**, **AS 2**, *and* **AS 3**. *Then for every $z \in \mathbf{U}_L$ and s near 0, the function $\mathbf{LM}f(s,z)$ has a pole at $s = 0$ of order at most $n(0') + 1$, and the function $\mathbf{LM}f(s,z)$ has the Laurent expansion*
$$\mathbf{LM}f(s,z) = \frac{R_{-n(0')-1}(0;z)}{s^{n(0')+1}} + \ldots + R_0(0; z) + R_1(0; z)s + \ldots$$
where, for each $j < 0$, $R_j(0; z) \in \mathbf{C}$ is a polynomial of degree $\leqq -\mathrm{Re}(p_0)$.

In the case when all $a_k \in \mathbf{Z}$, we define the **regularized product** to be the meromorphic function $D(z)$ such that
$$-\log D(z) = \mathrm{CT}_{s=0}\mathbf{LM}f(s,z) = R_0(0; z).$$
The regularized product is a particular meromorphic function of finite order that has zeros at the points $z = -\lambda_k$ with multiplicity a_k. Similarly, the regularized harmonic series is a particular meromorphic function that has simple poles at $z = -\lambda_k$ with residue a_k.

To compare with results that exist elsewhere in the literature, let us record the following formula. In the special case, define
$$\zeta_f(s,z) = \frac{1}{\Gamma(s)}\mathbf{LM}f(s,z).$$
If we assume that
$$f(t) = \sum_{k=1}^{\infty} a_k e^{-\lambda_k t}$$
satisfies the asymptotic conditions **AS 1**, **AS 2**, and **AS 3**, then we have the equality
$$\zeta_f(s,z) = \sum_{k=1}^{\infty} \frac{a_k}{(z + \lambda_k)^s}$$
for $\mathrm{Re}(z)$ and $\mathrm{Re}(s)$ sufficiently large. By Theorem 0.1 and Theorem 0.2, $\zeta_f(s,z)$ is holomorphic at $s = 0$ for $z \in \mathbf{U}_L$ and
$$\zeta'_f(0,z) = \mathrm{CT}_{s=0}\mathbf{LM}f(s;z) + \gamma R_{-1}(0;z).$$

From the definition of the Laplace-Mellin transform and Theorem 0.1, one has the differential equation
$$\partial_z \mathbf{LM}f(s,z) = -\mathbf{LM}f(s+1,z),$$

from which we conclude the relation

$$\partial_z R_0(0;z) = -R_0(1;z)$$

between the regularized harmonic series and the regularized product, when both functions are defined.

We say that a meromorphic function of finite order $G(z)$ is of **regularized product type** if

$$G(z) = Q(z)e^{P(z)} \prod_{j=1}^{n} D_j(\alpha_j z + \mu_j)^{k_j}$$

where:
  i) $Q(z)$ is a rational function,
  ii) $P(z)$ is a polynomial,
  iii) $D_j(z)$ is a regularized product associated to a theta function θ_j, and $\alpha_j, \mu_j \in \mathbf{C}$ with $\mathrm{Re}(\alpha_j) \geq 0$ and $k_j \in \mathbf{Z}$.

For simplicity, we shall write G as

$$G(z) = Q(z)e^{P(z)} D(z)$$

where

(1) $$D(z) = \prod_{j=1}^{n} D_j(\alpha_j z + \mu_j)^{k_j}$$

and say that $D(z)$ is a **quotient of regularized products**. The function $G(z)$ is said to be of **reduced order** M if $\deg P \leq M + 1$ and M is the largest integer that is $< -\mathrm{Re}(p_0)$ for each theta function θ_j. The results of Sect. 2 of [JoL 92a] show that such a function is of strict order $M+1$, as defined in [La 93]. As an example, the Lerch formula (see Sect. 2 of [JoL 92a]) states that the function $1/\Gamma(z)$ is of regularized product type. It correponds to the sequence $L = \mathbf{Z}_{\geq 0}$ with $a_k = 1$ for all k, $Q(z) = 0$ and $\deg P(z) = 0$. We note that $1/\Gamma(z)$ has reduced order $M = 0$.

If $D(z)$ is a regularized product, we say that the associated theta function θ is **meromorphic** if θ has a meromorphic extension to all $\mathbf{C}$. In the case of a quotient of regularized products as in (1), we say that the associated theta function is meromorphic if each theta function θ_j, corresponding to the regularized product $D_j(z)$, is meromorphic. In such a case, the theta function θ associated to $D(z)$ is written as

$$\theta(z) = \sum_{j=1}^{n} k_j \theta_j(z/\alpha_j) e^{-z\mu_j/\alpha_j},$$

in the sense of meromorphic functions.

1 The formal expression

As stated in the introduction, we are studying certain properties of the zeros and poles of a function Z, meromorphic and of finite order, which has a type of Euler product and functional equation. Recall that a meromorphic function is of **finite order** if it is a quotient of entire functions of finite order. Following classical practice (see, for example, [Bo 58]), we find it convenient to deal with two functions Z and $\tilde{Z}$. We shall assume the following general properties:

1. **Meromorphy.** The functions Z and $\tilde{Z}$ are meromorphic functions of finite order.

2. **Euler product.** There are sequences $\{\mathbf{q}\}$ and $\{\tilde{\mathbf{q}}\}$ of real numbers > 1 that depend on Z and $\tilde{Z}$, respectively, and that converge to infinity, such that for every $\mathbf{q}$ and $\tilde{\mathbf{q}}$, there exist complex numbers $c(\mathbf{q})$ and $c(\tilde{\mathbf{q}})$ such that for all $\operatorname{Re}(s) > \sigma_0$,

$$\log Z(s) = \sum_{\mathbf{q}} \frac{c(\mathbf{q})}{\mathbf{q}^s} \quad \text{and} \quad \log \tilde{Z}(s) = \sum_{\tilde{\mathbf{q}}} \frac{c(\tilde{\mathbf{q}})}{\tilde{\mathbf{q}}^s}. \tag{1}$$

The series are assumed to converge uniformly and absolutely in any half-plane of the form $\operatorname{Re}(s) \geq \sigma_0 + \varepsilon > \sigma_0$.

3. **Functional equation.** There are functions G and $\tilde{G}$, meromorphic and of finite order, such that

$$Z(s)G(s) = \tilde{Z}(\sigma_0 - s)\tilde{G}(\sigma_0 - s).$$

Remark. We note that our assumptions are minimal as far as Euler products and functional equations are concerned. In particular, they are much weaker than those made by Selberg [Sel 91] (see also [CoG 91]) because we do not assume anything like a Riemann hypothesis or Ramanujan-Petersson estimate concerning the coefficients $c(\mathbf{q})$. In addition, Selberg assumes that the fudge factors are concocted out of gamma functions, but we do not, so as to be able to apply the theorem to higher gamma functions or other functions which arise in the study of zeta functions of higher order. What does happen in practice is that the fudge factor is concocted out of regularized products.

The series (1) will be referred to as the **Euler sum** associated to $Z(s)$. Note that the Euler sum in (1) implies that all the zeros and poles of $Z(s)$ lie to the left of, and possibly on, the line $\operatorname{Re}(s) = \sigma_0$. The zeros or poles of $Z(s)$ that lie in the **critical strip** $0 \leq \operatorname{Re}(s) \leq \sigma_0$ will be called **primary**, and the others will be called **secondary**.

For any $\varrho \in \mathbf{C}$, define the function

$$v(\varrho) = \operatorname{ord}_\varrho Z.$$

The following result, which we call **Cramér's theorem** for our function Z, concerns the primary zeros and poles of the function $Z(s)$.

Theorem 1.1. *Assuming the notation above, let $a > 0$ and let $\mathscr{R}_a$ be the open infinite rectangle with vertices at the four points*

$$-a + i\infty, \quad -a, \quad \sigma_0 + a, \quad \sigma_0 + a + i\infty.$$

For simplicity, assume that G and $\tilde{G}$ have no zeros or poles on the vertical boundary components

$$(-a + i\infty, -a] \quad \text{and} \quad [\sigma_0 + a, \sigma_0 + a + i\infty)$$

of $\mathscr{R}_a$. Then the series

$$V_Z(z) = \sum_{\varrho \in \mathscr{R}_a} v(\varrho) e^{\varrho z}$$

converges uniformly and absolutely for all z in half-planes of the form $\text{Im}(z) \geq \eta > 0$. *Also, for any z with* $\text{Im}(z) > 0$, $V_Z(z)$ *can be written in the form*

$$2\pi i V_Z(z) = z e^{(\sigma_0 + a)z} \sum_{\mathbf{q}} \frac{c(\mathbf{q})}{\mathbf{q}^{\sigma_0 + a}(z - \log \mathbf{q})} - z e^{-az} \sum_{\tilde{\mathbf{q}}} \frac{c(\tilde{\mathbf{q}})}{\tilde{\mathbf{q}}^{\sigma_0 + a}(z + \log \tilde{\mathbf{q}})}$$

$$+ e^{\sigma_0 z} \int_{\sigma_0 + a - i\infty}^{\sigma_0 + a} e^{-sz} \tilde{G}'/\tilde{G}(s) ds + e^{\sigma_0 z} \int_{\sigma_0 + a - i\infty}^{\sigma_0 + a} e^{-sz} G'/G(\sigma_0 - s) ds$$

$$- z \int_{-a}^{\sigma_0 + a} e^{uz} \log|Z(u)| du - z \int_{-a}^{\sigma_0 + a} e^{uz} \arg(Z(u)) du$$

$$- e^{-az} \log Z(-a) + e^{-az} \log \tilde{Z}(\sigma_0 + a),$$

where the branch of $\log Z(-a)$ *is obtained by the analytic continuation of the Euler sum* (1) *along the top of the horizontal line segment* $[-a, \sigma_0 + a]$. *Finally:*

  (i) *The series over* $\mathbf{q}$ *and* $\tilde{\mathbf{q}}$ *define meromorphic functions of z with the obvious simple poles at* $\log \mathbf{q}$ *and* $-\log \tilde{\mathbf{q}}$, *with the obvious residues.*

  (ii) *The integrals involving Z are absolutely and uniformly convergent for z in any compact set, and define entire functions of z.*

  (iii) *The integrals involving G and $\tilde{G}$ converge absolutely and uniformly in half-planes of the form* $\text{Im}(z) \geq \eta > 0$.

The integrals on vertical lines are to be interpreted as follows. Let f be a meromorphic function of finite order. As we shall see in Lemma 2.1, for any vertical line $\text{Re}(s) = \sigma$, there is a sequence $\{T_m\}$ of real numbers converging to infinity, such that $f'/f(\sigma + iT_m)$ has polynomial growth, as a function of T_m. Then we define

$$\int_{\sigma_0 + a - i\infty}^{\sigma_0 + a} = \lim_{m \to \infty} \int_{\sigma_0 + a - iT_m}^{\sigma_0 + a}.$$

In Theorem 1.1, we note that all the terms except those involving G and $\tilde{G}$ are meromorphic in z. In fact, the two integrals with $\log|Z(u)|$ and $\arg(Z(u))$ define entire functions of z because $\log|Z(u)|$ is absolutely integrable over a finite interval (its singularities being at worse like $\log x$ near $x = 0$), and $\arg(Z(u))$ is piecewise continuous, so also absolutely integrable. Hence the function

$$F_Z(z) = 2\pi i V_Z(z) - e^{\sigma_0 z} \int_{\sigma_0 + a - i\infty}^{\sigma_0 + a} e^{-sz} [\tilde{G}'/\tilde{G}(s) + G'/G(\sigma_0 - s)] ds,$$

has a meromorphic continuation to all $z \in \mathbf{C}$ with poles at the points $\log \mathbf{q}$ and $-\log \tilde{\mathbf{q}}$. The residue of $F_Z(z)$ at $\log \mathbf{q}$ is obtained by substituting $\log \mathbf{q}$ for z in the $\mathbf{q}$ term, and hence

  the residue of F_Z at $\log \mathbf{q}$ is $c(\mathbf{q}) \log \mathbf{q}$.

Similarly,

  the residue of F_Z at $-\log \tilde{\mathbf{q}}$ is $c(\tilde{\mathbf{q}}) \log \tilde{\mathbf{q}} \tilde{\mathbf{q}}^{-\sigma_0}$.

Observe that the residues are independent of the choice of a, as they should be. The exponential coefficient e^{az} is such that it cancels with either $\mathbf{q}^a$ or $\tilde{\mathbf{q}}^a$ in the denominator when we make the appropriate substitution.

We are then led to study the terms which involve the fudge factors G and $\tilde{G}$. To further understand these terms, additional assumptions concerning the structure of G and $\tilde{G}$ are necessary. In particular, if the fudge factors G and $\tilde{G}$ are of regularized product type, one can analyze the integrals in Theorem 1.1 and determine the asymptotic behavior of V_Z as z approaches 0 with $\text{Im}(z) > 0$. To state the complete result, we need the following complex version of the asymptotic condition **AS 2**.

We say that f satisfies the **complex asymptotic condition** at 0 if the following condition holds:

AS 2C. For every $z_0 \in \mathbf{C} \backslash \mathbf{R}_{\geq 0}$ there exist sequences $\{p\}$ and $\{b_p\}$ as above such that for every q,

$$f(tz_0) - P_q(t) = O(t^{\text{Re}(q)} |\log t|^{m(q)}) \text{ for } t \to 0 \text{ and } t > 0.$$

Thus, the old asymptotic condition **AS 2** holds along every ray. We then have the following theorem, which improves the domain of meromorphy of V_Z and also gives us an asymptotic expansion at 0.

Theorem 1.2. *Assume that G and $\tilde{G}$ are of regularized product type. Then V_Z has a meromorphic continuation to the domain $\mathbf{C} \backslash \mathbf{R}_{\geq 0}$ and satisfies the complex asymptotic condition* **AS 2C**.

Corollary 1.3. *Assume that G and $\tilde{G}$ are of regularized product type, and let $\lambda'_k = \varrho_k/i$. Then the sequence $\Lambda' = \{\lambda'_k\}$ satisfies the convergence conditions* **DIR 1**, **DIR 2**, *and* **DIR 3**, *as well as the asymptotic conditions* **AS 1**, **AS 2**, *and* **AS 3**.

Proof. The conditions **DIR 1**, **DIR 2**, and **DIR 3** are immediate simply from the fact that Z has finite order and we are looking at zeros and poles going up in the critical strip. Then, as remarked at the end of [JoL 92a], Sect. 1, condition **AS 1** follows at once. As for **AS 2**, we apply Theorem 1.2 with $z = it$. Thus for this particular application, we look at the ray corresponding to the positive imaginary axis. Finally, the condition **AS 3** follows from the fact that Z is of finite order. □

For meromorphic theta functions, we have the following strengthening of Theorem 1.2.

Theorem 1.4. *Assume that G and $\tilde{G}$ are of regularized product type and that the theta functions θ_G and $\theta_{\tilde{G}}$ are meromorphic. Then there is a branch of $\log(-z)$ in $\mathbf{C} \backslash \mathbf{R}_{\geq 0}$ such that the function*

$$V_Z(z) - \theta_{\tilde{G}}(-z)e^{\sigma_0 z}\log(z) + \theta_G(z)\log(-z)$$

has a meromorphic continuation to $\mathbf{C}$. Further, the singularities of the extension are simple poles located at the points $\log \mathbf{q}$ and $-\log \tilde{\mathbf{q}}$, with additional poles that agree in location and order with the poles of θ_G and $\theta_{\tilde{G}}$.

We could consider the zeros and poles of Z that lie in the half-plane $\text{Im}(z) < 0$ and consider the corresponding function $V_Z(z)$ for values of z also in this half-plane. A formula corresponding to that in Theorem 1.1 exists, and one can establish results similar to those in Theorem 1.2, Corollary 1.3, and Theorem 1.4. By combining these results, we obtain the following theorem.

Theorem 1.5. *Let Z and $\tilde{Z}$ be meromorphic functions with Euler products and functional equations, as defined above. Assume that G and $\tilde{G}$ are of regularized product* type of reduced order M. Then Z and $\tilde{Z}$ are of regularized product type of reduced order M.

For a further discussion of this situation, see the ladder theorem at the end of Sect. 7.

We shall prove Theorem 1.2 and Theorem 1.4 in Sects. 3, 4, and 5, after the proof of Theorem 1.1 is given, which is the content of Sect. 2. The proof of Theorem 1.5 will be given after a summary of the results in Sect. 6. We conclude by presenting several examples in Sect. 7.

2 A contour integral and the proof of Theorem 1.1

This section is devoted to the proof of Theorem 1.1. Formally, the result follows by considering the contour integral

$$2\pi i V_Z(z) = \int_{-a+i\infty}^{-a} e^{sz} Z'/Z(s) ds + \int_{-a}^{\sigma_0+a} e^{sz} Z'/Z(s) ds + \int_{\sigma_0+a}^{\sigma_0+a+i\infty} e^{sz} Z'/Z(s) ds.$$

Convergence questions must be addressed to verify the formal calculations. We begin with the following general lemma about meromorphic functions of finite order.

Lemma 2.1. *Let f be a meromorphic function of order $< k$, and let $[\sigma_1, \sigma_2]$ be any finite segment of the real axis. For each sufficiently large positive integer m, there exists a real number T_m with $m \leq T_m \leq m+1$ having the following property. If S_m denotes the segment $\sigma + iT_m$ with $\sigma_1 \leq \sigma \leq \sigma_2$, then f has no zeros or poles on S_m and, for $m \to \infty$ and $s \in S_m$, we have*

$$|\log f(s)| = O(T_m^k)$$

and

$$|f'/f(s)| = O(T_m^{2k}).$$

The value of $\log f(s)$ above is obtained by analytic continuation starting at σ_2, going above the real axis to σ_1, up the right side of the vertical line $\mathrm{Re}(s) = \sigma_1$, and along the segment S_m.

Proof. Since f has order $< k$, the number of zeros and poles of f in the rectangle of height T above $[\sigma_1, \sigma_2]$ is $O(T^k)$. Hence for all m sufficiently large there is a number T_m, $m \leq T_m \leq m+1$, such that f has no zero or pole in the rectangle of width $1/T_m^k$ around the horizontal segment S_m. By the minimum modulus principle (see Chap. XIII, Theorem 3.4, of [La 93]) and the standard estimate for entire functions it follows that

$$-T_m^k \leq \log|f(s)| \leq T_m^k$$

for $s \in S_m$ and m sufficiently large. Furthermore,

(1) $$|\arg f(s)| = O(T_m^k).$$

The estimate (1) is a general fact about meromorphic functions, and we shall recall a proof below. Using this fact, for $s \in S_m$ we obtain

$$|\log f(s)| = |\log|f(s)| + \arg f(s)| = O(T_m^k),$$

which proves the first estimate.

As to the second, the function $\log f(s)$ is holomorphic on the rectangle of width $1/T_m^k$ around S_m. Around each point of S_m we consider Cauchy's formula applied to $\log f(s)$, with a circle centered at this point of radius $1/2T_m^k$. Then the derivative of $\log f$, which is f'/f, is bounded for $s \in S_m$ according to Cauchy's formula by

$$|f'/f(s)| \ll \|\log f\| \text{ (radius of the circle)}^{-1}$$
$$\ll T_m^{2k},$$

as was to be shown. □

We now recall how to prove (1). Without loss of generality, we suppose that f is entire, and we consider the log of each term in the Weierstrass product

$$f(s) = e^{P(s)} \prod \left(1 - \frac{s}{s_n}\right) e^{Q_k(s/s_n)}.$$

If $|s_n| \geq 2T$, say, then the log of each term can be taken according to the ordinary power series for the log near 1, and the argument of the product over such terms is bounded. For the finite number of remaining terms, of which there are at most $O(T^k)$, the argument of each factor $(1 - s/s_n)$ is bounded between -2π and 2π, as is the argument of $\exp Q_k(s/s_n)$. Summing over at most $O(T^k)$ terms gives the desired bound

$$\arg f(s) = O(|s|^k) \quad \text{for } |s| \to \infty.$$

In the above application, $\arg f(s)$ can be taken over either one of two definite paths, one of them, as described above, consists of the real segment $[\sigma_1, \sigma_2]$, the right side of the vertical ray $[\sigma_1, \sigma_1 + iT]$, and the horizontal segment $[\sigma_1 + iT, \sigma_2 + iT]$; or, alternately, the vertical segment $[\sigma_2, \sigma_2 + iT]$ and the horizontal segment $[\sigma_1 + iT, \sigma_2 + iT]$. The difference between these two evaluations of $\arg f(s)$ is $O(T^k)$ since the sum of these two evaluations is equal to the number of zeros inside the rectangle bounded by such lines.

Remark 1. It should be noted that Lemma 2.1 is a coarse estimate for an arbitrary meromorphic function of finite order, replacing much more refined estimates that exist for L-functions of number fields (see [La 70]). Here we need only these coarse estimates because we will integrate the functions considered in the lemma against a function that decreases exponentially on a vertical line.

We now begin our proof of Theorem 1.1. Choose an $\varepsilon > 0$ sufficiently small so that Z has no zeros or poles in the open rectangle with vertices

$$-a, \quad -a + i\varepsilon, \quad \sigma_0 + a + i\varepsilon, \quad \sigma_0 + a$$

or on the line segment $[-a + i\varepsilon, \sigma_0 + a + i\varepsilon]$. Note that the function Z may have zeros or poles on the horizontal line segment $[-a, \sigma_0 + a]$. For T sufficiently large, we shall study the contour integral

$$(2) \quad 2\pi i V_Z(z, \varepsilon; T) = \int_{-a+iT}^{-a+i\varepsilon} e^{sz} Z'/Z(s)ds + \int_{-a+i\varepsilon}^{\sigma_0+a+i\varepsilon} e^{sz} Z'/Z(s)ds$$
$$+ \int_{\sigma_0+a+i\varepsilon}^{\sigma_0+a+iT} e^{sz} Z'/Z(s)ds \int_{\sigma_0+a+iT}^{-a+iT} e^{sz} Z'/Z(s)ds.$$

We may assume that Z has no zeros or poles on the line segment connecting the points $-a + iT$ and $\sigma_0 + a + iT$, because we will pick $T = T_m$ for m sufficiently large. Let $\mathscr{B}_T$ denote the rectangle with vertices

$$-a + iT, \quad -a + i\varepsilon, \quad \sigma_0 + a + i\varepsilon, \quad \sigma_0 + a + iT.$$

By the residue theorem, we have

$$2\pi i V_Z(z, \varepsilon; T) = \sum_{\varrho \in \mathscr{B}_T} v(\varrho) e^{\varrho z}.$$

Theorem 1.1 will be established by studying each of the four integrals in (2). For simplicity of the coming discussion, let us call these integrals the left, bottom, right, and top integrals, respectively. We begin with the top integral which will be shown to be arbitrarily small upon by letting $T = T_m$ approach infinity.

Proposition 2.2. *Let T_m be as in Lemma 2.1. Then for $\mathrm{Im}(z) > 0$ we have*

$$\lim_{m \to \infty} \left[\int_{\sigma_0 + a + iT_m}^{-a + iT_m} e^{sz} Z'/Z(s) ds \right] = 0.$$

Proof. If $z = x + iy$ with $y > 0$, then

$$e^{sz} = O(e^{-yT_m}),$$

so, from Lemma 2.1, we have that the integral is bounded by $O(T_m^{2k} e^{-yT_m})$. □

From Lemma 2.1 we let

$$\lim_{m \to \infty} \int_{-a + iT_m}^{-a + i\varepsilon} e^{sz} Z'/Z(s) ds = \int_{-a + i\infty}^{-a + i\varepsilon} e^{sz} Z'/Z(s) ds$$

and

$$\lim_{m \to \infty} \int_{\sigma_0 + a + i\varepsilon}^{\sigma_0 + a + iT_m} e^{sz} Z'/Z(s) ds = \int_{\sigma_0 + a + i\varepsilon}^{\sigma_0 + a + i\infty} e^{sz} Z'/Z(s) ds.$$

Combining these equations with Proposition 2.2, we have the following Theorem.

Theorem 2.3. *With notation as above, we have*

$$2\pi i V_Z(z, \varepsilon) = \lim_{m \to \infty} 2\pi i V_Z(z, \varepsilon; T_m)$$

$$= \int_{-a + i\infty}^{-a + i\varepsilon} e^{sz} Z'/Z(s) ds + \int_{-a + i\varepsilon}^{\sigma_0 + a + i\varepsilon} e^{sz} Z'/Z(s) ds + \int_{\sigma_0 + a + i\varepsilon}^{\sigma_0 + a + i\infty} e^{sz} Z'/Z(s) ds.$$

As before, let us call the integrals in Theorem 2.3 the left, bottom and right integrals, respectively. By the above stated assumption on e, we have

$$V_Z(z, \varepsilon) = V_Z(z),$$

where $V_Z(z)$ is as defined in Theorem 1.1. To complete our proof of Theorem 1.1, we will compute the three integrals in Theorem 2.3 using the axioms of Euler product and functional equation. After these computations, we will let ε approach 0 to complete the proof.

First, let us use the functional equation to re-write the left integral as the sum of three integrals involving $\tilde{Z}$, G, and $\tilde{G}$. Specifically, we have

$$
(3) \quad \int_{-a+i\infty}^{-a+i\varepsilon} e^{sz} Z'/Z(s)\,ds = -\int_{\sigma_0+a-i\infty}^{\sigma_0+a-i\varepsilon} e^{(\sigma_0-s)z} Z'/Z(\sigma_0-s)\,ds
$$

$$
= e^{\sigma_0 z} \int_{\sigma_0+a-i\infty}^{\sigma_0+a-i\varepsilon} e^{-sz} \tilde{Z}'/\tilde{Z}(s)\,ds
$$

$$
(4) \quad + e^{\sigma_0 z} \int_{\sigma_0+a-i\infty}^{\sigma_0+a-i\varepsilon} e^{-sz} \tilde{G}'/\tilde{G}(s)\,ds
$$

$$
(5) \quad + e^{\sigma_0 z} \int_{\sigma_0+a-i\infty}^{\sigma_0+a-i\varepsilon} e^{-sz} G'/G(\sigma_0-s)\,ds.
$$

After we let $\varepsilon \to 0$, the integrals in (4) and (5) appear in the statement of Theorem 1.1. Note that letting $\varepsilon \to 0$ is justified since G and $\tilde{G}$ were assumed to be holomorphic and non-zero on the vertical lines of integration. As for (3), we can re-write this integral using the Euler sum of $\tilde{Z}$, yielding

$$
e^{\sigma_0 z} \int_{\sigma_0+a-i\infty}^{\sigma_0+a-i\varepsilon} e^{-sz} \tilde{Z}'/\tilde{Z}(s)\,ds
$$

$$
= e^{\sigma_0 z} e^{-sz} \log \tilde{Z}(s) \Big|_{\sigma_0+a-i\infty}^{\sigma_0+a-i\varepsilon} + z e^{\sigma_0 z} \int_{\sigma_0+a-i\infty}^{\sigma_0+a-i\varepsilon} e^{-sz} \log \tilde{Z}(s)\,ds
$$

$$
= e^{(-a-i\varepsilon)z} \log \tilde{Z}(\sigma_0 + a - i\varepsilon) + z e^{\sigma_0 z} \sum_{\tilde{q}} c(\tilde{q}) \int_{\sigma_0+a-i\infty}^{\sigma_0+a-i\varepsilon} e^{-sz} e^{-s \log \tilde{q}}\,ds
$$

$$
(6) \quad = e^{(-a-i\varepsilon)z} \log \tilde{Z}(\sigma_0 + a - i\varepsilon) - z e^{(-a-i\varepsilon)z} \sum_{\tilde{q}} \frac{c(\tilde{q})}{\tilde{q}^{\sigma_0+a-i\varepsilon}(z + \log \tilde{q})}.
$$

Both terms in line (6) appear in the statement of Theorem 1.1, after we let $\varepsilon \to 0$, which is justified by the Euler product condition and the fact that $a > 0$.

In the same manner as above, the right integral can be re-written using the Euler sum of Z, yielding

$$\int_{\sigma_0+a+i\varepsilon}^{\sigma_0+a+i\infty} e^{sz} Z'/Z(s) ds = e^{sz} \log Z(s) \Big|_{\sigma_0+a+i\varepsilon}^{\sigma_0+a+i\infty} - z \int_{\sigma_0+a+i\varepsilon}^{\sigma_0+a+i\infty} e^{sz} \log Z(s) ds$$

$$= -e^{(\sigma_0+a+i\varepsilon)z} \log Z(\sigma_0+a+i\varepsilon)$$

$$- z \int_{\sigma_0+a+i\varepsilon}^{\sigma_0+a+i\infty} e^{sz} \log Z(s) ds$$

$$= -e^{(\sigma_0+a+i\varepsilon)} \log Z(\sigma_0+a+i\varepsilon)$$

$$- z \sum_{\mathbf{q}} c(\mathbf{q}) \int_{\sigma_0+a+i\varepsilon}^{\sigma_0+a+i\infty} e^{sz} e^{-s \log \mathbf{q}} ds$$

$$= -e^{(\sigma_0+a+i\varepsilon)z} \log Z(\sigma_0+a+i\varepsilon)$$

(7) $$+ ze^{(\sigma_0+a+i\varepsilon)z} \sum_{\mathbf{q}} \frac{c(\mathbf{q})}{\mathbf{q}^{\sigma_0+a+i\varepsilon}(z-\log \mathbf{q})}.$$

The second term in line (7) appears in Theorem 1.1, after we let $\varepsilon \to 0$, which is valid by the Euler product condition. The first term in line (7) does not appear in Theorem 1.1 since it cancels with a term that appears in the evaluation of the bottom integral.

In the evaluation of the bottom integral, we see the importance of choosing $\varepsilon > 0$ before integrating by parts. By the choice of $\varepsilon > 0$, Z has no zeros or poles on the line segment $[-a+i\varepsilon, \sigma_0+a+i\varepsilon]$, so we have

$$\int_{-a+i\varepsilon}^{\sigma_0+a+i\varepsilon} e^{sz} Z'/Z(s) ds = e^{sz} \log Z(s) \Big|_{-a+i\varepsilon}^{\sigma_0+a+i\varepsilon} - z \int_{-a+i\varepsilon}^{\sigma_0+a+i\varepsilon} e^{sz} \log Z(s) ds$$

$$= e^{z(\sigma_0+a+i\varepsilon)} \log Z(\sigma_0+a+i\varepsilon) - e^{(-a+i\varepsilon)z} \log Z(-a+i\varepsilon)$$

$$- z \int_{-a+i\varepsilon}^{\sigma_0+a+i\varepsilon} e^{sz} \log Z(s) ds$$

(8) $$= e^{(\sigma_0+a+i\varepsilon)} \log Z(\sigma_0+a+i\varepsilon) - e^{(-a+i\varepsilon)z} \log Z(-a+i\varepsilon)$$

$$- z \int_{-a+i\varepsilon}^{\sigma_0+a+i\varepsilon} e^{sz} \log |Z(s)| ds$$

(9) $$- z \int_{-a+i\varepsilon}^{\sigma_0+a+i\varepsilon} e^{sz} \arg(Z(s)) ds.$$

Let $\varepsilon \to 0$ in (8) and combine equations (2) through (9) to complete the proof of Theorem 1.1. Note the cancellation of one term (7) with a term in (8). Also, as stated in Theorem 1.1, the value of $\log Z(-a)$ is obtained by the analytic continuation of the Euler sum of Z along the horizontal line segment $[\sigma_0+a+i\varepsilon, -a+i\varepsilon]$ then along the vertical line segment $[-a, -a+i\varepsilon]$, which is equivalent to the analytic continuation along the top of the horizontal line segment $[-a, \sigma_0+a]$ by the assumption of ε.

Remark 2. In the case that $Z(s)$ is real on the real axis, then $\arg(Z(s))$ is a step function on $[-a, \sigma_0 + s]$ and takes on values in $\mathbf{Z} \cdot \pi i$, except at the zeros and poles of Z, where the argument is undetermined. In this case the corresponding integral in (9) can be evaluated directly and trivially, as elementary integral.

Remark 3. When comparing our work with that of Cramér in [Cr 19], the reader should note that we have overcome a point of substantial technical difficulty that Cramér encountered when proving Theorem 1.1 for the Riemann zeta function $\zeta_{\mathbf{Q}}(s)$. By choosing a small positive a, we have avoided to consider the convergence of the Euler product of Z on the line $\mathrm{Re}(s) = \sigma_0$. Cramér used the fact that $\zeta_{\mathbf{Q}}(s)$ does not vanish on the vertical line $\mathrm{Re}(s) = 1$ as well as specific knowledge about the distribution of prime numbers, namely

$$\sum_{p \leq x} \frac{1}{p} = O(\log \log x) \quad \text{as} \quad x \to \infty$$

and the Landau theorem which states that the limit

$$\lim_{x \to \infty} \sum_{p \leq x} \frac{1}{p^{1+it}}$$

converges uniformly for t in compact subsets of $\mathbf{R} \setminus \{0\}$. By following Cramér's original proof exactly, we would have greatly increased the complexity of the axioms of meromorphy, Euler product, and functional equation.

3 Regularized products

We are now led to study the integrals

$$e^{\sigma_0 z} \int_{\sigma_0+a-i\infty}^{\sigma_0+a} e^{-sz} \tilde{G}'/\tilde{G}(s) ds \quad \text{and} \quad e^{\sigma_0 z} \int_{\sigma_0+a-i\infty}^{\sigma_0+a} e^{-sz} G'/G(\sigma_0 - s) ds,$$

and to determine conditions under which these integrals admit meromorphic extensions. For this, let us assume that G and $\tilde{G}$ can be written as

$$G(s) = Q(s) e^{P(s)} D(s) \phi(s) \quad \text{and} \quad \tilde{G}(s) = \tilde{Q}(s) e^{\tilde{P}(s)} \tilde{D}(s) \tilde{\phi}(s)$$

where $Q(s)$, $\tilde{Q}(s)$, $P(s)$, and $\tilde{P}(s)$ are polynomials, and $D(s)$ and $\tilde{D}(s)$ are quotients of regularized products. As discussed in the introduction, no assumptions will be made concerning the functions $\phi(s)$ and $\tilde{\phi}(s)$. As a result, we have

(1)
$$\int_{\sigma_0+a-i\infty}^{\sigma_0+a} e^{-sz} \tilde{G}'/\tilde{G}(s) ds$$

$$= \int_{\sigma_0+a-i\infty}^{\sigma_0+a} e^{-sz} \tilde{Q}'/\tilde{Q}(s) ds + \int_{\sigma_0+a-i\infty}^{\sigma_0+a} e^{-sz} \tilde{P}(s) ds$$

$$+ \int_{\sigma_0+a-i\infty}^{\sigma_0+a} e^{-sz} \tilde{D}'/\tilde{D}(s) ds + \int_{\sigma_0+a-i\infty}^{\sigma_0+a} e^{-sz} \tilde{\phi}'/\tilde{\phi}(s) ds.$$

and a similar expression exists for $\tilde{G}(\sigma_0 - s)$. The second integral on the right hand side of (1) is evaluated through the following elementary lemma.

Lemma 3.1. *For* $\mathrm{Im}(z) > 0$ *and any polynomial* P, *we have*

$$\int_{\sigma_0+a-i\infty}^{\sigma_0+a} e^{-sz}P(s)ds = P(-\partial_z)[-e^{-(\sigma_0+a)z}/z].$$

We will say nothing about the fourth integral which involves $\tilde{\phi}$. For the remainder of this section, we will consider general integrals involving regularized products (for definitions, see [JoL 92a]) and obtain results that apply to the third integral in (1). Since any polynomial can be viewed as the regularized product associated to a finite sequence, this discussion applies to the first integral in (1) as well as the third integral. Also, without loss of generality to the question being considered, we may assume that D_L itself is a regularized product rather than a quotient of regularized products.

Let $L = \{\lambda_k\}$ be a sequence giving rise to the zeta function

$$\zeta(s) = \sum_{k=1}^{\infty} a_k \lambda_k^{-s}$$

and theta function

$$\theta(t) = \sum_{k=1}^{\infty} a_k e^{-\lambda_k t}$$

where a_k are integers ≥ 0 called the multiplicity of λ_k. Assume these functions satisfy the convergence conditions **DIR 1**, **DIR 2**, and **DIR 3** and the asymptotic conditions **AS 1**, **AS 2**, and **AS 3**, as in Sect. 2 of [JoL 92a]. We then have the associated regularized product D_L, which is entire, with zeros at $\{-\lambda_k\}$. Let $u \in \mathbf{R}$ and consider the two integrals

$$(2) \quad I_u^-(L;z) = \int_{u-i\infty}^{u} e^{-sz} D_L'/D_L(s)ds \text{ and } I_u^+(L;z) = \int_u^{u+i\infty} e^{-sz} D_L'/D_L(s)ds.$$

These integrals are taken on rays (half-lines) vertically down and vertically up, respectively. Also, the integrals are written in such a way that they fit the application we have in mind, namely the study of the third integral in (1). Note that the integrals in (2) are both absolutely convergent for $\mathrm{Im}(z) > 0$.

Next we have to consider separately a finite number of exponential terms for the theta function. As usual, let

$$Q_N\theta_L(t) = \sum_{k=1}^{N} a_k e^{-\lambda_k t},$$

where a_k is the multiplicity of λ_k in L. Given $c > 0$, let N be chosen so large that

$$\theta_L(t) - Q_N\theta_L(t) = O(e^{-ct}) \quad \text{for } t \to \infty.$$

Let $L = L_0 \cup L_1$, where $L_0 = \{\lambda_0, \ldots, \lambda_{N-1}\}$ and $L_1 = \{\lambda_N, \lambda_{N+1}, \ldots\}$. We shall consider the product

$$D_L = D_{L_0} \cdot D_{L_1}.$$

Suppose for definitness that we deal with I_u^+ (the analysis for I_u^- is identical to that of I_u^+), and assume that D_L has no zeros on the ray $[u, u+i\infty)$. Then, for $\mathrm{Im}(z) > 0$, the following integrals are absolutely convergent, and we can write

$$(3) \quad I_u^+(L; z) = \int_u^{u+i\infty} e^{sz} D'_{L_0}/D_{L_0}(s)\,ds + \int_u^{u+i\infty} e^{sz} D'_{L_1}/D_{L_1}(s)\,ds$$

$$= I_u^+(L_0; z) + I_u^+(L_1; z).$$

We take care of the first integral in (3), which is a finite sum, with the value

$$I_u^+(L; z) = \sum_{k=0}^{N-1} \int_u^{u+i\infty} \frac{a_k e^{sz}}{s + \lambda_k}\,ds$$

$$= \sum_{k=0}^{N-1} a_k e^{-\lambda_k z} \int_{u+\lambda_k}^{u+\lambda_k+i\infty} e^{wz} \frac{dw}{w}.$$

Each one of these integrals is essentially an elementary integral. For a given complex number α and any z for which $\mathrm{Im}(z) > 0$, let

$$\mathrm{Ei}_\alpha(z) = \int_\alpha^{\alpha+i\infty} e^{wz} \frac{dw}{w}.$$

Then $\mathrm{Ei}'(z) = -e^{\alpha z}/z$, so

$$\mathrm{Ei}_\alpha(z) + \log(-z)$$

has an analytic continuation along every path in **C**. We may write $\mathrm{Ei}_\alpha(z)$ as the classical **exponential integral**

$$\mathrm{Ei}_\alpha(z) = -\int_1^z \frac{e^{\alpha \zeta}}{\zeta}\,d\zeta + c(\alpha),$$

where $c(\alpha)$ is the constant of integration.

In summary, we have proved the following proposition.

Proposition 3.2. *Let $u \in \mathbf{R}$ be such that D_L has no zero on the ray $[u, u+i\infty)$. Then for $\mathrm{Im}(z) > 0$,*

$$I_u^+(L_0; z) = \sum_{k=0}^{N-1} a_k e^{-\lambda_k z} \mathrm{Ei}_{u+\lambda_k}(z) + c(u+\lambda_k).$$

Furthermore, the function

$$I_u^+(L_0; z) + \sum_{k=0}^{N-1} a_k e^{-\lambda_k z} \log(-z)$$

has an analytic continuation to the entire complex plane.

We now assume that the sequence L_1 satisfies what we shall call the **convergence conditions**:

CONV 1. $|\theta_{L_1}(t)| = O(e^{-ct})$ for $t \to \infty$.

CONV 2. $c + u > 0$.

Then for $\text{Im}(z) > 0$, integrating by parts, we obtain:

$$I_u^+(L_1; z) = \int_u^{u+i\infty} e^{sz} \partial_s \log D(s) ds$$

$$= \frac{e^{uz}}{z} (\partial_s \log D)(u) - \frac{1}{z} \int_u^{u+i\infty} e^{sz} \partial_s^2 \log D(s) ds.$$

Iterating this procedure, we obtain inductively:

Lemma 3.3. *For every integer $r > N + 1$, there is a polynomial $P_{L_1, r-1}$ of degree $\leq r - 1$ such that for $\text{Im}(z) > 0$ we have*

$$I_u^+(L_1; z) = e^{uz} P_{L_1, r-1}(1/z) + (-1/z)^{r-1} \int_u^{u+i\infty} e^{sz} \partial_s^r \log D(s) ds.$$

We then select $r > N + 1$, so that $\xi_{L_1}(w + r, s)$ is holomorphic near $w = 0$, so then

$$\partial_s^r \log D_{L_1}(s) = (-1)^{r+1} \text{CT}_{w=0} \xi_{L_1}(w + r, s) = (-1)^{r+1} \xi_{L_1}(r, s).$$

Then referring to the formulas for the functions $T_r(z)$ from Sect. 2 of [JoL 92a], we obtain

(4) $\quad \partial_s^r \log D(s) = (-1)^{r+1} \Gamma(r) \sum \frac{a_k}{(s + \lambda_k)^r} = (-1)^{r+1} \int_0^\infty \theta_{L_1}(t) e^{-ts} t^r \frac{dt}{t}.$

We note that $\theta_{L_1}(t) t^r$ is bounded near the origin, and that

$$\theta_{L_1}(t) e^{-ts} = O(e^{-(c+u)t}) \quad \text{for } \text{Re}(s) = u \text{ and } t \to \infty.$$

Hence, by **CONV 1** and **CONV 2**, we shall have no convergence problem with the forthcoming double integral resulting from substituting (4) in the right side of Lemma 3.3.

Proposition 3.4. *Let $r > N + 1$. If L_1 satisfies the convergence conditions **CONV 1** and **CONV 2**, then for $\text{Im}(z) > 0$ we have*

$$e^{-uz} I_u^+(L_1; z) = P_{L_1, r-1}(1/z) + \frac{1}{z^{r-1}} \int_0^\infty \frac{\theta_{L_1}(t) e^{-ut}}{t - z} t^r \frac{dt}{t}.$$

The integral on the right is absolutely convergent for $z \notin \mathbf{R}_{\geq 0}$, and gives an analytic continuation of $I_u^+(L_1; z)$ to the domain $\mathbf{C} \backslash \mathbf{R}_{\geq 0}$.

Proof. The integral on the right of Proposition 3.3 becomes a double integral, and the order of integration can be reversed for $\text{Im}(z) > 0$, to yield

(5) $\quad \int_0^\infty \theta_{L_1}(t) t^r \int_u^{u+i\infty} e^{-s(t-z)} ds \frac{dt}{t} = \int_0^\infty \frac{\theta_{L_1}(t) e^{-u(t-z)}}{t - z} t^r \frac{dt}{t},$

whence the Proposition follows. □

We note that by following the above analysis for I_u^- rather than I_u^+, one is led to consider the integral

$$\int_0^\infty \frac{\theta_{L_1}(t)e^{-u(t+z)}}{t+z} t^r \frac{dt}{t},$$

rather than the integral in (5). Recall that we have already met integrals of this form in Sect. 5 of [JoL 92a], which, in general, we shall express as

$$Hf(z) = \int_0^\infty \frac{f(t)}{t-z} dt$$

for some function f. We are interested in the domain of analyticity of Hf, and its analytic continuation, which will involve analytic properties of f, allowing us to continue Hf across the positive real axis, and then determine the singularity in a neighborhood of the origin. Such integrals were considered by Hermite [Her 1881], as mentioned by Cramér [Cr 19], so we call Hf the **Hermite transform** of f.

Suppose that $f \in L^1([0,\infty))$. Then Hf is obviously analytic for z in the complement of the positive real axis, that is $z \notin \mathbf{R}_{\geq 0}$. In our investigation of Hf, we shall go further in two different directions. One direction involves making assumptions of meromorphy on the theta function θ_L; and the other involves using a result from [JoL 92a] along the line of Stirling's formula, to give an asymptotic expansion for $V_Z(z)$ in a neighborhood of 0 independently of these analytic assumptions. The next two sections are devoted to these two results and are logically independent. We start with the result having to do with meromorphic theta functions.

4 Meromorphy assumptions and their consequences

For sequences $L = \{\lambda_k\}$ and $A = \{a_k\}$ arising in practice, it is a relatively rare occurrence that the corresponding theta function

$$\theta_L(t) = \sum a_k e^{-\lambda_k t}$$

has a meromorphic continuation on $\mathbf{C}$. However, it is remarkable fact that in many known examples of a situation with an Euler product and functional equation

$$Z(s)G(s) = \tilde{Z}(\sigma_0 - s)\tilde{G}(\sigma_0 - s),$$

then the fudge factor G has the shape

$$G(s) = Q(s)e^{P(s)}D_L(s)\phi(s),$$

often without $\phi(s)$, and is such that the theta function θ_L has a meromorphic continuation. Thus, in a context when it becomes important to have such a continuation, we indeed have it! We shall give several examples in Sect. 7.

The goal of this section is to prove:

Theorem 4.1. *Let $L = \{\lambda_k\}$ be a sequence satisfying the three convergence conditions* **DIR 1**, **DIR 2**, *and* **DIR 3**, *and whose associated theta function satisfies the asymptotic conditions* **AS 1**, **AS 2**, *and* **AS 3**. *Assume θ_L has a meromorphic extension to $\mathbf{C}$. Let*

$u \in \mathbf{R}$ be such that no zero of D_L lies on the vertical ray $[u, u + i\infty)$. Let

$$I_u^+(L; z) = \int_u^{u+i\infty} e^{sz} D_L'/D_L(s) ds,$$

and decompose $L = L_0 \cup L_1$ as in Sect. 3, so that L_1 satisfies the convergence conditions as stated in Sect. 3. Thus

$$\theta_{L_0}(t) = Q_M \theta_L(t) = \sum_{k=0}^{N-1} a_k e^{-\lambda_k t} \quad \text{and} \quad \theta_{L_1}(t) = \sum_{k=M}^{\infty} a_k e^{-\lambda_k t}.$$

Let $f(z) = \theta_{L_1}(z)$ and $g(z) = \theta_{L_0}(z)$, so $f(z) + g(z) = \theta_L(z)$. Then

$$I_u^+(L; z) + f(z) \log(-z) + g(z) \log(-z) = I_u^+(L; z) + \log(-z) \theta_L(z)$$

has a meromorphic to $\mathbf{C}$.

The proof of Theorem 4.1 will come from a sequence of lemmas concerning the Hermite transform Hf. We shall start by assuming that $f \in L^1((0, \infty))$, and that f is analytic on $(0, \infty)$, i.e. analytic on some open set containing the positive real axis.

Let $\varepsilon > 0$ and $R > 0$, with ε being small and R being large. For $v > 0$, let $\mathscr{L}(\varepsilon, R; v)$ be the path given by:
- the horizontal line segment $[0, \varepsilon]$;
- the vertical line segment $[\varepsilon, \varepsilon + iv]$;
- the horizontal line segment $[\varepsilon + iv, R + iv]$;
- the vertical line segment $[R + iv, R]$;
- and the horizontal line segment $[R, \infty]$.

We can choose v so that f is analytic on the closed rectangle bounded by $\mathscr{L}(\varepsilon, R; v)$ and the interval $[\varepsilon, R]$. With this we define

$$H_{\varepsilon, R, v} f(z) = \int_{\mathscr{L}(\varepsilon, R; v)} \frac{f(\zeta)}{\zeta - z} d\zeta.$$

We compare Hf and $H_{\varepsilon, R, v} f$ in the following lemma.

Lemma 4.2. *Assume that $f \in L^1((0, \infty))$ and is analytic on $(0, \infty)$ with ε, R, v as above. Then*

$$H_{\varepsilon, R, v} f(z) = Hf(z) \quad \text{for } z \in \mathbf{C} \backslash (\mathbf{R}_{>0} \cup \mathscr{L}(\varepsilon, R; v)).$$

Furthermore, $H_{\varepsilon, R, v} f$ is analytic on $\mathbf{C} \backslash \mathscr{L}(\varepsilon, R; v)$, and thus gives the analytic continuation of Hf across the positive real axis from below.

Proof. Immediate from the assumptions, and the fact that the integral around the closed contour bounded by the rectangle with vertices at $\varepsilon, \varepsilon + iv, R + iv$, and R is equal to 0. □

It remains to determine the behavior of Hf in a neighborhood of 0, where Hf has a branch point. For this purpose, we split the integral

$$Hf(z) = \int_0^\delta \frac{f(t)}{t - z} dt + \int_\delta^\infty \frac{f(t)}{t - z} dt = H^{(0)} f(z) + H^{(\delta)} f(z).$$

The second integral on the right, $H^{(\delta)} f(z)$, is holomorphic for z in a neighborhood of 0. Hence the singularity comes from $H^{(0)} f(z)$, which we now determine, assuming that f is itself holomorphic in a neighborhood of 0.

Lemma 4.3. *Let $f(w)$ be anylitic for $|w| < \delta$ with some $\delta > 0$. Then*
$$H^{(0)}f(z) + f(z)\log(-z)$$
is analytic for z in some neighborhood of 0.

Proof. Write the convergent power series
$$f(t) = \sum_{k=0}^{\infty} c_k t^k \quad \text{for } |t| < \delta.$$

Then
$$H^{(0)}f(z) = \sum_{k=0}^{\infty} c_k \int_0^{\delta} \frac{t^k dt}{t-z}, \quad \text{and by writing } t = z + (t-z),$$
$$= \sum_{k=0}^{\infty} c_k \sum_{j=0}^{k} \binom{k}{j} z^j \int_0^{\delta} (t-z)^{k-j-1} dt.$$

We then have terms with $j = k$, terms with $j \neq k$ and with $j = 0$ and $k \neq 0$, and the other terms, thus giving rise to three disjoint sums which we denote by S_1, S_2, and S_3, respectively. We find:

(i) For the sum with $j = k$,
$$S_1(z) = \sum_{k=0}^{\infty} c_k z^k \int_0^{\delta} \frac{dt}{t-z} = f(z)[\log(\delta - z) - \log(-z)]$$
$$= -f(z)\log(-z) + \text{holomorphic at } z = 0,$$
as asserted above.

(ii) For the sum with $j = 0$ and $k \neq 0$,
$$S_2(z) = \sum_{k=1}^{\infty} c_k \int_0^{\delta} (t-z)^{k-1} dt,$$
which is holomorphic at 0.

(iii) For the remaining sum,
$$S_3(z) = \sum_{k=2}^{\infty} c_k \sum_{j=1}^{k-1} \binom{k}{j} z^j \int_0^{\delta} (t-z)^{k-j-1} dt,$$
which is analytic for z in some neighborhood of 0, and has the value 0 at $z = 0$. This concludes the proof of the lemma. □

We shall put together the information given by Lemmas 4.2 and 4.3 and the following general complex analysis result.

Lemma 4.4. (a) *Let f be meromorphic function on $\mathbf{C}$.*

(b) *Let H be analytic on $\mathbf{C}\setminus\mathbf{R}_{\geq 0}$, and such that H has an analytic continuation across the positive real axis.*

(c) *Assume that $H(z) + f(z)\log(-z)$ has an extension to an analytic function of z in some neighborhood of 0.*

Then $H(z) + f(z)\log(-z)$ has an extension to a meromorphic function on all of $\mathbf{C}$.

Proof. Let F be the meromorphic function on $\mathbf{C}\backslash\mathbf{R}_{\geq 0}$ such that

$$F(z) = H(z) + f(z)\log(-z).$$

By the first two conditions, we can continue F across the positive real axis, and we must show that this continuation is in fact equal to F just above this axis. By the third assumption, this continuation is equal to F in a neighborhood of 0. Hence it is equal to F is an open set above the real axis, thus proving the lemma. □

Theorem 4.5. *Assume:*
  (a) f *is meromorphic on* $\mathbf{C}$ *and holomorphic at* 0.
  (b) f, *when restricted to* $(0,\infty)$, *is integrable.*
Then $Hf(z) + f(z)\log(-z)$ *extends to a meromorphic function on* $\mathbf{C}$.

Remark. Hypotheses (a) and (b) of Theorem 4.5 imply the two hypotheses of Lemma 4.2, and therefore the hypotheses of Lemma 4.4 are satisfied, so that the conclusion of Theorem 4.5 follows. We shall apply Theorem 4.5 to the situation of Proposition 3.4, making a meromorphic assumption on θ_{L_1} and letting

$$f(\zeta) = \theta_{L_1}(\zeta)e^{-u\zeta}\zeta^{r-1}$$

and considering the integral

$$\frac{e^{uz}}{z^{r-1}}\int_0^\infty \frac{f(\zeta)}{t-\zeta}d\zeta.$$

We are now in the position to finish the proof of Theorem 4.1. We decompose the desired integral as a sum

$$I_u^+(L;z) = I_u^+(L_0;z) + I_u^+(L_1;z).$$

The term $I_u^+(L_0;z)$ is equal to $g(z)\log(-z) + h(z)$ with entire h, by Proposition 3.2. The theta function θ_{L_1} satisfies the convergence conditions **CONV 1** and **CONV 2** if we pick M large enough in Proposition 3.2. We can then apply Theorem 4.5 to conclude the proof of Theorem 4.1. □

5 Asymptotic developments near $z = 0$

We return to the general situation of Sect. 3, without the additional assumptions of meromorphy on the theta function. Recall from Sect. 3 that we are studying the integrals

(1) $\quad e^{\sigma_0 z}\displaystyle\int_{\sigma_0+a-i\infty}^{\sigma_0+a} e^{-sz}\tilde{G}'/\tilde{G}(s)ds\quad$ and $\quad e^{\sigma_0 z}\displaystyle\int_{\sigma_0+a-i\infty}^{\sigma_0+a} e^{-sz}G/G(\sigma_0-s)ds.$

Specifically, we want to determine when these integrals have asymptotic expansions as in **AS 2C**. Assume that one can write

(2) $\qquad G(s) = Q(s)e^{P(s)}D(s)\phi(s)\quad$ and $\quad \tilde{G}(s) = \tilde{Q}(s)e^{\tilde{P}(s)}\tilde{D}(s)\tilde{\phi}(s)$

where Q, $\tilde{Q}$, and $\tilde{P}$ are polynomials, and D and $\tilde{D}$ are quotients of regularized products. Again, no assumptions concerning ϕ and $\tilde{\phi}$ will be made. The following theorem, which is the main result of this section, studies the remaining factors in G and $\tilde{G}$.

Theorem 5.1. *With notation as above, the integrals*

$$\int_{\sigma_0+a-i\infty}^{\sigma_0+a} e^{-sz}\tilde{D}'/\tilde{D}(s)ds + \int_{\sigma_0+a-i\infty}^{\sigma_0+a} e^{-sz}\tilde{Q}'/\tilde{Q}(s)ds + \int_{\sigma_0+a-i\infty}^{\sigma_0+a} e^{-sz}\tilde{P}'(s)ds$$

and

$$\int_{\sigma_0+a-i\infty}^{\sigma_0+a} e^{-sz}D'/D(\sigma_0-s)ds + \int_{\sigma_0+a-i\infty}^{\sigma_0+a} e^{-sz}Q'/Q(\sigma_0-s)ds$$

$$+ \int_{\sigma_0+a-i\infty}^{\sigma_0+a} e^{-sz}P'(\sigma_0-s)ds$$

are defined for all z with $\mathrm{Im}(z) > 0$; they have meromorphic extensions to $\mathbf{C} \setminus \mathbf{R}_{\geq 0}$, and they satisfy the asymptotic condition **AS 2C**.

The proof of Theorem 5.1 will follow from a sequence of lemmas. Recall, again from Sect. 3, that the integrals in (1) were re-written using (2) and integration by parts. These calculations write the integrals (1) in terms of Hermite transforms and exponential integrals that can be easily verified to satisfy **AS 2C**. The following lemma records the observation that these elementary terms satisfy **AS 2C**.

Proposition 5.2. *The following functions satisfy* **AS 2C**:
  (i) $P(-\partial_z)[-e^{-(\sigma_0+a)z}/z]$ *as defined in Lemma 3.1.*
  (ii) $\sum_{k=0}^{M-1} a_k e^{-\lambda_k z} \mathrm{Ei}_{u+\lambda_k}(z)$ *as defined in Proposition 3.2.*
  (iii) $P_{L_1,r-1}(1/z)$ *as defined in Lemma 3.3.*

By Proposition 5.2 and Proposition 3.4, to show that the functions in Theorem 5.1 satisfy the asymptotic condition **AS 2C**, it suffices to show that the Hermite transform

$$(3) \qquad \int_0^\infty \frac{\theta_{L_1}(t)e^{-ut}}{t-z} t^r \frac{dt}{t}$$

satisfies **AS 2C**. Let us split the integral in (3) and write

$$(4) \qquad \int_0^\infty \frac{\theta_{L_1}(t)e^{-ut}}{t-z} t^r \frac{dt}{t} = \int_0^1 \frac{\theta_{L_1}(t)e^{-ut}}{t-z} t^r \frac{dt}{t} + \int_1^\infty \frac{\theta_{L_1}(t)e^{-ut}}{t-z} t^r \frac{dt}{t}$$

and consider the two integrals on the right hand side of (4) separately. The following proposition handles the second integral.

Proposition 5.3. *The function*

$$\int_1^\infty \frac{\theta_{L_1}(t)e^{-ut}}{t-z} t^r \frac{dt}{t}$$

is holomorphic near $z = 0$.

The proof of Proposition 5.3 is immediate since one can differentiate under the integral sign any number of times.

To finish the proof of Theorem 5.1, it suffices to prove:

Proposition 5.4. *The function of z given by*

$$\int_0^1 \frac{\theta_{L_1}(t)e^{-ut}}{t-z} t^r \frac{dt}{t}$$

satisfies **AS 2C**.

Proof. We shall split the first integral in the right side of (4) by using the fact that $\theta_{L_1}(t)$, hence $\theta_{L_1}(t)e^{-ut}t^r$, satisfies **AS 2**. More precisely, let us write

(5) $$\int_0^1 \frac{\theta_{L_1}(t)e^{-ut}}{t-z} t^r \frac{dt}{t} = \int_0^1 \frac{\theta_{L_1}(t) - P_q\theta_{L_1}(t)}{t-z} e^{-ut}t^r \frac{dt}{t}$$

$$+ \int_0^1 \frac{P_q\theta_{L_1}(t)}{t-z} e^{-ut}t^r \frac{dt}{t}.$$

The first integral on the right hand side of (5) is considered in the following lemma.

Lemma 5.5. *Let $m = [q] + r - 2$ where $[q] = [\mathrm{Re}(q)]$, by definition. Let $z_0 \in \mathbf{C}\backslash\mathbf{R}_{\geq 0}$. Then there are constants $c_0(q), \ldots, c_{m-1}(q) \in \mathbf{C}$ such that*

$$\int_0^1 \frac{\theta_{L_1}(t) - P_q\theta_{L_1}(t)}{t - z_0 y} e^{-ut}t^r \frac{dt}{t} = c_0(q) + \ldots + c_{m-1}(q)y^{m-1} + O(y^m)$$

for $y \to 0$.

Proof. Simply use Taylor's formula with remainder, as in Lemma 5.7 of [JoL 92a].

Next we consider the second integral in (5). For this, let us write

$$e^{uz} \int_0^1 \frac{P_q\theta_{L_1}(t)}{t-z} e^{-ut}t^r \frac{dt}{t} = \int_0^1 \left[\int_u^\infty e^{v(z-t)} P_q\theta_{L_1}(t)t^r dv\right] \frac{dt}{t}$$

$$= \int_u^\infty \left[\int_0^1 e^{v(z-t)} P_q\theta_{L_1}(t)t^r \frac{dt}{t}\right] dv$$

(6) $$= \int_u^\infty \left[e^{zv} \int_0^1 P_q\theta_{L_1}(t)e^{-tv}t^r \frac{dt}{t}\right] dv.$$

If the integral in (6) satisfies **AS 2C**, so does the second integral in (5), and the desired result will be established. Recall that (6) is precisely the type of integral considered in Sect. 5 of [JoL 92a]. Specifically, in the notation of [JoL 92a], (6) can be written as

(7) $$\int_u^\infty \left[e^{zv} \int_0^1 P_q\theta_{L_1}(t)e^{-tv}t^r \frac{dt}{t}\right] dv = \int_u^\infty e^{zv} J_3(r,v) dv$$

where

$$J_3(r,v) = \int_0^1 P_q\theta_{L_1}(t)e^{-tv}t^r \frac{dt}{t} dv.$$

To apply the results in Sect. 5 of [JoL 92a], we need to take the transform (7) of $J_3(r,v)$ over an interval $[a, \infty)$ with $a > 0$. If $u < 0$, note that the integral over the segment

$$\int_u^a e^{zv} \int_0^1 P_q \theta_{L_1}(t) e^{-tv} t^r \, \frac{dt}{t} \, dv$$

is entire in z, so to obtain an asymptotic expansion of (7), we are reduced to considering the case when $u > 0$.

In that case, we see that there are polynomials $B_p^{\#(r)}$ such that

$$\mathrm{CT}_{s=r} B_p(\partial_s) \left[\frac{\pi z^{s+p-1}}{\sin[\pi(s+p)]} \right] = B_p^{\#(r)}(\log z) z^{r+p-1}.$$

We may apply Lemma 5.3, Lemma 5.10, and Theorem 5.11 from Sect. 5 of [JoL 92a] to obtain:

Lemma 5.6. *Suppose $u > 0$. Then there is an entire function g_q such that*

$$e^{uz} \int_0^1 \frac{P_q \theta_{L_1}(t)}{t - z} e^{-ut} t^r \, \frac{dt}{t} = \sum_{\mathrm{Re}(p) < \mathrm{Re}(q)} B_p^{\#(r)}(\log z) z^{r+p-1} + g_q(z).$$

The function of z given by

$$z \mapsto \sum_{\mathrm{Re}(p) < \mathrm{Re}(q)} B_p^{\#(r)}(\log z) z^{r+p-1}$$

satisfies **AS 2C***, hence so does the integral in* (7)*.*

This completes the proof of Proposition 5.4.

To conclude the proof of Theorem 5.1, let us comment on the compatibility of the asymptotics presented in Lemma 5.5 and Lemma 5.6. For this, let us choose $\mathrm{Re}(q') > \mathrm{Re}(q)$ and write

$$\int_0^1 \frac{\theta_{L_1}(t) e^{-ut}}{t - z} t^r \, \frac{dt}{t} = \int_0^1 \frac{\theta_{L_1}(t) - P_q \theta_{L_1}(t)}{t - z} e^{-ut} t^r \, \frac{dt}{t} + \int_0^1 \frac{P_q \theta_{L_1}(t)}{t - z} e^{-ut} t^r \, \frac{dt}{t}$$

$$= \int_0^1 \frac{\theta_{L_1}(t) - P_{q'} \theta_{L_1}(t)}{t - z} e^{-ut} t^r \, \frac{dt}{t} + \int_0^1 \frac{P_{q'} \theta_{L_1}(t)}{t - z} e^{-ut} t^r \, \frac{dt}{t}.$$

If we choose $\mathrm{Re}(q') > \mathrm{Re}(q)$, one can apply Lemma 5.5 and Lemma 5.6 to obtain the (formal asymptotic) equality

$$(8) \quad \sum_{k \leq [q]+r-3} c_k(q) y^k + \sum_{\mathrm{Re}(p) < \mathrm{Re}(q)} B_p^{\#(r)}(\log z) z^{r+p-1}$$

$$+ g_q(z_0 y) + O(y^{[q]+r-2})$$

$$= \sum_{k \leq [q']+r-3} c_k(q') y^k + \sum_{\mathrm{Re}(p) < \mathrm{Re}(q')} B_p^{\#(r)}(\log z) z^{r+p-1}$$

$$+ g_{q'}(z_0 y) + O(y^{[q']+r-2}).$$

Since the asymptotics of $\theta_{L_1}(t)$ are consistent when taking $\text{Re}(q') > \text{Re}(q)$, the terms of the form
$$\sum_{\text{Re}(p) < \text{Re}(q')} B_p^{\#(r)} (\log z) z^{r+p-1}$$
in (8) are consistent. So, even though the asymptotics in Lemma 5.5 involving the coefficients $c_k(q)$ and that of Lemma 5.6 involving g_q individually change when increasing $\text{Re}(q)$ to $\text{Re}(q')$, Eq. (8) shows that the sum is consistent. This completes the proof of Theorem 5.1.

6 Cramér's theorem

We have now completed the proof of Theorem 1.1, Theorem 1.2, and Theorem 1.4. It may be useful to the reader to go briefly through the main steps before proving Theorem 1.5.

We integrate $Z'/Z(s)$ against the rapidly decreasing function e^{sz} [with $\text{Im}(z) > 0$] around the standard rectangle slightly above the real segment of the critical strip. Then passing to the limit, we get $V_Z(z)$ by the residue theorem. By simple estimates, the integral over the top segment vanishes at infinity, so our integral over the rectangle breaks up into three integrals: over the left vertical ray, the bottom segment, and the right vertical ray. The integral over the bottom segment presents no difficulty since we are only interested in the analytic behavior of V_Z. The two integrals over the vertical rays are handled similarly. By the functional equation, the integral of Z'/Z over the left vertical ray is transmuted into a sum of integrals of $\tilde{Z}'/\tilde{Z}$, $\tilde{G}'/\tilde{G}$, and G'/G over the right vertical ray. The integrals of $\tilde{Z}'/\tilde{Z}$ and Z'/Z over the right vertical rays give rise to the sums containing the "prime powers" **q** and **q̃**. This leaves the integrals involving the logarithmic derivatives of the fudge factors G and $\tilde{G}$.

Assuming that G and $\tilde{G}$ themselves split into elementary factors (rational functions and exponential of polynomials), and regularized products, we are reduced to determining the analytic behavior of integrals
$$\int e^{sz} D'_L/D_L(s) ds$$
over vertical rays. For these integrals, we have to either make meromorphy assumptions on the theta function or to use analytic results which were considered previously in connection with Stirling's formula for regularized products. Putting everything together yields the final results.

We are not concerned here with exact formulas, but with qualitative questions, namely with meromorphic continuation and whether the convergence conditions for Dirichlet series **DIR 1**, **DIR 2**, and **DIR 3** and the asymptotic assumptions for theta functions **AS 1**, **AS 2**, and **AS 3** are satisfied. The first three are essentially trivially satisfied, and the main problem lies with the asymptotic condition **AS 2**. Both Cramér's theorem and its proof emphasize aspects of the integral around the rectangle besides the aspects encountered in the other two main results of analytic number theory, namely the explicit formulas and the counting of zeros, which are quantitative. Observe that in these cases we integrate Z'/Z against three types of functions:

– For Cramér's theorem, the function e^{sz} which is very smooth and decreases exponentially at infinity.
– For the explicit formulas, a function $\Phi(s)$ which is a Mellin transform of a not-so-smooth function, and decreases rather slowly at infinity.
– For the counting of zeros, the constant function 1, not decreasing at all.

Each case reflects a different property of the function Z, although these properties are not independent. In subsequent papers of this series, we shall deal with the explicit formulas and the counting of zeros in the same context of regularized products that we have considered all along.

We conclude this section by proving Theorem 1.5. Let

$$\Lambda_+ = \{\varrho/i \in \mathbf{C} | Z(\varrho) = 0, \ -a \leqq \operatorname{Re}(\varrho) \leqq \sigma_0 + a, \ \operatorname{Im}(\varrho) > 0\}$$

and

$$\Lambda_- = \{\varrho/(-i) \in \mathbf{C} | Z(\varrho) = 0, \ -a \leqq \operatorname{Re}(\varrho) \leqq \sigma_0 + a, \ \operatorname{Im}(\varrho) < 0\}.$$

By Corollary 1.3, the theta function θ_{Λ_+} associated to the sequence Λ_+ satisfies the asymptotic conditions **AS 1**, **AS 2**, and **AS 3**. Similarly, by considering the contour integral in the lower half plane, one would show that the theta function θ_{Λ_-} associated to the sequence Λ_- satisfies the asymptotic conditions **AS 1**, **AS 2**, and **AS 3**. Theorem 1.6 of [JoL 92a] then implies that the functions

$$D_+(z) = \exp(-\mathrm{CT}_{s=0}\mathbf{LM}\theta_{\Lambda_+}(s,z))$$

and

$$D_-(z) = \exp(-\mathrm{CT}_{s=0}\mathbf{LM}\theta_{\Lambda_-}(s,z))$$

are meromorphic functions of finite order with

$$D_+(z) = 0 \quad \text{if and only if } -z \in \Lambda_+,$$

and

$$D_-(z) = 0 \quad \text{if and only if } -z \in \Lambda_-.$$

Since G and $\tilde{G}$ are of regularized product type, the functional equation and Euler product for Z and $\tilde{Z}$ imply that the zeros and poles of $Z(s)$ in the half plane $\operatorname{Re}(s) < -a$ coincide in order and location with those of G and $\tilde{G}$. Hence, there is a function D_G of regularized product type, indeed formed from the regularized products that exist in the decomposition of G and $\tilde{G}$, such that $Z \cdot D_G$ has no zeros or poles in the half-planes $\operatorname{Re}(s) < -a$ and $\operatorname{Re}(s) > \sigma_0 + a$. With all this, we conclude that the function

$$Z(s) \cdot \frac{D_G(s)}{D_+(-s/i)D_-(s/i)}$$

is of finite order and has no zeros and poles. Therefore, there exists a polynomial $P_Z(s)$ such that

$$Z(s) = \frac{D_+(-s/i)D_-(s/i)}{D_G(s)} \cdot e^{P_Z(s)},$$

thus showing that $Z(s)$ is of regularized product type.

It remains to show that Z is of reduced order M, given that G and $\tilde{G}$ are of reduced order M. First, let us show that $\deg P_Z \leqq M + 1$. For this, we consider the expression

$$Z'/Z(s) = iD'_+/D_+(-s/i) - iD'_-/D_-(s/i) - D'_G/D_G(s) + P'_Z(s)$$

for $s = x \in \mathbf{R}$ with $x \to \infty$. By Theorem 1.1, Lemma 1.2, and Corollary 2.2 of the appendix to [JoL 93], we have, for some integer m, the asymptotic relations

$$D'_+/D_+(-x/i) = O(x^M(\log x)^m) \quad \text{for } |x| \to \infty,$$

and
$$D'_-/D_-(x/(-i)) = O(x^M(\log x)^m) \quad \text{for } x \to \infty.$$
From the general Gauss formula (Theorem 4.1 of [JoL 92a]) with $w = x - 1$ and $z = 1$, we obtain the asymptotic relation
$$D'_G/D_G(x) = O(x^M(\log x)^m) \quad \text{for } x \to \infty.$$
From the Euler sum, we have
$$Z'/Z(x) = o(1) \quad \text{for } x \to \infty.$$
Therefore, by combining the above, we obtain the asymptotic relation
$$P'_Z(x) = O(x^M(\log x)^m) \quad \text{for } x \to \infty,$$
thus showing that P_Z is a polynomial of degree $\leq M + 1$.

By assumption, D_G is of regularized product type and of reduced order M, so it remains to verify that D_+ and D_- are of reduced order M. We give the argument for D_+ with the argument for D_- being identical. As before, let us write a function g of restricted classical type as
$$G(s) = Q(s)e^{P(s)}D(s)$$
where Q is a rational functioin, P is a polynomial of degree $\leq M + 1$, and D is the quotient of regularized products of reduced order M. By setting $z = it$ in Cramér's theorem, we conclude that the asymptotics of $\theta_{\Lambda_+}(t)$ as $t \to 0$ are given by the asymptotics of
$$\int_{\sigma_0+a-i\infty}^{\sigma_0+a} e^{-s \cdot it} D'/D(s) ds + \int_{\sigma_0+a-i\infty}^{\sigma_0+a} e^{-s \cdot it} Q'/Q(s) ds + \int_{\sigma_0+a-i\infty}^{\sigma_0+a} e^{-s \cdot it} P'(s) ds$$
as $t \to 0$. Since $\deg P' \leq M$, Proposition 5.2(i) and 5.2(ii) show that the second and third integrals are $O(t^{-(M+1)})$ as $t \to 0$. By linearity, we may assume that D is a single regularized product and we use Proposition 3.2, Proposition 3.4 and the result of Sect. 5 to determine the asymptotics of the first integral. Specifically, if we take $r = M + 2$ in Proposition 3.4 and Lemma 5.6, and use the inequality $-\text{Re}(p_0) \leq M + 1$, we see that the first integral is $O(t^{-(M+1)}|\log t|^m)$ for some integer m. Combining these bounds, we conclude that the theta function $\theta_{\Lambda_+}(t)$ has the asymptotic behavior
$$\theta_{\Lambda_+}(t) = O(t^{-(M+1)}|\log t|^m) \quad \text{as } t \to 0,$$
thus showing that D_+ is of reduced order M and completing the proof of theorem 1.5.

7 Examples

To conclude this paper we will present several examples of functions with Euler product and functional equation to which we can apply Cramér's theorem. The discussion will be kept brief and various references to the literature will be given. Our list of examples is by no means exhaustive, and in subsequent papers we shall give further specific applications of Cramér's theorem.

Throughout this discussion we will use the following interpretation of the Lerch formula from Sect. 2 of [JoL 92a]. Let L be a sequence that satisfies the three convergence conditions **DIR 1**, **DIR 2**, **DIR 3**, and whose associated theta function satisfies the three asymptotic conditions **AS 1**, **AS 2**, and **AS 3**. Then there is a polynomial P_L such that the Weierstrass product associated to L and the regularized product associated to L differ by the multiplicative factor $\exp(P_L)$.

Number fields

In the most classical case of L-functions associated to number fields K (Dirichlet, Hecke, Artin), the function Z is the L-function itself, and the fudge factors G and $\tilde{G}$ are functions of the form

$$[s(s-1)]^{\delta_L} A^s \prod \Gamma(\alpha_i s + \mu_i)$$

where $\alpha_i > 0$, δ_L equals 0 or 1, and A is a constant that depends on the degree, the absolute value of the discriminant, and the number of complex embeddings of the number field K (see p. 254 of [La 70]). Recall that up to an additive linear factor, $\log \Gamma(s)$ is the logarithm of the regularized product with associated theta function

$$\theta_{\mathbf{Z}_{\geq 0}}(t) = \sum_{n \geq 0} e^{-nt} = \frac{1}{1 - e^{-t}},$$

which evidently has a meromorphic extension to $\mathbf{C}$. From this one can show that for $\alpha > 0$ and $\mu \in \mathbf{C}$, up to an additive linear factor, the function $\log \Gamma(\alpha s + \mu)$ is the logarithm of the regularized product with associated theta function

$$\theta_{\alpha,\mu}(t) = \sum_{n \geq 0} e^{-(n+\mu)t/\alpha} = \frac{e^{-\mu t/\alpha}}{1 - e^{-t/\alpha}}.$$

Hence, up to an additive linear factor, $\log \prod \Gamma(\alpha_i s + \mu_i)$ is the logarithm of the regularized product with associated theta function

$$\theta_L(t) = \sum \theta_{\alpha_i,\mu_i}(t).$$

As in the proof of Theorem 1.5, one considers the Cramér theorem when applied to both the upper and lower halves of the critical strip, and combines these results with the Lerch formula to show that the L-functions associated to number fields, are regularized products, up to multiplicative factors of the form $\exp(Q(s))$ for some polynomial Q, hence are of regularized product type. Further, Theorem 5.1 computes the asymptotics of the associated theta function near zero. In the case when $K = \mathbf{Q}$, we obtain the classical Cramér theorem [Cr 19] which implies that the Riemann zeta function $\zeta_\mathbf{Q}$ is of regularized product type. We note that Cramér's theorem was previously extended to Dirichlet L-series in [Ka 90] and independently at the same time as our paper to Hecke L-series in [Il 92].

Varieties

The zeta functions of varieties over numbers fields are conjectured to have a meromorphic continuation and functional equation, whose fudge factors G and $\tilde{G}$ are gamma factors. For the conjectured precise determination of these factors, see [Ser 69]. The Cramér theorem would apply to these zeta functions as well.

Automorphic Forms

The L-functions associated to automorphic forms have Euler products and functional equations. There is no question here of giving a summary of the literature, and we merely refer the reader to the following books and surveys which we find useful: Bump [Bu 89], Gelbart-Shahidi [GeS 88], and Shimura [Sh 71].

The Selberg zeta function for compact Riemann surfaces

Let X denote a compact Riemann surface with hyperbolic metric and canonical sheaf $\mathcal{K}$. Let π denote a finite-dimensional unitary representation of the fundamental group of X, and let j denote any half-integer. The Selberg zeta function ([He 76] or [Sel 56]) associated to the vector sheaf

$$\mathcal{E} = \mathcal{K}^j \otimes \mathcal{C}(\pi)$$

has an Euler product and functional equation, so Cramér's theorem applies. The Euler product is over the set of primitive geodesics and the fudge factors G and $\tilde{G}$ are of the form

$$e^{P(s)} \prod \Gamma_2(\alpha_i^{(2)} s + \mu_i^{(2)}) \prod \Gamma_1(\alpha_j^{(1)} s + \mu_j^{(1)})^{-1}$$

(see [DP 86], [JoL 92a], [Sa 87], [Wi 92]). Here P is a polynomial of degree 2, $\alpha_i^{(k)} > 0$, and Γ_1 is the classical gamma function. The function Γ_2 is the Barnes double gamma function whose role in the study of determinants of Laplacians has been studied in, for example, [Ku 91b] and [Va 88]. For our purpose, we need to know only that $\log \Gamma_2(s)$ is, up to an additive quadratic factor, a regularized product with associated theta function

$$\theta_2(t) = (-\partial_t)\left[\frac{1}{1 - e^{-t}}\right] = \frac{e^{-t}}{(1 - e^{-t})^2},$$

which is meromorphic. With this, the Lerch formula and Cramér's theorem, when applied to the upper and lower halves of the critical strip, imply that the Selberg zeta function associated to $\mathcal{E}$ is of regularized product type.

More generally, one meets meromorphic theta functions by considering the sequence $L = \mathbf{Z}_{\geq 0}$ when the integer n has multiplicity $p(n)$, where p is a polynomial. Then the associated theta function has the form

$$\theta(t) = \sum_{n \geq 0} p(n) e^{-nt} = p(-\partial_t)\left[\frac{1}{1 - e^{-t}}\right].$$

The Selberg zeta function for non-compact Riemann surfaces

If X is a non-compact hyperbolic Riemann surface of finite volume with h cusps, there is, as in the compact case, a Selberg zeta function with Euler product and functional equation. However, in the non-compact case, the fudge factors G and $\tilde{G}$ are of the form

$$e^{P(s)} \prod \Gamma_2(\alpha_i^{(2)} s + \mu_i^{(2)}) \prod \Gamma_1(\alpha_j^{(1)} s + \mu_j^{(1)})^{-1} \cdot A(s),$$

where the function $\phi(s) = A(s)/A(1-s)$, which does not appear in the compact case, is the determinant of the constant terms in the Fourier expansions in the cusps of the Eisenstein series (i.e., the scattering determinant; see the discussion starting on p. 498 of [He 83], or see p. 49 of [Sel 56]). In general, very little is known about the scattering matrix, but at least, we can prove:

Theorem 7.1. *Let Z be the Selberg zeta function of a non-compact hyperbolic Riemann surface of finite volume, and let ϕ be the determinant of the scattering matrix as above. Then both ϕ and Z are of regularized product type.*

Proof. Associated to each cusp P_j is an Eisenstein series $E_j(s,z)$. Let $\phi_{i,j}(s)$ be the constant term of $E_j(s,z)$ in a Fourier series expansion in the cusp P_i, and let

$$\phi(s) = \det[\phi_{i,j}(s)].$$

In [He 83] following Selberg [Se 56] it is shown that ϕ satisfies the functional equation

$$\phi(s)\phi(1-s) = 1.$$

It is also shown that there exist constants, $a, b \in \mathbf{R}^+$ and sequences $\{\mathbf{r}\}$ and $\{c(\mathbf{r})\}$, with $\mathbf{r} > 1$ tending to infinity, such that

$$\phi(s) = G(s)L(s)$$

where

$$G(s) = \left[\frac{\pi^{s-1}\Gamma_1(s-1/2)}{\pi^s \Gamma_1(s)}\right]^h e^{as+b}$$

and $L(s)$ has a Dirichlet series of the form

$$L(s) = 1 + \sum_{\mathbf{r}} \frac{c(\mathbf{r})}{\mathbf{r}^s}.$$

So, L has an Euler sum (i.e. a Dirichlet series for $\log L(s)$), and

$$L(s)G(s) = \tilde{L}(1-s)\tilde{G}(1-s)$$

where $\tilde{L} = 1/L$ and $\tilde{G} = 1/L$. Hence, Cramér's theorem applies to show that L, hence ϕ, is of regularized product type of reduced order $M = 0$.

The Selberg zeta function Z associated to X has an Euler product and (non-symmetric) functional equation with fudge factors factors involving Γ_1, Γ_2 and ϕ. Since these three functions are of regularized product type, the Cramér theorem applies to show that Z is of regularized product type, thus concluding the proof of the theorem. □

Remark. If X is realized as a quotient of the hyperbolic upper half plane $\mathfrak{h}$ by a congruence subgroup, the function ϕ has been computed and is given in terms of classical gamma functions and Dirichlet L-functions with character (see [He 83] or [Hu 84]). When the congruence subgroup is equal to the discrete group $PSL(2,\mathbf{Z})$ and Z is the Selberg zeta function associated to the sheaf $\mathscr{O}$, one has

$$A(s) = \xi_\mathbf{Q}(2s) = \pi^{-s}\Gamma(s)\zeta_\mathbf{Q}(2s)$$

(see [He 83], [Kub 73], and [Sel 56]). By Theorem 7.1, we conclude that the Selberg zeta function associated to the trivial sheaf $\mathscr{O}$ over $PSL(2,\mathbf{Z})\backslash\mathfrak{h}$ is of regularized product type (see, for example, [Ko 91a] and [Ko 92]). As a corollary, it follows, without using the trace formula, that the Kurokawa theta function [Kur 88] satisfies **AS 2**.

In addition, by Theorem 7.1, we can apply all the results of [JoL 92a]. In particular, for congruence subgroups we recover known results amounting to our Lerch formula in Efrat [Ef 88] and [Ef 91], and Koyama [Ko 92]. They also attempted a proof for the non-congruence subgroup case, using the trace formula, without success. The Cramér theorem thus provides a more effective tool, bypassing the trace formula at this step. In the present state of the literature, however, the trace formula is still used to prove the functional equation axiom.

The Selberg zeta function for rank one symmetric spaces

Gangolli [Ga 77] and Gangolli-Warner [GW 80] have defined and studied Selberg zeta functions for finite volume quotients of symmetric spaces of rank one. In this setting the Selberg zeta function has an Euler product and functional equation. For compact quotients, the fudge factors G and $\tilde{G}$ are explicitly computed and are such that their logarithmic derivatives are quotients of gamma functions (see p. 1 and 39 of [Ga 77] or p. 32 of [GW 80]). This allows one to compute the zeros and poles of G and $\tilde{G}$, from which one can show that G and $\tilde{G}$ are of the form

$$Q(s)e^{P(s)}D_L(s)$$

where Q and P are polynomials, and D_L is the quotient of regularized products with associated meromorphic theta function. In the non-compact case, there is an extra factor $\phi(s)$ which, as in the case of $PSL(2, \mathbf{R})$, has the representation-theoretic interpretation of being the determinant of an intertwining operator (see Theorem 4.4 of [GW 80] as well as [Ko 92b]).

Millson-Shintani zeta function and the η-invariant

In [Mi 78] Millson (following Shintani) defines a Selberg-type zeta for a compact $4n - 1$ dimensional Riemannian manifold of constant negative curvature (see also [MoS 89]). This zeta function has an Euler product and functional equation due to Shintani, of the form

$$Z(s)Z(4n - 2 - s) = C,$$

so the fudge factors G and $\tilde{G}$ are constant. As a result, Cramér's theorem shows that the theta function formed from the zeros in the upper half of the critical strip of this zeta function is itself a meromorphic theta function.

The ladder theorem

Let $\Lambda = \{\lambda_k\}$ be a sequence satisfying the three convergence conditions **DIR 1**, **DIR 2,**, and **DIR 3**, and whose associated theta function satisfies the three asymptotic conditions **AS 1**, **AS 2**, and **AS 3**. We form the regularized products $D_\Lambda(z)$. Suppose there is a G, Z, $\tilde{G}$, and $\tilde{Z}$ with Euler product and functional equation as in Sect. 1, and assume that

$$D_\Lambda(s(s - \sigma_0)) = Z(s)G(s).$$

Further, assume G and $\tilde{G}$ are of regularized product type. Put

$$\Lambda' = \{\lambda'_k\} = \{\varrho_k/i\},$$

where ϱ_k ranges over the zeros of Z in the upper half of the critical strip. Then, by the general Cramér's theorem, Λ' satisfies the three convergence conditions and the associated theta function satisfies the three asymptotic conditions. Indeed, the function $D_\Lambda(s(s - \sigma_0))$ vanishes when

$$s(s - \sigma_0) + \lambda_k = 0,$$

or

$$s = \frac{\sigma_0}{2} \pm i\sqrt{\lambda_k - \frac{\sigma_0^2}{4}}.$$

Theorem 7.8 of [JoL 92a] applies to show that the sequence $\{\varrho_k/i\}$ where

$$\varrho_k = \frac{\sigma_0}{2} + i\sqrt{\lambda_k - \frac{\sigma_0^2}{4}}$$

satisfies the convergence conditions and the asymptotic conditions.

References

[Bo 58] BOCHNER, S.: On Riemann's functional equation with multiple gamma factors. Ann. Math. **67**, 29–41 (1958)

[Bu 89] BUMP, D.: The Rankin-Selberg method: a survey. In: Number theory, trace formula, and discrete groups, pp. 49–109. Aubert, K.E. et al. (eds.) London: Academic Press 1989

[CoG 91] CONREY, J.B., GHOSH, A.: On the Selberg class of Dirichlet series: small weights. Preprint (1991)

[Cr 19] CRAMÉR, H.: Studien über die Nullstellen der Riemannschen Zetafunktion. Math. Z. **4**, 104–130 (1919)

[De 92] DENINGER, C.: Local L-factors of motives and regularized products. Invent. Math. **107**, 135–150 (1992)

[DP 86] D'HOKER, E., PHONG, D.: On determinants of Laplacians on Riemann surfaces. Commun. Math. Phys. **105**, 537–545 (1986)

[Ef 88] EFRAT, I.: Determinants of Laplacians on surfaces of finite volume. Commun. Math. Phys. **119**, 443–451 (1988)

[Ef 91] EFRAT, I.: Erratum: Determinants of Laplacians on surfaces of finite volume. Commun. Math. Phys. **138**, 607 (1991)

[Ga 77] GANGOLLI, R.: Zeta functions of Selberg's type for compact space forms of symmetric space of rank one. Ill. J. Math. **21**, 1–42 (1977)

[GW 80] GANGOLLI, R., WARNER, G.: Zeta functions of Selberg's type for some non-compact quotients of symmetric spaces of rank one. Nagoya Math. J. **78**, 1–44 (1980)

[GeS 88] GELBART, S., SHAHIDI, F.: Analytic properties of automorphic L-functions. San Diego: Academic Press 1988

[He 76] HEJHAL, D.A.: The Selberg trace formula for $PSL(2, \mathbf{R})$, vol. 1. (Lect. Notes in Math., vol. 548. Berlin, Heidelberg, New York: Springer 1976

[He 83] HEJHAL, D.A.: The Selberg trace formula for $PSL(2, \mathbf{R})$, vol. 2. (Lect. Notes Math., vol. 1001. Berlin Heidelberg, New York: 1983

[Her 1881] HERMITE, C.: Sur quelques points de la théorie des fonctions. J. Crelle **91**, 48–75 (1881) (Collected papers, vol. IV, pp. 48–75) Paris: Gauthier-Villiars 1917

[Hu 84] HUXLEY, M.N.: Scattering matrices for congruence subgroups. In: Modular forms, pp. 157–196. Rankin, R.A. (ed.) New York: Wiley 1984

[Il 92] ILLIES, G.: Regularisierte Produkte und Heckesche L-Reihen. Preprint (1992)

[In 32] INGHAM, A.E.: The distribution of prime numbers. Cambridge: Cambridge University Press 1932

[JoL 92a] JORGENSON, J., LANG, S.: Complex analytic properties of regularized products. Yale University Preprint (1992), to appear Lect. Notes in Math.

[JoL 92b] JORGENSON, J., LANG, S.: A Parseval formula for functions with an asymptotic expansion at the origin. Yale University Preprint (1992), to appear Lect. Notes in Math.

[JoL 93] JORGENSON, J., LANG, S.: Explicit formulas and regularized products. Yale University Preprint (1993)

[Ka 90] KACZOROWSKI, J.: The k-functions in multiplicative number theory. I. Acta Arith. **56**, 195–211 (1990)

[Ko 91a] KOYAMA, S.: Determinant expression of Selberg zeta functions. I. Trans. Am. Math. Soc. **324**, 149–168 (1991)

[Ko 91b] KOYAMA, S.: Determinant expression of Selberg zeta functions. III. Proc. Am. Math. Soc. **113**, 303–311 (1991)
[Ko 92] KOYAMA, S.: Determinant expression of Selberg zeta functions. II. Trans. Am. Math. Soc. **329**, 755–772 (1992)
[Kub 73] KUBOTA, T.: Elementary theory of Eisenstein series. New York: Wiley 1973
[Kur 88] KUROKAWA, N.: Parabolic components of zeta functions. Proc. Japan Acad. Ser. A **64**, 21–24 (1988)
[Kur 91] KUROKAWA, N.: Multiple sine functions and Selberg zeta functions. Proc. Japan Acad. Ser. A **67**, 61–64 (1991)
[Kur 92] KUROKAWA, N.: Gamma factors and Plancherel measures. Proc. Japan Acad. Ser. A **68**, 256–260 (1992)
[La 70] LANG, S.: Algebraic number theory. Menlo Park, CA.: Addison-Wesley 1970, reprinted as Grad. Texts Math., vol. 110. Berlin Heidelberg New York: Springer 1986
[La 93] LANG, S.: Complex analysis. (Grad. Texts Math., vol 103. Berlin Heidelberg New York: Springer 1985, 3rd. edition 1993
[Mi 78] MILLSON, J.: Closed geodesics and the η-invariant. Ann. Math. **108**, 1–39 (1978)
[MoS 89] MOSCOVICI, H., STANTON, R.J.: Eta invariants of Dirac operators on locally symmetric manifolds. Invent. Math. **95**, 629–666 (1989)
[Sa 87] SARNAK, P.: Determinants of Laplacians. Commun. Math. Phys. **110**, 113–120 (1987)
[Sel 56] SELBERG, A.: Harmonic analysis and discontinuous groups in weakly symmetric Riemannian spaces with applications to Dirichlet series. J. Indian Math. Soc. B *20*, 47–87 (1956) (Collected papers volume I, pp. 423–463. Berlin Heidelberg New York: Springer 1989
[Sel 91] SELBERG, A.: Old and new conjectures and results about a class of Dirichlet series. Collected papers volume II, pp. 47–63. Berlin Heidelberg New York: Springer 1991
[Ser 69] SERRE, J.-P.: Facteurs locaux des fonctions zeta des variétés algébriques. Sem. Delange-Poitou-Pisot 1969–70, Exposé 19
[Sh 71] SHIMURA, G.: Introduction to the arithmetic theory of automorphic functions. Princeton: Princeton University Press 1971
[Va 88] VARDI, I.: Determinants of Laplacians and multiple gamma functions. SISM J. Math. Anal. **19**, 493–507 (1988)
[Vi 79] VIGNÉRAS, M.-F.: L'équation fonctionelle de la fonction zéta de Selberg du groupoe modulaire $SL(2, \mathbf{Z})$. Astérisque **61**, 235–249 (1979)
[Wi 92] WILLIAMS, F.L.: A factorization of the Selberg zeta function attached to a rank 1 space form. Manuscr. Math. **77**, 17–39 (1992)

Lecture Notes in Mathematics   1564

Editors:
A. Dold, Heidelberg
B. Eckmann, Zürich
F. Takens, Groningen

Subseries:
Mathematisches Institut der Universität
und Max-Planck-Institut für Mathematik,
Bonn - vol. 18

Advisor:
F. Hirzebruch

Jay Jorgenson Serge Lang

Basic Analysis of Regularized Series and Products

Springer-Verlag
Berlin Heidelberg New York
London Paris Tokyo
Hong Kong Barcelona
Budapest

Authors

Jay A. Jorgenson
Serge Lang
Department of Mathematics
Yale University
Box 2155 Yale Station
New Haven, CT 06520, USA

Mathematics Subject Classification (1991): 11M35, 11M41, 11M99, 30B50, 30D15, 35P99, 35S99, 39B99, 42A99

(Authors' Note: there is no MSC number for regularized products, but there should be.)

ISBN 3-540-57488-3 Springer-Verlag Berlin Heidelberg New York
ISBN 0-387-57488-3 Springer-Verlag New York Berlin Heidelberg

This work is subject to copyright. All rights are reserved, whether the whole or part of the material is concerned, specifically the rights of translation, reprinting, re-use of illustrations, recitation, broadcasting, reproduction on microfilms or in any other way, and storage in data banks. Duplication of this publication or parts thereof is permitted only under the provisions of the German Copyright Law of September 9, 1965, in its current version, and permission for use must always be obtained from Springer-Verlag. Violations are liable for prosecution under the German Copyright Law.

© Springer-Verlag Berlin Heidelberg 1993
Printed in Germany

2146/3140-54321 - Printed on acid-free paper

Foreword

The two papers contained in this volume provide results on which a series of subsequent papers will be based, starting with [JoL 92b], [JoL 92d] and [JoL 93]. Each of the two papers contains an introduction dealing at greater length with the mathematics involved.

The two papers were first submitted in 1992 for publication in *J. reine angew. Math.* A referee emitted the opinion: "While such generalized products are of interest, they are not of such central interest as to justify a series of long papers in expensive journals." The referee was cautious, stating that this "view is subjective", and adding that he "will leave to the judgement of the editors whether to pass on this recommendation to the authors". The recommendation, in addition not to publish "in expensive journals", urged us to publish a monograph instead. In any case, the editors took full responsibility for the opinion about the publication of our series "in expensive journals". We disagree very strongly with this opinion. In fact, one of the applications of the complex analytic properties of regularized products contained in our first paper is to a generalization of Cramér's theorem, which we prove in great generality, and which appears in *Math. Annalen* [JoL 92b]. The referee for *Math. Ann.* characterized this result as "important and basic in the field".

Our papers were written in a self-contained way, to provide a suitable background for an open-ended series. Thus we always considered the possible alternative to put them in a Springer Lecture Note, and we are very grateful to the SLN editors and Springer for publishing them.

Acknowledgement: During the preparation of these papers, the first author received support from the NSF Postdoctoral Fellowship DMS-89-05661 and from NSF grant DMS-93-07023. The second author benefited from his visits at the Max Planck Institut in Bonn.

Part I

Some Complex Analytic Properties of Regularized Products and Series

Content

Introduction 1
1. Laplace-Mellin Transforms 9
2. Laurent Expansion at $s = 0$,
   Weierstrass Product, and Lerch Formula 30
3. Expressions at $s = 1$ 38
4. Gauss Formula 48
5. Stirling Formula 53
6. Hankel Formula 65
7. Mellin Inversion Formula 73
   Table of Notation 85

Part II

A Parseval Formula for Functions with a Singular Asymptotic Expansion at the Origin

Content

Introduction 91
1. A Theorem on Fourier Integrals 93
2. A Parseval Formula 105
3. The General Parseval Formula 109
4. The Parseval Formula for $I_w(a + it)$ 114

Bibliography 119

Part I

Some Complex Analytic Properties of Regularized Products and Series

Introduction

We shall describe how parts of analytic number theory and parts of the spectral theory of certain operators (differential, pseudo-differential, elliptic, etc.) are being merged under a more general analytic theory of regularized products of certain sequences satisfying a few basic axioms. The most basic examples consist of the sequence of natural numbers, the sequence of zeros with positive imaginary part of the Riemann zeta function, and the sequence of eigenvalues, say of a positive Laplacian on a compact manifold. The resulting theory applies to the zeta and L-functions of number theory, or representation theory and modular forms, to Selberg-like zeta functions in spectral theory, and to the theory of regularized determinants familiar in physics and other parts of mathematics.

Let $\{\lambda_k\}$ be a sequence of distinct complex numbers, tending to infinity in a sector contained in the right half plane. We always put $\lambda_0 = 0$ and $\lambda_k \neq 0$ for $k \geq 1$. We are also given a sequence $\{a_k\}$ of complex numbers. We assume the routine conditions that the Dirichlet series

$$\sum_{k=1}^{\infty} \frac{a_k}{\lambda_k^{\sigma}} \quad \text{and} \quad \sum_{k=1}^{\infty} \frac{1}{\lambda_k^{\sigma}}$$

converge absolutely for some $\sigma > 0$. If $a_k \in \mathbf{Z}_{\geq 0}$ for all k, we view a_k as a multiplicity of λ_k, and we call this **the spectral case**. We may form other functions, namely:

The **theta series** $\theta(t) = \sum_{k=0}^{\infty} a_k e^{-\lambda_k t}$ for $t > 0$;

The **zeta function** $\zeta(s) = \sum_{k=1}^{\infty} a_k \lambda_k^{-s}$;

The **Hurwitz zeta function** $\zeta(s, z) = \sum_{k=0}^{\infty} a_k (z + \lambda_k)^{-s}$;

The **xi function** $\xi(s, z) = \Gamma(s)\zeta(s, z)$, which can be written as the **Laplace-Mellin transform** of the theta function, that is

$$\xi(s,z) = \int_0^\infty \theta(t) e^{-zt} t^s \frac{dt}{t} = \mathbf{LM}\theta(s,z).$$

For the sequence $\{\lambda_k\}$ with $a_k = 1$ for all k, consider the derivative

$$\zeta'(s) = \sum_{k=1}^\infty \frac{-\log \lambda_k}{\lambda_k^s}.$$

Putting $s = 0$ formally, as Euler would do (cf. [Ha 49]), we find

$$\zeta'(0) = \sum_{k=1}^\infty -\log \lambda_k.$$

Therefore if the zeta function has an analytic continuation at $s = 0$, then

$$\exp(-\zeta'(0)) = \prod_{k=1}^\infty \lambda_k$$

may be viewed as giving a value for the meaningless infinite product on the right. Similarly, using the sequence $\{\lambda_k + z\}$ instead of $\{\lambda_k\}$, we would obtain a value for the meaningless infinite product

$$\mathbf{D}(z) = \exp(-\zeta'(0,z)) = \prod_{k=1}^\infty (\lambda_k + z),$$

where the derivative here is the partial derivative with respect to the variable s. To make sense of this procedure in the spectral case, under certain conditions one shows that the sequence $\{\lambda_k\}$ also determines:

The **regularized product**

$$D(z) = e^{P(z)} E(z),$$

where P is a normalizing polynomial, and $E(z)$ is a standard Weierstrass product having zeros at the numbers $-\lambda_k$ with multiplicity a_k. The degree of P and the order of the Weierstrass product will be characterized explicitly below in terms of the sequence $\{\lambda_k\}$ and appropriate conditions.

We keep in mind the following four basic examples of the spectral case.

Example 1. The gamma function. Let $\lambda_k = k$ range over the natural numbers. Then the zeta function is the Riemann zeta function $\zeta_{\mathbf{Q}}$, and the Hurwitz zeta function is the classical one (whence the name we have given in the general case). The theta function is simply

$$\theta(t) = \sum_{k=0}^{\infty} e^{-kt} = \frac{1}{1-e^{-t}}.$$

The corresponding Weierstrass product is that of the gamma function.

Example 2. The Dedekind zeta function. Let F be an algebraic number field. The **Dedekind zeta function** is defined for $\mathrm{Re}(s) > 1$ by the series

$$\zeta_F(s) = \sum \mathbf{N}\mathfrak{a}^{-s}$$

where $\mathfrak{a}$ ranges over the (non-zero) ideals of the ring of algebraic integers of F, and $\mathbf{N}\mathfrak{a}$ is the absolute norm, in other words, the index $\mathbf{N}\mathfrak{a} = (\mathfrak{o} : \mathfrak{a})$. Then

$$\theta(t) = \sum_{k=0}^{\infty} a_k e^{-kt}$$

where a_k is the number of ideals $\mathfrak{a}$ such that $\mathbf{N}\mathfrak{a} = k$. This theta function is different from the one which occurs in Hecke's classical proof of the functional equation of the Dedekind zeta functions (cf. [La 70], Chapter XIII). Of course, this example extends in a natural way to Dirichlet, Hecke, Artin, and other L-series classically associated to number fields. In these extensions, a_k is usually not an integer.

Zeta functions arising from representation theory and the theory of automorphic functions constitute an extension of the present example, but we omit here further mention of them to avoid having to elaborate on their more complicated definitions.

Example 3. Regularized determinant of an operator. In this case, we let $\{\lambda_k\}$ be the sequence of eigenvalues of an operator. In the most classical case, the operator is the positive Laplacian on a compact Riemannian manifold, but other much more complicated examples also arise naturally, involving possibly non-compact manifolds or pseudo differential operators. Suitably normalized, the function $D(z)$ is viewed as a regularized determinant (generalizing the characteristic polynomial in finite dimensions).

Example 4. Zeros of the zeta function. Let $\{\rho_k\}$ range over the zeros of the Riemann zeta function with positive imaginary part. Let a_k be the multiplicity of ρ_k, conjecturally equal to 1. Put $\lambda'_k = \rho_k/i$. The sequence $\{\lambda'_k\}$ is thus obtained by rotating the vertical strip to the right, so that it becomes a horizontal strip. A theorem of Cramér [Cr 19] gives a meromorphic continuation (with a logarithmic singularity at the origin) for the function

$$2\pi i V(z) = \sum a_k e^{\rho_k z} = \sum a_k e^{-\lambda'_k t}$$

which amounts to a theta function in this case (after the change of variables $z = it$).

This fourth example generalizes as follows. Suppose given a sequence $\{\lambda_k\}$ such that the corresponding zeta function has an Euler product and functional equation whose fudge factors are of regularized product type. (These notions will be defined quite generally in [JoL 92b].) We are then led to consider the sequence $\{\lambda'_k\}$ defined as above. From §7, one sees that a regularized product exists for the sequence $\{\lambda_k\}$. We will show in [JoL 92b] that a regularized product also exists for the sequence $\{\lambda'_k\}$. Passing from $\{\lambda_k\}$ to $\{\lambda'_k\}$ will be called **climbing the ladder** in the hierarchy of regularized products.

Basic Formulas. The example of the gamma function provides a basic table of properties which can be formulated and proved under some additional basic conditions which we shall list in a moment. The table includes:

The multiplication formula
The Lerch formula
The (other) Gauss formula
The Stirling formula
The Hankel formula
The Mellin inversion formula
The Parseval formula.

The multiplication formula may be viewed as a special case of the Artin formalism treated in [JoL 92d]. The Parseval formula, which determines the Fourier transform of Γ'/Γ as a distribution on a vertical line will be addressed in the context of a general result in Fourier analysis in [JoL 92c]. Here we show that the other formulas can be expressed and proved in a general context, under certain axioms (covering all four examples and many more complicated analogues). We shall find systematically how the simple expression $1/(1 - e^{-t})$ is replaced by theta functions throughout the formalism developed in this part. More generally, whenever the above expression occurs in mathematics, one should be on the lookout for a similar more general structure involving a theta function associated to a sequence having a regularized product.

The Asymptotic Expansion Axiom. The main axiom is a certain asymptotic expansion of the theta function at the origin, given as **AS 2** in §1; namely, we assume that there exists a sequence of complex numbers $\{p\}$ whose real parts tend to infinity, and polynomials B_p such that

$$\theta(t) \sim \sum_p B_p(\log t) t^p.$$

The presence of log terms is essential for some applications.

In Example 1, the asymptotic expansion of the theta function is immediate, since $\theta(t) = 1/(1 - e^{-t})$. In Example 2, this expansion follows from the consideration of §7 of the present part. In both Examples 1 and 2, B_p is constant for all p, so we say that there are no log terms.

In the spectral theory of Example 3, Minakshisundaram-Pleijel [MP 49] introduced the zeta function formed with the sequence of eigenvalues λ_k, and Ray-Singer introduced the so-called analytic torsion [RS 73], namely $\zeta'(0)$. Voros [Vo 87] and Cartier-Voros [CaV 90] gave further examples and results, dealing with a sequence of numbers whose real part tends to infinity. We have found their axiomatization concerning the corresponding theta function useful. However, both articles [Vo 87] and [CaV 90] leave some basic questions open in laying down the foundations of regularized products. Voros himself states: "In the present work, we shall not be concerned with rigorous proofs, which certainly imply additional regularity properties for the sequence $\{\lambda_k\}$." Furthermore, Cartier-Voros have only certain specific and special applications in mind (the Poisson summation formula and the Selberg trace formula in the case of compact Riemann surfaces). Because of our more general asymptotic expansion for the theta function, the theory becomes applicable to arbitrary compact manifolds with arbitrary Riemannian metrics and elliptic pseudo-differential operators where the log terms appear starting with [DuG 75], and continuing with [BrS 85], [Gr 86] and [Ku 88] for the spectral theory. In this case, the theta function is the trace of the heat kernel, and its asymptotic expansion is proved as a consequence of an asymptotic expansion for the heat kernel itself.

In Example 4 for the Riemann zeta function, the asymptotic expansion follows as a corollary of Cramér's theorem. A log term appears in the expansion. The generalization in [JoL 92b] involves some extra work. Some of the arguments used to prove the asymptotic expansion **AS 2** are given in §5 of the present part (especially Theorem 5.11), because they are directly related to those used to prove Stirling's formula. Indeed, the Stirling formula gives an asymptotic expansion for the log of a regularized product at infinity; in [JoL 92b] we require in addition an asymptotic expansion for the Laplace transform of the log of the regularized product in a neighborhood of zero.

Normalization of the Weierstrass Product by the Lerch Formula. We may now return to describe more accurately our normalization of the Weierstrass product. When the Hurwitz zeta function $\zeta(s,z)$ is holomorphic at $s = 0$, and all numbers a_k are positive integers, there is a unique entire functions $\mathbf{D}(z)$ whose zeros are the numbers $-\lambda_k$ with multiplicities a_k, with a normalized Weierstrass product such that the **Lerch formula** is valid, namely

$$\log \mathbf{D}(z) = -\zeta'(0, z),$$

where the derivative on the right is with respect to the variable s.

As to the Weierstrass order of $\mathbf{D}$, let p_0 be a leading exponent in the asymptotic expansion for $\theta(t)$, i.e. $\operatorname{Re}(p_0) \leq \operatorname{Re}(p)$ for all p such that $B_p \neq 0$. Let M be the largest integer $< -\operatorname{Re}(p_0)$. Then $M + 1$ is the order of $\mathbf{D}$.

In the general case with the log terms present in the asymptotic expansion, the Hurwitz zeta function $\zeta(s,z)$ may be meromorphic at $s = 0$ instead of being holomorphic, and we show in §3 how to make the appropriate definitions so that a similar formula is valid.

In Example 1, this formula is the classical Lerch formula. In Example 2, and various generalizations to L-functions of various types, the formula is new as far as we know. In Example 3, the formula occurs in many special cases of the theory of analytic torsion of Ray-Singer and in Voros [Vo 87], formula (4.1). In Example 4, the formula specializes to a formula discovered by Deninger for the Riemann zeta function $\zeta_\mathbf{Q}$ (see [De 92], Theorem 3.3).

Applications. Aside from developing a formalism which we find interesting for its own sake, we also give systematically fundamental analytic results which are used in the subsequent series of parts, including not only the generalization of Cramér's theorem mentioned above, but for instance our formulation of general explicit formulas analogous to those of analytic number theory (see [JoL 93]). These particular applications deal with cases when the zeta function has an Euler product and functional equation. Such cases may arise from a regularized product by a change of variables $z = s(s-1)$. The Selberg zeta function itself falls in this category. However, so far the Euler product does not play a role. A premature change of variables $z = s(s-1)$ obscures the basic properties of the regularized product and Dirichlet series which do not depend on the Euler product.

Although, as we have pointed out, some special cases of our formulas are known, many others are new. Our results and formulas concerning regularized products are proved in sufficient generality to apply in several areas of mathematics where zeta functions occur, e.g. analytic number theory, representation theory, spectral theory, ergodic theory and dynamical systems, etc. For example, our Lerch formula is seen to apply to Selberg type zeta functions not only for Riemann surfaces but for certain higher dimensional manifolds as well.

Furthermore, our general principle of climbing the ladder of regularized products applies to the scattering determinant associated to a non-compact hyperbolic Riemann surface of finite volume. In [JoL 92b] we shall use results of Selberg to show that the scattering determinant satisfies our axioms, and hence is of regularized product type. As a second application of our Cramér's theorem, we then conclude that the Selberg zeta function in the non-compact case is also of regularized product type. These facts were not known previously.

Our theory also applies to Ruelle type zeta functions arising in ergodic theory and dynamical systems (for example, see [Fr 86] and references given in that part).

Therefore, we feel that it is timely to deal systematically with the theory of regularized products, which we find central in mathematics.

For the convenience of the reader, a table of notation is included at the end of this part.

§1. Laplace-Mellin Transforms

We first recall some standard results concerning Laplace-Mellin transforms. The **Mellin transform** of a measurable function f on $(0, \infty)$ is defined by

$$\mathbf{M}f(s) = \int_0^\infty f(t) t^s \frac{dt}{t}.$$

The **Laplace transform** is defined to be

$$\mathbf{L}f(z) = \int_0^\infty f(t) e^{-zt} \frac{dt}{t}.$$

The **Laplace-Mellin transform** combines both, with the definition

$$\mathbf{LM}f(s, z) = \int_0^\infty f(t) e^{-zt} t^s \frac{dt}{t}.$$

We now worry about the convergence conditions. The next lemma is standard and elementary.

Lemma 1.1. *Let I be an interval of real numbers, possibly infinite. Let U be an open set of complex numbers, and let $f = f(t, z)$ be a continuous function on $I \times U$. Assume:*

(a) *For each compact subset K of U the integral*

$$\int_I f(t, z) \, dt$$

is uniformly convergent for $z \in K$.

(b) *For each t the function $z \mapsto f(t, z)$ is holomorphic.*

Let

$$F(z) = \int_I f(t, z) \, dt.$$

Then the second partial $\partial_2 f$ satisfies the same two conditions as f, the function F is holomorphic on U, and

$$F'(z) = \int_I \partial_2 f(t, z) \, dt.$$

For a proof, see [La 85], Chapter XII, §1. We then have immediately:

Lemma 1.2. Special Case. *Let z be such that $\mathrm{Re}(z) > 0$ and let $b_p \in \mathbf{C}$. Then for $\mathrm{Re}(s) > 0$ we have*

$$\int_0^\infty b_p e^{-zt} t^s \frac{dt}{t} = b_p \frac{\Gamma(s)}{z^s},$$

the integral being absolutely convergent, uniformly for

$$\mathrm{Re}(z) \geq \delta_1 > 0 \quad \text{and} \quad \mathrm{Re}(s) \geq \delta_2 > 0.$$

General Case. *For any polynomial B, let $B(\partial_s)$ be the associated constant coefficient differential operator. For $\mathrm{Re}(z) > 0$, $p \in \mathbf{C}$ and $\mathrm{Re}(s + p) > 0$ we have*

$$\int_0^\infty e^{-zt} B(\log t) t^{s+p} \frac{dt}{t} = B(\partial_s) \left[\frac{\Gamma(s+p)}{z^{s+p}} \right].$$

The proof of Lemma 1.2 follows directly from an interchange of differentiation and integration, which is valid for z and s in the above stated region.

At this point let us record several very useful formulas. For any z with $\mathrm{Re}(z) > 0$ and any s with $\mathrm{Re}(s) > 0$,

$$\int_1^\infty e^{-zt} t^s \frac{dt}{t} = \frac{1}{z^s} \int_{1/z}^\infty e^{-u} u^s \frac{du}{u} = \frac{\Gamma(s)}{z^s} - \int_0^1 e^{-zt} t^s \frac{dt}{t}$$

where the path of integration in the second integral is such that u/z is real. By expanding e^{-zt} in a power series about the origin, we have

$$\int_0^1 e^{-zt} t^s \frac{dt}{t} = \sum_{k=0}^\infty \frac{(-z)^k}{k!} \frac{1}{s+k},$$

which shows that the given integral can be meromorphically continued to all $s \in \mathbf{C}$ and all $z \in \mathbf{C}$. The integration by parts formula

$$s \int_0^1 e^{-zt} t^s \frac{dt}{t} = e^{-z} + z \int_0^1 e^{-zt} t^{s+1} \frac{dt}{t},$$

also provides a meromorphic continuation of the given integral.

The next lemma shows how an asymptotic expression for a function $f(t)$ near $t = 0$ gives a meromorphic continuation and an asymptotic expansion at infinity for the Laplace-Mellin transform of $f(t)$. We first consider a special case.

Lemma 1.3. Special Case. *Let f be piecewise continuous on $(0, \infty)$. Assume:*

(a) $f(t)$ *is bounded for $t \to \infty$.*

(b) $f(t) = b_p t^p + O(t^{\mathrm{Re}(q)})$ *for some $b_p \in \mathbf{C}$, $p, q \in \mathbf{C}$ such that $\mathrm{Re}(p) < \mathrm{Re}(q)$, and $t \to 0$.*

Then for $\mathrm{Re}(s) = \sigma > -\mathrm{Re}(p)$ and $\mathrm{Re}(z) > 0$ the Laplace-Mellin integral

$$\mathbf{LM}f(s, z) = \int_0^\infty f(t) e^{-zt} \, t^s \, \frac{dt}{t}$$

converges absolutely, and for $\mathrm{Re}(s) > -\mathrm{Re}(q)$ the function $\mathbf{LM}f$ has a meromorphic continuation such that

$$\mathbf{LM}f(s, z) = b_p \frac{\Gamma(s+p)}{z^{s+p}} + g(s, z)$$

where for fixed z, $s \mapsto g(s, z)$ is holomorphic for $\mathrm{Re}(s) > -\mathrm{Re}(q)$. The only possible poles in s of $\mathbf{LM}f$ when $\mathrm{Re}(z) > 0$ and when $\mathrm{Re}(s + q) > 0$ are at $s = -p - n$ with $n \in \mathbf{Z}_{\geq 0}$. All poles are simple, and the residue at $s = -p$ is b_p.

Proof. We decompose the integral into a sum:

$$\mathbf{LM}f(s, z) = \int_0^\infty (f(t) - b_p t^p) e^{-zt} \, t^s \, \frac{dt}{t} + \int_0^\infty e^{-zt} b_p t^{s+p} \, \frac{dt}{t}$$

$$= \int_0^1 (f(t) - b_p t^p) e^{-zt} \, t^s \, \frac{dt}{t}$$

$$+ \int_1^\infty (f(t) - b_p t^p) e^{-zt} \, t^s \, \frac{dt}{t} + b_p \frac{\Gamma(s+p)}{z^{s+p}}.$$

55

The second integral, from 1 to ∞, is entire in s, and converges uniformly for all z such that $\text{Re}(z) \geq \delta > 0$, and for all s such that $\text{Re}(s)$ is in a finite interval of $\mathbf{R}$. Also, we have

$$b_p \frac{\Gamma(s+p)}{z^{s+p}} = \frac{b_p}{s+p} - b_p(\gamma + \log z) + O(s+p),$$

since

$$\Gamma(s) = \frac{1}{s} - \gamma + O(s).$$

So there remains to analyze the first integral, from 0 to 1. More generally, let $q \in \mathbf{C}$, and let g be piecewise continuous on $(0, 1]$ satisfying

$$g(t) = O(t^{\text{Re}(q)}) \quad \text{for } t \to 0.$$

Then the integral

$$I_1(s, z) = \int_0^1 g(t) e^{-zt} t^s \frac{dt}{t}.$$

is obviously holomorphic in $z \in \mathbf{C}$ and $\text{Re}(s) > -\text{Re}(q)$, by Lemma 1.1. This concludes the proof of Lemma 1.3. $\square$

We are now going to show the effect of introducing logarithmic terms. For $p \in \mathbf{C}$, we let B_p denote a polynomial with complex coefficients and we put

$$b_p(t) = B_p(\log t).$$

We then let $B(\partial_s)$ be the corresponding constant coefficient partial differential operator.

Lemma 1.3. General Case. *Let f be piecewise continuous on $(0, \infty)$. Assume:*

(a) $f(t)$ *is bounded for* $t \to \infty$.

(b) $f(t) = b_p(t)t^p + O(t^{\text{Re}(q)}|\log t|^m)$ *for some function*

$$b_p(t) = B_p(\log t) \in \mathbf{C}[\log t],$$

such that $\text{Re}(p) < \text{Re}(q)$, $m \in \mathbf{Z}_{\geq 0}$, *and* $t \to 0$.

Then for $\text{Re}(s) = \sigma > -\text{Re}(p)$ and $\text{Re}(z) > 0$ the Laplace-Mellin integral

$$\mathbf{LM}f(s,z) = \int_0^\infty f(t) e^{-zt}\, t^s\, \frac{dt}{t}$$

converges absolutely, and for $\text{Re}(s) > -\text{Re}(q)$ the function $\mathbf{LM}f$ has a meromorphic continuation such that

$$\mathbf{LM}f(s,z) = B_p(\partial_s)\left[\frac{\Gamma(s+p)}{z^{s+p}}\right] + g(s,z)$$

where for fixed z, $s \mapsto g(s,z)$ is holomorphic for $\text{Re}(s) > -\text{Re}(q)$. The only possible singularity of $\mathbf{LM}f$ when $\text{Re}(z) > 0$ and $\text{Re}(s+q) > 0$ are poles of order at most $\deg B_p + 1$ at $s = -p-n$ with $n \in \mathbf{Z}_{\geq 0}$.

The proof is the same as in the special case invoking Lemma 1.1.

In brief, the presence of the logarithmic term $b_p(t)$ introduces a pole at $s = -p$ of order $\deg B_p + 1$, and Lemma 1.3 explicitly describes the polar part of the expansion of the Laplace-Mellin integral near $s = -p$.

The preceding lemmas give us information for

$$g(t) = O(t^{\text{Re}(q)} |\log t|^m).$$

We end our sequence of lemmas by describing more precisely the behavior due to the term $b_p(t) t^p$.

Lemma 1.4. *Let f be piecewise continuous on $(0, \infty)$. Assume:*

(a) *There is some $c \in \mathbf{R}$ such that $f(t) = O(e^{ct})$ for $t \to \infty$.*

(b) $f(t) = b_p(t) t^p + O(t^{\text{Re}(q)} |\log t|^m)$ *with* $b_p(t) = B_p(\log t)$, $\text{Re}(p) < \text{Re}(q)$, $m \in \mathbf{Z}_{\geq 0}$ *and* $t \to 0$.

Then $\mathbf{LM}f(s,z)$ is meromorphic in each variable for

$$\text{Re}(z) > c \quad \text{and} \quad \text{Re}(s) > -\text{Re}(q)$$

except possibly for poles at $s = -(p+n)$ with $n \in \mathbf{Z}_{\geq 0}$ of order at most $\deg B_p + 1$.

Proof. We split the integral:

$$\int_0^\infty f(t)e^{-zt}t^s \frac{dt}{t} = I_1(s,p,z) + I_2(s,p,z) + I_3(s,z)$$

where

(1) $$I_1(s,p,z) = \int_0^1 (f(t) - b_p(t)t^p)e^{-zt}t^s \frac{dt}{t},$$

(2) $$I_2(s,p,z) = \int_0^1 B_p(\log t)e^{-zt}t^{s+p} \frac{dt}{t},$$

(3) $$I_3(s,z) = \int_1^\infty f(t)e^{-zt}t^s \frac{dt}{t}.$$

The specified poles in s are going to come only from I_2. That is:

$I_1(s,p,z)$ is holomorphic for $z \in \mathbf{C}$ and $\mathrm{Re}(s) > -\mathrm{Re}(q)$.

$I_2(s,p,z)$ has a meromorphic continuation given by expanding e^{-zt} in its Taylor series and integrating term by term to get

$$I_2(s,p,z) = \sum_{n=0}^\infty (-1)^n \frac{z^n}{n!} B_p(\partial_s) \left[\frac{1}{n+p+s}\right].$$

$I_3(s,z)$ is holomorphic for $\mathrm{Re}(z) > c$ and all $s \in \mathbf{C}$. The theorem follows. □

We now consider an infinite sequence $L = \{\lambda_k\}$ of distinct complex numbers satisfying:

DIR 1. For every positive real number c, there is only a finite number of k such that $\mathrm{Re}(\lambda_k) \leq c$.

We use the convention that $\lambda_0 = 0$ and $\lambda_k \neq 0$ for $k \geq 1$. Under condition **DIR 1** we delete from the complex plane $\mathbf{C}$ the horizontal half lines going from $-\infty$ to $-\lambda_k$ for each k, together, when necessary, the horizontal half line going from $-\infty$ to 0. We define the open set:

$\mathbf{U}_L =$ the complement of the above half lines in $\mathbf{C}$.

If all λ_k are real and positive, then we note that $\mathbf{U}_L$ is simply $\mathbf{C}$ minus the negative real axis $\mathbf{R}_{\leq 0}$.

We also suppose given a sequence $A = \{a_k\}$ of distinct complex numbers. With L and A, we form the **asymptotic exponential polynomials** for integers $N \geq 1$:

$$Q_N(t) = a_0 + \sum_{k=1}^{N-1} a_k e^{-\lambda_k t}.$$

Throughout we shall also write

$$a_k = a(\lambda_k).$$

Similarly, we are given a sequence of complex numbers

$$\{p\} = \{p_0, \ldots, p_j, \ldots\}$$

with

$$\mathrm{Re}(p_0) \leq \mathrm{Re}(p_1) \leq \cdots \leq \mathrm{Re}(p_j) \leq \cdots$$

increasing to infinity. To every p in this sequence, we associate a polynomial B_p and, as before, we set

$$b_p(t) = B_p(\log t).$$

We then define the **asymptotic polynomials at 0** to be

$$P_q(t) = \sum_{\mathrm{Re}(p) < \mathrm{Re}(q)} b_p(t) t^p.$$

In many, perhaps most, applications the exponents p are real. Because there are significant cases when the exponents are not necessarily real, we lay the foundations in appropriate generality. We define

$$m(q) = \max \deg B_p \quad \text{for} \quad \mathrm{Re}(p) = \mathrm{Re}(q).$$

59

Let $\mathbf{C}\langle T\rangle$ be the algebra of polynomials in T^p with arbitrary complex powers $p \in \mathbf{C}$. Then, with this notation,

$$P_q(t) \in \mathbf{C}[\log t]\langle t\rangle.$$

We let f be a piecewise continuous function on $(0, \infty)$ satisfying the following **asymptotic conditions** at infinity and zero.

AS 1. Given a positive number C and $t_0 > 0$, there exists N and $K > 0$ such that

$$|f(t) - Q_N(t)| \leq K e^{-Ct} \text{ for } t \geq t_0.$$

AS 2. For every q, we have

$$f(t) - P_q(t) = O_q(t^{\mathrm{Re}(q)} |\log t|^{m(q)}) \text{ for } t \to 0,$$

where, as indicated, the implied constant depends on q.

Often we will write **AS 2** as

$$f(t) \sim \sum_p b_p(t) t^p.$$

Also, we will write $P_q(t) = P_q f(t)$ to denote the dependence on f. The crucial condition is **AS 2** and, in practice, is the most difficult to verify.

Let p_0 be an exponent of the asymptotic expansion **AS 2** with smallest (negative) value of $\mathrm{Re}(p_0)$. Let M be the largest integer $< -\mathrm{Re}(p_0)$. We call M the **reduced order** of the sequence $\{\lambda_k\}$. For instance, the sequence $\mathbf{Z}_{\geq 0}$ has reduced order 0.

The case when all a_k are non-negative integers will be called the **spectral case**. In such a situation one can view the coefficients a_k as determining a multiplicity in which the element λ_k appears in the sequence L.

The case when all the polynomials B_p are either 0 or constants, denoted by b_p, will be called the **special case**. In such a situation P_q is a polynomial with complex exponents and without the log terms.

Define $n(q)$ to be

$$n(q) = \max_{\text{Re}(p)<\text{Re}(q)} \deg B_p.$$

Throughout we will use p' to denote an element in the sequence $\{p\}$ with next largest real part for which $B_{p'}$ is not zero. In particular, this means that

$$n(q') = \max_{\text{Re}(p)\leq\text{Re}(q)} \deg B_p.$$

Theorem 1.5. *Let f satisfy **AS 1** and **AS 2**. Then $\mathbf{LM}f$ has a meromorphic continuation for $s \in \mathbf{C}$ and $z \in \mathbf{U}_L$. For each z, the function $s \mapsto \mathbf{LM}f(s,z)$ has poles only at the points $-(p+n)$ with $b_p \neq 0$ in the asymptotic expansion of f at 0. A pole at $-(p+n)$ has order at most $n(p)+1$. In the special case when the asymptotic expansion at 0 has no log terms, the poles are simple.*

Proof. We first do the analytic continuation in z, for $\text{Re}(s)$ large. We subtract an exponential polynomial Q_N from $f(t)$, using the asymptotic axiom **AS 1**, to write

$$\int_0^\infty f(t)e^{-zt}t^s\frac{dt}{t} = \int_0^\infty [f(t)-Q_N(t)]e^{-zt}t^s\frac{dt}{t} + \int_0^\infty Q_N(t)e^{-zt}t^s\frac{dt}{t}$$

$$= \int_0^\infty [f(t)-Q_N(t)]e^{-zt}t^s\frac{dt}{t}$$

$$+ a_0\int_0^\infty e^{-zt}t^s\frac{dt}{t} + \sum_{k=1}^{N-1} a_k \int_0^\infty e^{-\lambda_k t}e^{-zt}t^s\frac{dt}{t}$$

$$= \int_0^\infty [f(t)-Q_N(t)]e^{-zt}t^s\frac{dt}{t}$$

$$+ a_0\frac{\Gamma(s)}{z^s} + \sum_{k=1}^{N-1} a_k \frac{\Gamma(s+p)}{(z+\lambda_k)^{s+p}}.$$

61

It is immediate that the terms involving the gamma function have the above stated meromorphy properties. In particular, note that the appearance of the term

$$\sum_{k=1}^{N-1} a_k \frac{\Gamma(s+p)}{(z+\lambda_k)^{s+p}}$$

requires us to restrict z to $\mathbf{U}_L$. So, at this time, it remains to study the integral involving $f - Q_N$. Since f satisfies **AS 2**, then so does $f - Q_N$, and the integral

$$\int_0^\infty [f(t) - Q_N(t)] e^{-zt} t^s \frac{dt}{t}$$

is absolutely convergent for $\text{Re}(s) > -\text{Re}(p_0)$ and $\text{Re}(z) > -C$ if

$$f(t) - Q_N(t) = O(e^{-Ct}) \text{ for } t \to \infty.$$

By taking N sufficiently large, one can make C arbitrarily large, so this process shows how to meromorphically continue $\mathbf{LM}f(s,z)$ as a function of z. Next we show how to continue meromorphically in s.

For this we can apply Lemma 1.4 to $f - Q_N$, which shows that for $\text{Re}(z) > -C$ the function

$$s \mapsto \mathbf{LM}f(s,z)$$

is meromorphic in $\mathbf{C}$. In addition, Lemma 1.4 shows that the only possible poles are as described in the statement of the theorem. Note that the set of these poles is discrete, because the values $p+n$ tend to infinity. This completes the proof of the theorem. □

We shall use a systematic notation for the coefficients of the Laurent expansion of $\mathbf{LM}f(s,z)$ near $s = s_0$. Namely we let $R_j(s_0; z)$ be the coefficient of $(s - s_0)^j$, so that

$$\mathbf{LM}_f(s,z) = \sum R_j(s_0; z)(s - s_0)^j.$$

We shall be particularly interested when $s_0 = 0$ or $s_0 = 1$. Also, when necessary, we will express the dependence of the coefficients on the function f by writing

$$R_{j,f}(s_0; z) = R_j(s_0; z).$$

Theorem 1.6. *Let f satisfy* **AS 1** *and* **AS 2**. *Then for every $z \in \mathbf{U}_L$ and s near 0, the function $\mathbf{LM}f(s,z)$ has a pole at $s=0$ of order at most $n(0')+1$, and the function $\mathbf{LM}f(s,z)$ has the Laurent expansion*

$$\mathbf{LM}f(s,z) = \frac{R_{-n(0')-1}(0;z)}{s^{n(0')+1}} + \cdots + R_0(0;z) + R_1(0;z)s + \ldots$$

where, for each $j < 0$, $R_j(0;z) \in \mathbf{C}$ is a polynomial of degree $\leq -\operatorname{Re}(p_0)$.

Proof. For any $C > 0$, choose N as in **AS 1** so that $f - Q_N$ is bounded as $t \to \infty$. Since

$$\mathbf{LM}f(s,z) = \mathbf{LM}[f - Q_N](s,z) + \mathbf{LM}Q_N(s,z)$$
$$= \mathbf{LM}[f - Q_N](s,z) + a_0 \frac{\Gamma(s)}{z^s} + \sum_{k=1}^{N-1} a_k \frac{\Gamma(s+p)}{(z+\lambda_k)^{s+p}},$$

it suffices to prove the theorem for $f - Q_N$, or, equivalently, assume that $f(t)$ is bounded as $t \to \infty$.

Let

$$P_{0'}(t) = \sum_{\operatorname{Re}(p) \leq 0} b_p(t) t^p,$$

so we include $\operatorname{Re}(p) = 0$ in the sum. We decompose the integral into a sum:

$$\mathbf{LM}f(s,z) = \int_0^\infty f(t) e^{-zt} t^s \frac{dt}{t}$$
$$= \int_0^\infty [f(t) - P_{0'}(t)] e^{-zt} t^s \frac{dt}{t}$$
$$+ \int_1^\infty f(t) e^{-zt} t^s \frac{dt}{t} + \int_0^1 P_{0'}(t) e^{-zt} t^s \frac{dt}{t}.$$

The first two integrals are holomorphic at $s = 0$ and $\operatorname{Re}(z) > 0$.

By Lemma 1.2, the third integral is simply

$$\sum_{\mathrm{Re}(p)\leq 0} B_p(\partial_s) \int_0^1 e^{-zt} t^{s+p} \frac{dt}{t}$$

$$= \sum_{\mathrm{Re}(p)\leq 0} \sum_{k=0}^{\infty} \frac{(-z)^k}{k!} B_p(\partial_s) \left[\frac{1}{s+p+k}\right]$$

(4) $$= \sum_{\mathrm{Re}(p)+k=0} \frac{(-z)^k}{k!} B_p(\partial_s) \left[\frac{1}{s}\right] + h_0(z) + O(s),$$

where $h_0(z)$ is entire in z. This proves the theorem, and (4) gives us an explicit determination of the polynomials $R_j(0; z)$ for $j < 0$. $\square$

The constant term $R_0(s_0; z)$ is so important that we give it a special notation; namely, for a meromorphic function $G(s)$ we let

$$\mathrm{CT}_{s=s_0} G(s) = \text{constant term in the Laurent expansion}$$

of $G(s)$ at $s = s_0$.

That is,

$$\mathrm{CT}_{s=0} \mathbf{LM} f(s, z) = R_0(0; z).$$

Corollary 1.7. *Define*

$$\zeta_f(s, z) = \frac{1}{\Gamma(s)} \mathbf{LM} f(s, z) \quad \text{and} \quad \xi_f(s, z) = \mathbf{LM} f(s, z).$$

Then, in the special case, $\zeta_f(s, z)$ is holomorphic at $s = 0$ for $z \in \mathbf{U}_L$ and

$$\zeta_f'(0, z) = \mathrm{CT}_{s=0} \xi_f(s; z) + \gamma R_{-1, f}(0; z).$$

Proof. In the special case $n(0') = 0$, so $\mathbf{LM} f(s, z)$ has a pole at $s = 0$ of order at most 1. Since

$$\frac{1}{\Gamma(s)} = s + \gamma s^2 + O(s^3),$$

we have

$$\zeta_f(s, z) = \frac{1}{\Gamma(s)} \xi_f(s, z)$$

$$= R_{-1, f}(0; z) + (\mathrm{CT}_{s=0} \xi_f(s, z) + \gamma R_{-1, f}(0; z))s + O(s^2),$$

from which the corollary follows. $\square$

Theorem 1.8. *The function* $\mathbf{LM}f$ *satisfies the equation*

$$\partial_z \mathbf{LM}f(s,z) = -\mathbf{LM}f(s+1,z), \quad \text{for } s \in \mathbf{C} \text{ and } z \in \mathbf{U}_L.$$

Proof. This is immediate by differentiating under the integral sign for $\mathrm{Re}(s)$ and $\mathrm{Re}(z)$ sufficiently large, and follows otherwise by analytic continuation. □

Corollary 1.9. *For any integer j, we have*

$$\partial_z R_j(s_0; z) = -R_j(s_0 + 1, z).$$

In particular, for the constant terms, we have

$$\partial_z R_0(0; z) = -R_0(1, z).$$

Corollary 1.10. *The Mellin transform*

$$\xi_f(s) = \mathbf{M}f(s),$$

has a meromorphic continuation to $s \in \mathbf{C}$ whose only possible poles are at $s = -p$ such that $b_p \neq 0$.

Proof. Write $f = f_0 + f_1$ with $f_0 = Q_N$ and $f_1 = f - Q_N$ with N sufficiently large so that we can apply **AS 1** to f_1 with $C > 0$. Then $\mathbf{M}f_0$ is entire in s. As for f_1, we have, for any q,

$$\mathbf{M}f_1(s) = \int_0^1 [f_1(t) - P_q(t)] t^s \frac{dt}{t}$$

(5)
$$+ \sum_{\mathrm{Re}(p) < \mathrm{Re}(q)} B_p(\partial_s) \left[\frac{1}{s+p}\right] + \int_1^\infty f_1(t) t^s \frac{dt}{t}.$$

The first integral in (5) is holomorphic for $\mathrm{Re}(s) > -\mathrm{Re}(q)$, by Lemma 1.4. By the construction of f_1, the second integral in (5) is entire in s. The sum in (6) is meromorphic for all $s \in \mathbf{C}$ with possible poles at $s = -p$. With all this, the proof is complete. □

22

Given the sequences A and L as defined in **AS 1**, one can consider:

a Dirichlet series

$$\zeta_{L,A}(s) = \zeta(s) = \sum_{k=1}^{\infty} a_k \lambda_k^{-s},$$

a theta series

$$\theta_{L,A}(t) = \theta(t) = a_0 + \sum_{k=1}^{\infty} a_k e^{-\lambda_k t},$$

a reduced theta series

$$\theta_{L,A}^{(1)}(t) = \theta^{(1)}(t) = \sum_{k=1}^{\infty} a_k e^{-\lambda_k t},$$

and, more generally, for each positive integer N,

a truncated theta series

$$\theta_{L,A}^{(N)}(t) = \theta^{(N)}(t) = \sum_{k=N}^{\infty} a_k e^{-\lambda_k t}.$$

We shall assume throughout that the theta series converges absolutely for $t > 0$. From **DIR 1** it follows that the convergence of the theta series is uniform for $t \geq \delta > 0$ for every δ. We shall apply the above results to $f = \theta$ with the associated sequence of exponential polynomials $Q_N \theta$ being the natural ones, namely

$$Q_N \theta(t) = a_0 + \sum_{k=1}^{N-1} a_k e^{-\lambda_k t}.$$

Note that the above notation leads to the formula

$$\theta(t) - Q_N \theta(t) = \theta^{(N)}(t)$$

The absolute convergence of the theta series $\theta(t)$ describes a type of convergence of $\theta^{(N)}$ near infinity that is uniform for all N. The following condition describes a type of uniformity of the asymptotics of $\theta(t)$ near $t = 0$.

AS 3. Given $\delta > 0$, there exists an $\alpha > 0$ and a constant $C > 0$ such that for all N and $0 < t \leq \delta$ we have

$$\left|\theta^{(N)}(t)\right| = |\theta(t) - Q_N(t)| \leq C/t^\alpha.$$

We shall see that the three conditions **AS 1**, **AS 2** and **AS 3** on the theta series correspond to conditions on the zeta function describing the growth of the sequence $\{\lambda_k\}$ and also the sequence $\{a_k\}$. The condition **DIR 1** simply states that the sequence $\{\lambda_k\}$ converges to infinity, in some weak sense. The following condition gives a slightly stronger convergence requirement.

DIR 2.

(a) The Dirichlet series

$$\sum_k \frac{a_k}{\lambda_k^\sigma}$$

converges absolutely for some real σ. Equivalently, we can say that there exists some $\sigma_0 \in \mathbf{R}_{>0}$ such that

$$|a_k| = O(|\lambda_k|^{\sigma_0}) \text{ for } k \to \infty.$$

(b) The Dirichlet series

$$\sum_k \frac{1}{\lambda_k^\sigma}$$

converges absolutely for some real σ. Specifically, let σ_1 be a real number for which

$$\sum_k \frac{1}{|\lambda_k|^{\sigma_1}} < \infty.$$

Theorem 1.11. *Assume that the theta series*

$$\theta(t) = \sum_{k=1}^\infty a_k e^{-\lambda_k t}$$

satisfies **AS 1**, **AS 2** and **AS 3**, and assume that $\text{Re}(\lambda_k) > 0$ for all $k > 0$. Then for $\text{Re}(s) > \alpha$,

$$\xi(s) = \mathbf{M}\theta(s) = \Gamma(s) \sum_{k=1}^{\infty} \frac{a_k}{\lambda_k^s}$$

in the sense that the series on the right converges absolutely, to $\mathbf{M}\theta(s)$. In particular, the convergence condition **DIR 2(a)** is satisfied for the Dirichlet series

$$\sum_{k=1}^{\infty} \frac{a_k}{\lambda_k^s}.$$

Proof. For $\sigma = \text{Re}(s) > \alpha$, we have

$$\xi(s) - \Gamma(s) \sum_{k=1}^{N-1} \frac{a_k}{\lambda_k^s} = \int_0^1 [\theta(t) - Q_N\theta(t)] t^s \frac{dt}{t} + \int_1^{\infty} [\theta(t) - Q_N\theta(t)] t^s \frac{dt}{t}.$$

From **AS 1** we have

$$\left| \int_1^{\infty} [\theta(t) - Q_N\theta(t)] t^s \frac{dt}{t} \right| \leq K \int_1^{\infty} e^{-Ct} t^\sigma \frac{dt}{t},$$

which goes to zero as $C \to \infty$. Similarly, by **AS 3** we have that

$$\left| \int_0^1 [\theta(t) - Q_N\theta(t)] t^s \frac{dt}{t} \right| \leq C \int_0^1 t^{\sigma - \alpha} \frac{dt}{t},$$

which is uniformly bounded if $\sigma > \alpha$. Therefore, we have that for $\sigma \geq \alpha + \epsilon > \alpha$, the difference

$$\xi(s) - \Gamma(s) \sum_{k=1}^{N-1} \frac{a_k}{\lambda_k^s}$$

is uniformly bounded for all N. By letting N approach ∞, we can interchange limit and integral, by dominated convergence, to show that for $\sigma \geq \alpha + \epsilon > \alpha$,

$$\lim_{N \to \infty} \left[\xi(s) - \Gamma(s) \sum_{k=1}^{N-1} \frac{a_k}{\lambda_k^s} \right] = 0,$$

which completes the proof of the theorem. □

Remark 1. If $-\mathrm{Re}(p_0)$ is an integer, then $M+2$ is the smallest integer m for which the Dirichlet series

$$\sum_{1}^{\infty} \frac{|a_k|}{|\lambda_k|^m} < \infty$$

converges. If $-\mathrm{Re}(p_0)$ is not an integer, then $M+1$ is the smallest such m. In any event, we let m_0 be the smallest such m. The exponent $\mathrm{Re}(p_0)$, which comes from **AS 2**, is not independent of the integer m_0, which comes from **DIR 2**. In fact

(6) $$m_0 - 1 \leq -\mathrm{Re}(p_0) < m_0.$$

Indeed

$$\xi(s) = \Gamma(s) \sum_{k=1}^{\infty} a_k \lambda_k^{-s}$$

$$= \sum_{\mathrm{Re}(p) < \mathrm{Re}(q)} \int_0^1 b_p(t) t^{p+s} \frac{dt}{t} + \int_1^{\infty} [\theta(t) - P_q(t)] t^s \frac{dt}{t}.$$

The second integral on the right is holomorphic for

$$\mathrm{Re}(s) > -\mathrm{Re}(q).$$

The first integral on the right has its first pole at $s + p_0 = 0$, so at $-p_0$, whence $m_0 > -\mathrm{Re}(p_0)$. The first inequality in (6) follows from the minimality of m_0 since $\mathrm{Re}(p_0) \leq 0$.

The following condition on the sequence L requires that, beyond what is stated in the convergence condition **DIR 1**, the sequence λ_k approaches infinity in a sector.

DIR 3. There is a fixed $\epsilon > 0$ such that for all k sufficiently large, we have

$$-\frac{\pi}{2} + \epsilon \leq \arg(\lambda_k) \leq \frac{\pi}{2} - \epsilon.$$

Equivalently, there exists positive constants C_1 and C_2 such that for all k sufficiently large,

$$C_1|\lambda_k| \leq \operatorname{Re}(\lambda_k) \leq C_2|\lambda_k|.$$

Theorem 1.12. *Let (L, A) be sequences for which the associated Dirichlet series*

$$\sum_{k=1}^{\infty} \frac{a_k}{\lambda_k^s}$$

satisfies the three convergence conditions **DIR 1**, **DIR 2** *and* **DIR 3**. *Then the theta series*

$$\theta(t) = \sum_{k=1}^{\infty} a_k e^{-\lambda_k t}$$

satisfies **AS 1** *and* **AS 3**.

Proof. Let us first show how **AS 3** follows. Directly from **DIR 2** and **DIR 3** we have, for some constants c_1 and c_2, the inequalities

$$\left|\theta(t) - Q_N\theta(t)\right| \leq \left|\sum_{k=N}^{\infty} a_k e^{-\lambda_k t}\right|$$

$$\leq \sum_{k=N}^{\infty} |a_k| e^{-\operatorname{Re}(\lambda_k)t}$$

$$\leq c_1 \sum_{k=N}^{\infty} |\lambda_k|^{\sigma_0} e^{-c_2|\lambda_k|t}.$$

Note that for any $x \geq 0$, there is a constant $c = c(\sigma_0 + \sigma_1)$ such that

$$x^{\sigma_0 + \sigma_1} e^{-x} \leq c.$$

Let us apply this inequality to $x = c_2|\lambda_k|t$ and then sum for $k \geq N$ to obtain, for any $t > 0$,

$$\left|\theta(t) - Q_N\theta(t)\right| \leq c_1 \sum_{k=N}^{\infty} |\lambda_k|^{\sigma_0} e^{-c_2|\lambda_k|t}$$

(7)
$$\leq c_1 \cdot c(c_2 t)^{-\sigma_0 - \sigma_1} \sum_{k=N}^{\infty} |\lambda_k|^{-\sigma_1}.$$

Now if we let
$$\alpha = \sigma_0 + \sigma_1,$$

and
$$C = c_1 c(c_2)^{-\alpha} \sum_{k=1}^{\infty} |\lambda_k|^{-\sigma_1},$$

then (7) becomes

$$\left|\theta(t) - Q_N\theta(t)\right| \leq C/t^\alpha,$$

which establishes the asymptotic condition **AS 3**.

In order to establish **AS 1**, we need some preliminary calculations. First, note that by choosing $c_3 < c_2$, we can write

$$\left|\theta(t) - Q_N\theta(t)\right| \leq \left|\sum_{k=N}^{\infty} a_k e^{-\lambda_k t}\right| \leq c_1' \sum_{k=N}^{\infty} e^{-c_3|\lambda_k|t}.$$

Now let
$$C_N = \min\{c_3|\lambda_k|\} \quad \text{for } k \geq N,$$

and

(8)
$$t_0^{(N)} = \max\left\{\frac{\log|\lambda_k|^{\sigma_1}}{c_3|\lambda_k| - \frac{1}{2}C_N}\right\} \quad \text{for } k \geq N.$$

By **DIR 1** and **DIR 3** we have

$$\lim_{N \to \infty} C_N = \infty,$$

and since
$$\frac{1}{2}C_N \leq \frac{1}{2}c_3|\lambda_k| \quad \text{for} \quad k \geq N,$$
we can write
$$c_3|\lambda_k| - \frac{1}{2}C_N \geq c_3|\lambda_k| - \frac{1}{2}c_3|\lambda_k| = \frac{1}{2}c_3|\lambda_k| > 0.$$
Therefore,
(9) $$\frac{\log|\lambda_k|\sigma_1}{c_3|\lambda_k| - \frac{1}{2}C_N} \leq \frac{\log|\lambda_k|}{|\lambda_k|} \cdot \frac{2\sigma_1}{c_3} \quad \text{for} \quad k \geq N.$$

By combining (8) and (9) we conclude that
$$t_0^{(N)} \leq \max\left\{\frac{\log|\lambda_k|}{|\lambda_k|}\right\} \cdot \frac{2\sigma_1}{c_3} \quad \text{for} \quad k \geq N,$$
so, in particular,
$$\lim_{N \to \infty} t_0^{(N)} = 0.$$
Therefore, there exists $t_0 < \infty$ such that
$$t_0^{(N)} \leq t_0 \quad \text{for all} \quad N.$$
Note that for $t > t_0$ and any $k \geq N$ we have
$$(c_3|\lambda_k| - \frac{1}{2}C_N)t \geq (c_3|\lambda_k| - \frac{1}{2}C_N)t_0$$
$$\geq (c_3|\lambda_k| - \frac{1}{2}C_N)t_0^{(N)}$$
$$\geq \frac{c_3|\lambda_k| - \frac{1}{2}C_N}{c_3|\lambda_k| - \frac{1}{2}C_N} \log|\lambda_k|\sigma_1$$
$$\geq \log|\lambda_k|\sigma_1,$$
Therefore, for $t > t_0$,
$$e^{-(c_3|\lambda_k| - \frac{1}{2}C_N)t} \leq |\lambda_k|^{-\sigma_1},$$

which gives the bound

$$\left|\theta(t) - Q_N\theta(t)\right| \le c_1' e^{-\frac{1}{2}C_N t} \sum_{k=N}^{\infty} e^{-(c_3|\lambda_k| - \frac{1}{2}C_N)t}$$

$$\le c_1' e^{-\frac{1}{2}C_N t} \sum_{k=N}^{\infty} |\lambda_k|^{-\sigma_1},$$

which shows that **AS 1** holds, and completes the proof of the theorem. □

Remark 2. The convergence condition **DIR 1** is assumed as part of the asymptotic condition **AS 1**. Theorem 1.11 asserts that if the theta series

$$\theta(t) = \sum_{k=1}^{\infty} a_k e^{-\lambda_k t}$$

satisfies the asymptotic conditions **AS 1**, **AS 2** and **AS 3**, then

$$\xi(s) = \mathbf{M}\theta(s) = \Gamma(s) \sum_{k=1}^{\infty} \frac{a_k}{\lambda_k^s}.$$

Further, the sequences (L, A) satisfies the convergence condition **DIR 2(a)** and, by Corollary 1.10, $\xi(s)$ has a meromorphic continuation to all $s \in \mathbf{C}$. Theorem 1.12 states that if the two sequences (L, A) satisfy the three convergence conditions **DIR 1**, **DIR 2** and **DIR 3**, then the corresponding theta series satisfies the asymptotic conditions **AS 1** and **AS 3**. As previously stated, the condition **AS 2** is quite delicate and, in practice, is the most difficult to verify. In §7 we will show, under additional meromorphy and growth condition hypothesis on ξ, **AS 2** follows from the three convergence conditions **DIR 1**, **DIR 2** and **DIR 3**.

§2. Laurent expansion at $s=0$, Weierstrass product, and the Lerch formula

In this section, we consider the case when a_k is a non-negative integer for all k, which we define to be **the spectral case**. The corresponding Dirichlet series is then also called spectral, or **the spectral zeta function**. We develop some ideas of Voros [Vo 87], especially about his formula (4.1) which we formulate in a general context below as Theorem 2.1. Our arguments are somewhat different from those of Voros, who makes "no pretence of rigour". Basically, we want to make sense out of an infinite product of a sequence of complex numbers $L = \{\lambda_k\}$, counted with multiplicities $A = \{a_k\}$, which satisfies certain conditions. The results in §1 and the above definitions establish a line of investigation via a zeta function. In this section we indicate the line along Weierstrass products, and we show how the two approaches connect.

We recall the construction of a Weierstrass product. Let λ be a non-zero complex number, let m be an integer $\geq m_0$, and let

$$E_m(z,\lambda) = \left(1 - \frac{z}{\lambda}\right) \exp\left(\frac{z}{\lambda} + \frac{1}{2}\left(\frac{z}{\lambda}\right)^2 + \ldots + \frac{1}{m-1}\left(\frac{z}{\lambda}\right)^{m-1}\right),$$

or also

$$\log E_m(z,\lambda) = -\sum_{n=m}^{\infty} \frac{1}{n}\left(\frac{z}{\lambda}\right)^n.$$

We shall work under the assumptions **DIR 2** and **DIR 3**, and, for $m \geq m_0$, we define the **Weierstrass product**

$$D_{m,L}(z) = z^{a_0} \prod_{k=1}^{\infty} E_m(z, -\lambda_k)^{a_k}.$$

By the elementary theory of Weierstrass products, $D_{m,L}(z)$ is an entire function of strict order $\leq m$, that is

$$\log |D_{m,L}(z)| = O(|z|^m) \quad \text{for} \quad |z| \to \infty.$$

We use the adjective "strict" to avoid putting an ε in the exponent on the right hand side.

Let $D(z)$ be an entire function of strict order $\leq m$ with the same zeros as $D_{m,L}(z)$, counting multiplicities. Then there exists a polynomial $P_D(z)$ of degree $\leq m$ such that

$$D(z) = e^{P_D(z)} D_{m,L}(z).$$

We shall describe conditions that determine the polynomial P_D uniquely.

Suppose the three convergence conditions **DIR 1**, **DIR 2** and **DIR 3** are satisfied, so we have the spectral zeta function and the numbers

$$\zeta(n) = \sum_{k=1}^{\infty} \frac{a_k}{\lambda_k^n} \quad \text{for} \quad n \geq m.$$

We then have the power series of $D(z)$ at the origin coming from the expansion

$$(1) \quad \log D(z) = a_0 \log z + P_D(z) + \sum_{n=m}^{\infty} (-1)^{n-1} \zeta(n) \frac{z^n}{n}$$

$$= a_0 \log z + c_0 + \sum_{n=1}^{\infty} (-1)^{n-1} c_n \frac{z^n}{n}$$

where the coefficients c_n satisfy

$$c_n = \zeta(n) \quad \text{for} \quad n \geq m+1.$$

If $D(0) \neq 0$, meaning $a_0 = 0$, then $D(0) = e^{c_0}$, and we have

$$(2) \quad -\log[D(z)/D(0)] = \sum_{n=1}^{\infty} (-1)^n c_n \frac{z^n}{n}.$$

From this we see that the coefficients of $P_D(z)$ can be determined from the expansion (2) and the numbers

$$\zeta(m_0), \ldots, \zeta(m).$$

Since $D_{m_0,L}(z)$ is an entire function, the logarithmic derivative

$$\frac{d}{dz} \log D_{m_0,L}(z) = D'_{m_0,L}/D_{m_0,L}(z)$$

is meromorphic, and further derivatives $(d/dz)^r \log D_{m_0,L}(z)$ are also meromorphic, immediately expressible as sums that are of Mittag-Leffler type as follows. For an integer $r \geq 1$ let us define

$$T_{r,L}(z) = T_r(z) = \frac{(-1)^{r-1}}{\Gamma(r)} \left(\frac{d}{dz}\right)^r \log D_{m_0,L}(z).$$

We have trivially

$$\frac{d}{dz} T_r(z) = -r T_{r+1}(z).$$

Since the logarithmic derivative transforms products to sums, as part of the standard elementary theory of Weierstrass products we obtain for each $r \geq 1$ the expansion:

(3)
$$T_r(z) = \begin{cases} \dfrac{a_0}{z^r} + \sum_{k=1}^{\infty} \left[\dfrac{a_k}{(z+\lambda_k)^r} - \sum_{n=0}^{N} \binom{-r}{n} \dfrac{a_k z^n}{\lambda_k^{n+r}} \right]; & r < m_0 \\ \dfrac{a_0}{z^r} + \sum_{k=1}^{\infty} \dfrac{a_k}{(z+\lambda_k)^r}; & r \geq m_0 \end{cases}$$

where $N = m_0 - r - 1$. In (3) we have the usual binomial coefficient

$$\binom{-r}{n} = \frac{\Pi_n(r)}{n!}$$

where

$$\Pi_n(r) = (-1)^n r(r+1)\ldots(r+n-1).$$

Note that for $n > 1$ we have

$$\Pi_n(r) = -r \Pi_{n-1}(r+1).$$

Let us define $\Pi_n(r)$ for negative n through this recursive relation.

Because of the absolute convergence of the theta series $\theta(t)$, we have, by Theorem 1.8, for all z such that $\text{Re}(z + \lambda_k) > 0$ for all k, and any $r \geq m_0$,

$$T_r(z) = \frac{a_0}{z^r} + \sum_{k=1}^{\infty} \frac{a_k}{(z+\lambda_k)^r} = \frac{1}{\Gamma(r)} \int_0^{\infty} \theta(t) e^{-zt} t^r \frac{dt}{t}.$$

If we let

$$\theta_z(t) = e^{-zt}\theta(t) = a_0 e^{-zt} + \sum_{k=1}^{\infty} a_k e^{-(z+\lambda_k)t},$$

we can write $T_r(z)$ as a Mellin transform:

$$T_r(z) = \frac{1}{\Gamma(r)} \int_0^{\infty} \theta_z(t) t^r \frac{dt}{t}.$$

The convergence of the integral and sum is uniform for $\mathrm{Re}(z)$ sufficiently large.

We shall apply Theorem 1.8 with $f = \theta$ and $\mathbf{LM}f = \xi$ to obtain the following theorem, which we call the **Lerch formula**.

Theorem 2.1. *In the spectral case, assume that the theta function θ satisfies the asymptotic conditions* **AS 1**, **AS 2** *and* **AS 3** *with the natural sequence of exponential polynomials (4). Then there exists a unique polynomial $P_L(z)$ of degree $\leq m_0 - 1$ such that if we define*

$$D_L(z) = e^{P_L(z)} D_{m_0, L}(z),$$

then for all $z \in \mathbf{C}$ with $\mathrm{Re}(z)$ sufficiently large, we have

$$D_L(z) = \exp(-\mathrm{CT}_{s=0}\xi(s,z)).$$

Hence $\exp(-\mathrm{CT}_{s=0}\xi(s,z))$ has an analytic continuation to all $z \in \mathbf{C}$ to the entire function $D_L(z)$. In particular,

$$D_L'/D_L(z) = -\partial_z \mathrm{CT}_{s=0}\xi(s,z) = \mathrm{CT}_{s=1}\xi(s,z) = R_0(1;z).$$

Proof. If we count $\lambda_0 = 0$ with multiplicity a_0, we find for $\mathrm{Re}(s)$ large and $r \geq m_0$,

$$(\partial_z)^r \xi(s,z) = \Gamma(s) \sum_{k=0}^{\infty} (-s)(-s-1)\ldots(-s-r+1) \frac{a_k}{(z+\lambda_k)^{s+r}}$$

$$= \Gamma(s) \Pi_r(s) \sum_{k=0}^{\infty} \frac{a_k}{(z+\lambda_k)^{s+r}}$$

where, as before,
$$\Pi_r(s) = (-1)^r s(s+1)\ldots(s+r-1).$$

We now look at the constant term in a Laurent expansion at $s = 0$. If $r > m_0$, then the series

$$\sum_{k=0}^{\infty} \frac{a_k}{(z+\lambda_k)^r}$$

converges. At $s = 0$ the gamma function has a first order pole with residue 1. Note that the $\Pi_r(0) = 0$. So, Theorem 1.6 gives us a bound $m_0 - 1$ for the degrees of the polynomials in z occuring as coefficients in the negative powers in s of the Laurent expansion of $\xi(s,z)$. From this, we conclude that $(\partial_z)^r \xi(s,z)$ is holomorphic near $s = 0$. With all this, we can set $s = 0$ and obtain the equality

$$(\partial_z)^r \mathrm{CT}_{s=0} \xi(s,z) = (-1)^r \Gamma(r) \sum_{k=0}^{\infty} \frac{a_k}{(z+\lambda_k)^r},$$

from which we obtain, using (3),

$$(\partial_z)^r \left[\log D_{m_0}(z) + \mathrm{CT}_{s=0} \xi(s,z)\right] = 0.$$

Hence, there is a polynomial $P_L(z)$ of degree $\leq r - 1$ such that

$$\log D_{m_0, L}(z) + P_L(z) = -\mathrm{CT}_{s=0} \xi(s,z)$$

for $\mathrm{Re}(z)$ sufficiently large, which completes the proof of the theorem upon setting $r = m_0$. □

We call $D(z) = D_L(z)$ the **regularized product** associated to the sequence $L + z$. In particular, $D(0)$ is the regularized product of the sequence L.

Remark 1. Assume L can be written as the disjoint union of the sequences L' and L'' where L' satisfies the convergence conditions **DIR 1**, **DIR 2** and **DIR 3**, and $\theta_{L'}(t)$ satisfies the asymptotic conditions **AS 1**, **AS 2** and **AS 3**. Then L'' and $\theta_{L''}(t)$ necessarily satisfy these conditions and

$$\theta(t) = \theta_{L'}(t) + \theta_{L''}(t).$$

From this, we immediately have

$$\xi(s,z) = \xi_{L'}(s,z) + \xi_{L''}(s,z)$$

and

$$D_L(z) = D_{L'}(z) D_{L''}(z).$$

A particular example of such a decomposition is the case when L' is a finite subset of L, in which case $D_{L'}(z) = \prod[(z + \lambda_k)e^{\gamma}]$; see Remark 2 below.

It now becomes of interest to determine in some fashion the coefficients of the polynomial $P_L(z)$, and for this it suffices to determine $P_L^{(r)}(0)$. We define the **reduced sequence** L_0 to be the sequence which is the same as L except that we delete $\lambda_0 = 0$. Then

$$\theta_{L_0}(t) = \theta(t) - a_0$$

and, as discussed in Remark 1, we have

$$D_{m_0, L_0}(z) = z^{-a_0} D_{m_0, L}(z).$$

Theorem 2.2. *The polynomials P_L and P_{L_0}, as defined in Theorem 2.1, are equal. If we let ∂_2 be the partial derivative with respect to the second variable, then*

$$P_{L_0}^{(r)}(0) = -\partial_2^r \mathrm{CT}_{s=0} \xi(s,0)$$

for $0 \le r \le m_0 - 1$. That is

$$P_{L_0}(z) = -\sum_{k=0}^{m_0-1} \partial_2^k \mathrm{CT}_{s=0} \xi(s,0) \frac{z^k}{k!}.$$

Proof. From Theorem 2.1 let us write

$$P_{L_0}(z) = -\log D_{m_0, L_0}(z) - \mathrm{CT}_{s=0} \xi(s,0).$$

79

The canonical product of $D_{m_0,L_0}(z)$ is such that the Taylor series of $\log D_{m_0,L_0}(z)$ at the origin begins with powers of z which are at least z^{m_0}. Hence,

$$\left(\frac{\partial}{\partial z}\right)^r \log D_{m_0,L_0}(z)\bigg|_{z=0} = 0 \quad \text{for } 0 \leq r \leq m_0 - 1.$$

This proves the theorem. □

Remark 2. The normalization $D_L(z)$ that we have given is the most convenient one for the formalism we are developing. One may also define the **characteristic determinant $\mathbf{D}_L(z)$** by the condition

$$-\log \mathbf{D}_L(z) = R_{1,\zeta}(0;z) = \mathrm{CT}_{s=0}\left[s^{-1}\zeta(s,z)\right] = \mathrm{CT}_{s=0}\left[\frac{\xi(s,z)}{s\Gamma(s)}\right],$$

so $R_{1,\zeta}(0;z)$ is the coefficient of s in the Laurent expansion of $\zeta(s,z)$ at $s=0$. In the special case we have that

$$R_{1,\zeta}(0;z) = \zeta'(0,z).$$

Note that $\log \mathbf{D}_L(z)$ and $\log D_L(z)$ differ by an obvious polynomial in z coming from the Laurent expansion of $\Gamma(s)$ at $s = 0$. For example, if L is a finite sequence, then $\mathbf{D}_L(z) = \prod(z + \lambda_k)$. Also, let us record the formula

$$\zeta'(0,z) = R_{0,\xi}(0;z) + \gamma R_{-1,\xi}(0;z)$$
$$= \mathrm{CT}_{s=0}\xi(s,z) + \gamma R_{-1,\xi}(0;z),$$

which, again, holds only in the special case.

Example 1. Theorems 2.1 and 2.2 provide a general setting for the some classical formulas. Voros [Vo 87] gives examples, including the simplest case which concerns the sequence $\lambda_k = k$ and the gamma function (see his Example c). Theorems 2.1 and 2.2 contain as a special case the classical Lerch formula, which states that

$$\log \mathbf{D}(z) = -\zeta'_\mathbf{Q}(0,z) \quad \text{if } \mathbf{D}(z) = \sqrt{2\pi}/\Gamma(z)$$

and

$$\xi_\mathbf{Q}(s,z) = \Gamma(s)\sum_{n=0}^{\infty} \frac{1}{(z+n)^s} = \Gamma(s)\zeta_\mathbf{Q}(s,z).$$

Indeed, we have the immediate relation

$$\log \mathbf{D}(z) = \log D(z) + (\gamma + 1)z - \gamma/2$$

coming from the definitions and the expansion

$$\Gamma(s) = \frac{1}{s} - \gamma + O(s)$$

at $s = 0$.

Example 2. Let $\zeta_\mathbf{Q}$ be the Riemann zeta function, and let $\{\rho_k\}$ be the sequence of zeros in the critical strip with $\mathrm{Im}(\rho_k) > 0$. Let $\lambda'_k = \rho_k/i$. It is a corollary of a theorem of Cramér [Cr 19] that the theta function

$$\theta(t) = \sum a_k e^{-\lambda'_k t}$$

satisfies **AS 2**. The Lerch formula for the sequence $\{\lambda'_k\}$ (with multiplicities a_k) then specializes to a formula discovered by Deninger [De 92], Theorem 3.3. In [JoL 92b] we extend Cramér's theorem to a wide class of functions having an Euler product and functional equation, including the L-series of a number field (Hecke and Artin), L-functions arising in representation theory and modular forms, Selberg-type zeta funtions and L-functions for Riemann surfaces and certain higher dimensional manifolds, etc. The present section therefore applies to these functions as well, and thus the Lerch formula is valid for them. Note that Deninger's method did not give him the analytic continuation for the expressions in his formula, but such continuation occurs naturally in our approach.

§3. Expressions at $s = 1$.

In this section we return to general Dirichlet series, as in §1, meaning we are given sequences (L, A) whose Dirichlet series that satisfy the convergence conditions **DIR 1**, **DIR 2** and **DIR 3**, with associated theta series $\theta_{L,A} = \theta$ that satisfies the three asymptotic conditions **AS 1**, **AS 2**, and **AS 3**. With this, we will study the xi-function

$$\xi(s,z) = \mathbf{LM}\theta(s,z) = \int_0^\infty \theta(t) e^{-zt} t^s \frac{dt}{t}$$

near $s = 1$. Referring to **AS 2**, let us define the **principal part of the theta function** $P_0 \theta(t)$ by

(1) $$P_0 \theta(t) = \sum_{\mathrm{Re}(p)<0} b_p(t) t^p.$$

Recall that, by definition, $m(0) = \max \deg B_p$ for $\mathrm{Re}(p) = 0$ so

(2) $$\theta(t) - P_0 \theta(t) = \begin{cases} O(1) & \text{in the special case} \\ O(|\log t|^{m(0)}) & \text{in the general case,} \end{cases}$$

as t approaches zero.

Theorem 3.1. *Let C and N be related as in **AS 1**. Then the function*

$$\xi(s,z) - \Gamma(s) \sum_{k=0}^{N-1} \frac{a_k}{(\lambda_k + z)^s} - \int_0^1 P_0 \theta(t) e^{-zt} t^s \frac{dt}{t}$$

has a holomorphic continuation to the region

$$\{\mathrm{Re}(s) > 0\} \times \{\mathrm{Re}(z) > -C\}.$$

Proof. Let us write

$$\xi(s,z) - \Gamma(s) \sum_{k=0}^{N-1} \frac{a_k}{(\lambda_k + z)^s} - \int_0^1 P_0\theta(t) e^{-zt} t^{s+p} \frac{dt}{t}$$

$$= \int_0^\infty [\theta(t) - Q_N\theta(t)] e^{-zt} t^s \frac{dt}{t} - \int_0^1 P_0\theta(t) e^{-zt} t^s \frac{dt}{t}$$

(3) $$= \int_0^1 [\theta(t) - P_0\theta(t)] e^{-zt} t^s \frac{dt}{t}$$

(4) $$+ \int_1^\infty [\theta(t) - Q_N\theta(t)] e^{-zt} t^s \frac{dt}{t}$$

(5) $$- \int_0^1 Q_N\theta(t) e^{-zt} t^s \frac{dt}{t}.$$

Using **AS 2**, we have that the integral in (3) is holomorphic for all $z \in \mathbf{C}$ and $\mathrm{Re}(s) > 0$, as is the integral in (5), and the integral in (4) is holomorphic for all $s \in \mathbf{C}$ and $\mathrm{Re}(z) > -C$. Combining this, the stated claim has been proved. □

Corollary 3.2. *The function*

$$\xi(s,z) - \int_0^1 P_0\theta(t) e^{-zt} t^s \frac{dt}{t}$$

is meromorphic at $s = 1$ for all z with singularities that are simple poles at $z = -\lambda_k$. Also, the residue at $z = -\lambda_k$ is equal to a_k.

Proof. Immediate from the proof of Theorem 3.1 by taking C, consequently N, sufficiently large. □

We now study the **singular term**

$$\int_0^1 P_0\theta(t) e^{-zt} t^s \frac{dt}{t} = \sum_{\mathrm{Re}(p)<0} \int_0^1 b_p(t) e^{-zt} t^s \frac{dt}{t}$$

83

that appears in Corollary 3.2. To do so, we expand e^{-zt} in a power series and apply the following lemma.

Lemma 3.3. *Given p, there is an entire function $h_p(z)$ such that as s approaches 1, such that:*

(a) *Special Case:*

$$\int_0^1 b_p e^{-zt} t^{s+p} \frac{dt}{t} = \sum_{k=0}^{\infty} \frac{(-z)^k}{k!} b_p \cdot \frac{1}{s+p+k}$$

$$= \begin{cases} \frac{(-z)^{-p-1}}{(-p-1)!} b_p \cdot \frac{1}{s-1} + h_p(z) + O(s-1), & p \in \mathbf{Z}_{<0} \\ h_p(z) + O(s-1), & p \notin \mathbf{Z}_{<0}. \end{cases}$$

(b) *General Case:*

$$\int_0^1 b_p(t) e^{-zt} t^{s+p} \frac{dt}{t} = \sum_{k=0}^{\infty} \frac{(-z)^k}{k!} B_p(\partial_s) \left[\frac{1}{s+p+k} \right]$$

$$= \begin{cases} \frac{(-z)^{-p-1}}{(-p-1)!} B_p(\partial_s) \left[\frac{1}{s-1} \right] + h_p(z) + O(s-1), & p \in \mathbf{Z}_{<0} \\ h_p(z) + O(s-1), & p \notin \mathbf{Z}_{<0}. \end{cases}$$

From Lemma 3.3, we can assert the existence of an entire function $h(z)$ such that the singular term can be written as

$$\int_0^1 P_0\theta(t) e^{-zt} t^s \frac{dt}{t} = \sum_{p+k=-1} \frac{(-z)^k}{k!} B_p(\partial_s) \left[\frac{1}{s-1} \right] + h(z) + O(s-1).$$

The following theorem, which is the main result of this section, then follows directly from Theorem 3.1 and Lemma 3.3.

Theorem 3.4. *Near $s = 1$, the Hurwitz xi function $\xi(s, z)$ has the expansion*

$$\xi(s,z) = \frac{R_{-n(1)-1}(1;z)}{(s-1)^{n(1)+1}} + \cdots + \frac{R_{-1}(1;z)}{(s-1)} + R_0(1;z) + O(s-1),$$

where:

(a) *For $j < 0$, $R_j(1;z)$ is a polynomial of degree $< -\mathrm{Re}(p_0)$; in fact, the polar part of $\xi(s,z)$ near $s = 1$ is expressed by*

$$\frac{R_{-n(1)-1}(1;z)}{(s-1)^{n(1)+1}} + \cdots + \frac{R_{-1}(1;z)}{(s-1)} = \sum_{p+k=-1} \frac{(-z)^k}{k!} B_p(\partial_s) \left[\frac{1}{s-1}\right];$$

(b) $R_0(1;z) = \mathrm{CT}_{s=1}\xi(s,z)$ *is a meromorphic function in z for all $z \in \mathbf{C}$ whose singularities are simple poles at $z = -\lambda_k$ with residue equal to a_k. Furthermore,*

$$\mathrm{CT}_{s=1}\xi(s,z) = -\partial_z \mathrm{CT}_{s=0}\xi(s,z).$$

In the special case, the expansion of $\xi(s,z)$ near $s = 1$ simplifies to

$$\xi(s,z) = \frac{R_{-1}(1;z)}{s-1} + R_0(1;z) + O(s-1),$$

with

$$R_{-1}(1;z) = \sum_{p+k=-1} b_p \frac{(-z)^k}{k!}.$$

Let (L, A) be sequences of complex numbers that satisfy the three convergence conditions **DIR 1**, **DIR 2** and **DIR 3** and such that the associated theta function

$$\theta_{L,A}(t) = \theta(t) = \sum_{k=1}^{\infty} a_k e^{-\lambda_k t}$$

satisfies the three asymptotic conditions **AS 1**, **AS 2** and **AS 3**. We define the **regularized harmonic series** $R(z)$ associated to (L, A) to be

$$R_{L,A}(z) = R(z) = \mathrm{CT}_{s=1}\xi(s,z) = -\partial_z \mathrm{CT}_{s=0}\xi(s,z).$$

85

Remark 1. Theorem 3.4 states that the regularized harmonic series associated to (L, A) is a meromorphic function in z whose singularities are simple poles at $z = -\lambda_k$ and corresponding residues equal to a_k. Further, by **DIR 2** and **DIR 3**, Theorem 1.8 and Theorem 1.12, we have, for any integer $n \geq m_0$, the expression

$$\partial_z^n R(z) = (-1)^n \mathrm{CT}_{s=1} \xi(s+n, z) = (-1)^n \Gamma(n) \sum_{k=1}^{\infty} \frac{a_k}{(z+\lambda_k)^n}.$$

If all numbers a_k are integers, then one can assert the existence of a meromorphic function $D(z)$, unique up to constant factor, which satisfies the relation

(6) $$D'/D(z) = R(z).$$

The Spectral Case. In this case, with $a_k \in \mathbf{Z}_{\geq 0}$ for all k, Theorem 3.4 asserts the existence of an entire function $D(z)$, unique up to constant factor, such that

(7) $$D'/D(z) = \mathrm{CT}_{s=1} \xi(s, z).$$

Further, we have, by the Lerch formula (Theorem 2.1),

The Basic Identity:

$$R(z) = \mathrm{CT}_{s=1} \xi_L(s, z) = -\partial_z \mathrm{CT}_{s=0} \xi_L(s, z) = D'_L/D_L(z).$$

We shall normalize the constant factor in (7) so that

$$D(z) = D_L(z).$$

Next we shall give another type of expression for the singular term, which also gives an expression for $\mathrm{CT}_{s=1} \xi_L(s, z)$ leading into the Gauss formula of the next section.

Consider the function

(8) $$F_q(s, z) = \int_0^{\infty} [\theta(t) - P_q \theta(t)] e^{-zt} t^s \frac{dt}{t}$$

and set $F = F_0$, which is especially important among the Laplace-Mellin transforms of the functions $\theta - P_q \theta$. Following the results and techniques in §1, we shall study an analytic continuation of $F_q(s, z)$ and then compare $F_q(s, z)$ with $\xi(s, z)$.

Theorem 3.5. *The function $F_q(s,z)$ has a meromorphic continuation to the region*
$$\{\text{Re}(s) > -\text{Re}(q)\} \times \mathbf{U}_L.$$

Proof. Note that we can write (8) as

$$F_q(s,z) = \qquad\qquad\qquad\qquad\text{region of meromorphy}$$

(9) $\quad\displaystyle\int_0^1 [\theta(t) - P_q\theta(t)] \, e^{-zt} t^s \frac{dt}{t} \qquad \text{Re}(s) > -\text{Re}(q), \text{ all } z$

(10) $\quad + \displaystyle\int_1^\infty [\theta(t) - Q_N\theta(t)] \, e^{-zt} t^s \frac{dt}{t} \qquad \text{all } s, \text{Re}(z) > -C.$

(11) $\quad + \displaystyle\int_1^\infty [Q_N\theta(t) - P_q\theta(t)] \, e^{-zt} t^s \frac{dt}{t} \qquad \text{all } s, z \in \mathbf{U}_L.$

For (9), the assertion of meromorphy on the right follows from condition **AS 2** and Lemma 1.3; for (10) the assertion follows by Lemma 1.4, picking N and C as in **AS 1**. As for (11), the integral is a sum of integrals of elementary functions which we now recall. First we have

(12) $\quad\displaystyle\int_1^\infty e^{-\lambda_k t} e^{-zt} t^s \frac{dt}{t} = \frac{\Gamma(s)}{(z+\lambda_k)^s} - \int_0^1 e^{-\lambda_k t} e^{-zt} t^s \frac{dt}{t},$

so that

$$\int_1^\infty Q_N\theta(t) e^{-zt} t^s \frac{dt}{t} = \sum_{k=1}^\infty \left[\frac{\Gamma(s)}{(z+\lambda_k)^s} - \int_0^1 e^{-\lambda_k t} e^{-zt} t^s \frac{dt}{t} \right].$$

For the special case, when b_p is constant, we have

$$\int_1^\infty b_p e^{-zt} t^{s+p} \frac{dt}{t} = b_p \frac{\Gamma(s+p)}{z^{s+p}} - b_p \int_0^1 e^{-zt} t^{s+p} \frac{dt}{t}$$

(13) $\qquad\qquad = b_p \dfrac{\Gamma(s+p)}{z^{s+p}} - \displaystyle\sum_{k=0}^\infty \frac{(-z)^k}{k!} b_p \cdot \frac{1}{s+p+k},$

and for the general case we have the expansion replacing b_p by $B_p(\partial_s)$, namely:

$$\int_1^\infty b_p(t) e^{-zt} t^{s+p} \frac{dt}{t} = B_p(\partial_s) \left[\frac{\Gamma(s+p)}{z^{s+p}}\right] - \int_0^1 b_p(t) e^{-zt} t^{s+p} \frac{dt}{t}$$

(14) $$= B_p(\partial_s) \left[\frac{\Gamma(s+p)}{z^{s+p}}\right] - \sum_{k=0}^\infty \frac{(-z)^k}{k!} B_p(\partial_s) \left[\frac{1}{s+p+k}\right].$$

From this, the conclusion stated above follows since the integral in (11) is simply a sum of terms of the form given in (12), (13) and (14). □

Theorem 3.6. *For any q and for all (s,z) in the region of meromorphy, we have:*

a) *In the special case,*

$$\xi(s,z) = F_q(s,z) + \sum_{\mathrm{Re}(p)<\mathrm{Re}(q)} b_p \frac{\Gamma(s+p)}{z^{s+p}};$$

b) *in the general case,*

$$\xi(s,z) = F_q(s,z) + \sum_{\mathrm{Re}(p)<\mathrm{Re}(q)} B_p(\partial_s) \left[\frac{\Gamma(s+p)}{z^{s+p}}\right].$$

Proof. For $\mathrm{Re}(s)$ large, one can interchange the sum and integral in the definition of $F_q(s,z)$ and use that

$$\int_0^\infty P_q \theta_L(t) e^{-zt} t^s \frac{dt}{t} = \sum_{\mathrm{Re}(p)<\mathrm{Re}(q)} B_p(\partial_s) \left[\frac{\Gamma(s+p)}{z^{s+p}}\right].$$

The rest follows from the definition of $\xi(s,z)$ and analytic continuation. □

By taking $q = 0$ in Theorem 3.6 and writing $F_0(s,z) = F(s,z)$, we obtain the following corollary.

Corollary 3.7. *The constant term of $\xi(s,z)$ at $s=1$ is given by*

$$\mathrm{CT}_{s=1}\xi(s,z) = F(1,z) + \sum_{\mathrm{Re}(p)<0} \mathrm{CT}_{s=1} B_p(\partial_s)\left[\frac{\Gamma(s+p)}{z^{s+p}}\right]$$

where

$$F(1,z) = \int_0^\infty [\theta(t) - P_0\theta(t)]\, e^{-zt} dt.$$

The constant term on the right is obtained simply by multiplying the Laurent expansions of $\Gamma(s+p)$ and z^{-s-p} and is a universal expression, meaning an expression that depends solely only on B_p for $\mathrm{Re}(p) < 0$. A direct calculation shows that there exists a polynomial $\tilde{B}_p$ of degree at most $\deg B_p + 1$ such that

$$\mathrm{CT}_{s=1} B_p(\partial_s)\left[\frac{\Gamma(s+p)}{z^{s+p}}\right] = z^{-p-1}\tilde{B}_p(\log z).$$

The possible pole of $\Gamma(s+p)$ at $s=1$ accounts for the possibility of $\deg \tilde{B}_p$ exceeding $\deg B_p$. For convenience of the reader, we present these calculations explicitly in the special case.

Let $p \in \mathbf{C}$ and write

$$\Gamma(s+p) = \sum_{k=-1}^\infty \frac{c_k(p+1)}{(s-1)^k}$$

and

$$z^{-p-s} = z^{-p-1} e^{(s-1)(-\log z)} = z^{-p-1} \sum_{l=0}^\infty \frac{(-\log z)^l}{l!}(s-1)^l.$$

Then

$$\frac{\Gamma(s+p)}{z^{s+p}} = z^{-1-p} \sum_{n=-1}^\infty \left[\sum_{k+l=n} \frac{c_k(p+1)(-\log z)^l}{l!}\right](s-1)^n,$$

from which we obtain the equation

$$\mathrm{CT}_{s=1}\left[\frac{\Gamma(s+p)}{z^{s+p}}\right] = [c_0(p+1) - c_{-1}(p+1)\log z]z^{-1-p}.$$

Note that if $p \notin \mathbf{Z}_{<0}$, then $c_{-1}(p+1) = 0$ and $c_0(p+1) = \Gamma(p+1)$.

Corollary 3.8. In the special case, $R_0(1;z)$ has the integral expression

$$CT_{s=1}\xi(s,z) = F(1,z) + \sum_{Re(p)<0} b_p \left[c_0(p+1) - c_{-1}(p+1) \log z \right] z^{-1-p}$$

where

$$F(1,z) = \int_0^\infty [\theta(t) - P_0\theta(t)] e^{-zt} dt.$$

Example 1. Suppose we are in the spectral case and also that $\theta(t)$ is such that the principal part $P_0\theta(t)$ is simply b_{-1}/t. Then the equations in Theorem 3.6 become

$$\xi(s,z) = \frac{b_{-1}}{s-1} + CT_{s=1}\xi(s,z) + O(s-1)$$

where

$$CT_{s=1}\xi(s,z) = \int_0^\infty \left[\theta(t) - \frac{b_{-1}}{t} \right] e^{-zt} dt - b_{-1} \log z.$$

Using that

$$\int_0^\infty [e^{-zt} - e^{-t}] \frac{dt}{t} = -\log z,$$

we can write

$$CT_{s=1}\xi(s,z) = \int_0^\infty \left[\theta(t)e^{-zt} - \frac{b_{-1}e^{-t}}{t} \right] dt.$$

In this case, the integral expression for $R_0(1;z)$ reminds one of the Gauss formula for the logarithmic derivative of the gamma function, which will be studied in the next section.

Example 2. The regularized harmonic series is simply related to the classical Selberg zeta function $Z_X(s)$ associated to a compact hyperbolic Riemann surface X. For convenience, let us briefly recall the definition of the Selberg zeta function.

Let $\theta_X(t)$ be the trace of the heat kernel corresponding to the hyperbolic Laplacian that acts on C^∞ functions on X; hence, $\theta_X(t)$ is the theta function associated to the sequence L of eigenvalues of the Laplacian. The sequence A counts multiplicities of the eigenvalues. It is well known that

$$P_0 \theta(t) = \frac{b_{-1}}{t},$$

where b_{-1} is a constant that depends solely on the genus g of X, namely

$$b_{-1} = 2\pi(2g - 2).$$

The logarithmic derivative of the Selberg zeta function is defined by the equation

$$Z'_X/Z_X(s) = (2s-1) \int_0^\infty [\theta_X(t) - b_{-1}k(t)] e^{-s(s-1)t} dt$$

where $k(t)$ is a universal function, independent of X (see [Sa 87]). From Theorem 3.6, the definition of the Selberg zeta function, and the example above, we have the relation

$$(2s-1)D'_L/D_L(s(s-1)) - Z'_X/Z_X(s)$$

$$= (2s-1)b_{-1} \left(\int_0^\infty \left[k(t) - \frac{1}{t} \right] e^{-s(s-1)t} dt - \log(s(s-1)) \right)$$

$$= (2s-1)b_{-1} \left(\int_0^\infty \left[k(t) e^{-s(s-1)t} - \frac{e^{-t}}{t} \right] dt \right).$$

This relation was used in [JoL 92d] to prove that the Selberg zeta function satisifies Artin's formalism ([La 70]).

§4. Gauss Formula

Next we show that a classical formula of Gauss for Γ'/Γ can be formulated and proved more generally for the regularized harmonic series. As before, we let $P_0\theta$ denote the principal part of an asymptotic expansion at 0. Define

$$\theta_z(t) = e^{-zt}\theta(t).$$

If $\theta(t)$ satisfies the three asymptotic conditions, then it is immediate that $e^{-zt}\theta(t)$ also satisfies the three asymptotic conditions. As before, if the principal part of the theta function is

$$P_0\theta(t) = \sum_{\mathrm{Re}(p)<0} b_p(t)t^p,$$

then

(1) $$P_0\theta_z(t) = \sum_{\mathrm{Re}(p)+k<0} \frac{(-z)^k}{k!} b_p(t)t^{p+k}.$$

Note that for any complex w, we have

$$\xi_{L+z,A}(s,w) = \xi_z(s,w) = \xi(s, z+w).$$

Recall from Theorem 1.8 that

$$-\frac{\partial}{\partial z}\xi(s, z+w) = \xi(s+1, z+w),$$

so, in particular, we have

$$R(z+w) = -\partial_z \mathrm{CT}_{s=0}\xi(s, z+w) = \mathrm{CT}_{s=1}\xi(s, z+w).$$

In the spectral case, we have, by the Lerch formula,

$$R(z+w) = D'_L/D_L(z+w).$$

Finally, recall that $\mathbf{C}\langle T\rangle$ is the algebra of polynomials in T^p with arbitrary complex powers $p \in \mathbf{C}$.

With all this, we can follow the development leading to Corollary 3.8 and state the following theorem, which we call the **general Gauss formula**.

Theorem 4.1. *There is a polynomial $S_w(z)$ of degree in z*

$$\deg_z S_w < -\mathrm{Re}(p_0)$$

with coefficients in $\mathbf{C}[\log w]\langle w\rangle$ such that for any $w \in \mathbf{C}$ with $\mathrm{Re}(w) > 0$ and $\mathrm{Re}(w) > \max_k\{-\mathrm{Re}(\lambda_k + z)\}$,

$$R(z+w) = \int_0^\infty [\theta_z(t) - P_0\theta_z(t)]\, e^{-wt} dt + S_w(z)$$

In the special case, $S_w(z) \in \mathbf{C}\langle w\rangle[z] + \mathbf{C}\langle w\rangle \log w[z]$.

Proof. From Theorem 1.8, we have for sufficiently large $\mathrm{Re}(s)$ and $\mathrm{Re}(z)$, while viewing w as fixed, the equalities

$$-\partial_z \xi(s, z+w) = \xi(s+1, z+w)$$
$$= \xi_z(s+1, w)$$
$$= \int_0^\infty [\theta_z(t) - P_0\theta_z(t)]\, e^{-wt} t^{s+1} \frac{dt}{t}$$
$$+ \int_0^\infty P_0\theta_z(t) e^{-wt} t^{s+1} \frac{dt}{t}.$$

To compute the constant term in the expansion at $s = 0$, we can substitute $s = 0$ in the Laplace-Mellin integral of $\theta_z - P_0\theta_z$, thus getting the desired integral on the right hand side of the formula in the theorem. As for the integral of $P_0\theta_z$ we can use the expression (1) for the principal part of θ_z to get

$$\int_0^\infty P_0\theta_z(t) e^{-wt} t^s dt = \sum_{\mathrm{Re}(p)+k<0} \frac{(-z)^k}{k!} B_p(\partial_s) \left[\frac{\Gamma(s+p+k+1)}{w^{s+p+k+1}}\right].$$

In fact, we obtain an explicit formula for $S_w(z)$, namely

$$S_w(z) = \sum_{\mathrm{Re}(p)+k<0} \frac{(-z)^k}{k!} \mathrm{CT}_{s=0} B_p(\partial_s) \left[\frac{\Gamma(s+p+k+1)}{w^{s+p+k+1}}\right].$$

This expression shows that $S_w(z)$ is a polynomial in z, of degree $< -\mathrm{Re}(p_0)$, with coefficients in $\mathbf{C}\langle w\rangle[\log w]$. Recall that in the special case, all B_p are constants, and, for any p, the expression

$$\mathrm{CT}_{s=0} B_p(\partial_s) \left[\frac{\Gamma(s+p+k+1)}{w^{s+p+k+1}}\right]$$

lies in $\mathbf{C}\langle w\rangle + \mathbf{C}\langle w\rangle \log w$. With this, the proof of the theorem is complete. $\square$

Remark. A direct calculation shows that there exists a polynomial B_p^* of degree at most $\deg B_p + 1$ such that

$$\mathrm{CT}_{s=0} B_p(\partial_s) \left[\frac{\Gamma(s+p)}{z^{s+p}}\right] = z^{-p} B_p^*(\log z).$$

The possible pole of $\Gamma(s+p)$ at $s = 0$ accounts for the possibility of $\deg B_p^*$ exceeding $\deg B_p$. Using the relation

$$\frac{\Gamma(s+p+k+1)}{w^{s+p+k+1}} = \frac{(s)\cdots(s+k)\Gamma(s+p)}{w^{k+1}\cdot w^{s+p}}$$

one can hope to express $S_w(z)$ in terms of the polynomials B_p^*. However, even in the special case, such an expression is quite involved.

Corollary 4.2. *For fixed $w \in \mathbf{C}$ with $\mathrm{Re}(w) > 0$ and*

$$\mathrm{Re}(w) > \max_k\{-\mathrm{Re}(\lambda_k)\},$$

and for $z \in \mathbf{C}$ with $\mathrm{Re}(z)$ sufficiently large, the integral

$$I_w(z) = \int_0^\infty [\theta_z(t) - P_0\theta_z(t)]\, e^{-wt}\, dt,$$

as a function of z, is holomorphic in z and has a meromorphic continuation to all $z \in \mathbf{C}$ with poles at the points $\lambda_k + w$ in $L + w$ and corresponding residues equal to a_k.

Proof. Immediate from Theorem 3.4(b) and Theorem 4.1. $\square$

In the case when $a_k \in \mathbf{Z}$ for all k, this allows us to write the integral $I_w(z)$ from Corollary 4.2 as

$$I_w = H'_w/H_w$$

where H_w is a meromorphic function on $\mathbf{C}$, uniquely defined up to a constant factor. Following Theorem 2.1, we define $S_w^\#(z)$ to be the integral of $S_w(z)$ with zero constant term, so $S_w(z) = \partial_z S_w^\#(z)$. Define $H_w(z)$ by the relation

$$D(z+w) = e^{S_w^\#(z)} H_w(z).$$

Then $H_w(z)$ is the unique meromorphic function of z such that

$$H'_w/H_w(z) = I_w(z)$$

and

$$H_w(0) = D(w).$$

Furthermore, we have

$$D'_L/D_L(z+w) = H'_w(z)/H_w(z) + S_w(z) = I_w(z) + S_w(z).$$

In the spectral case, meaning $a_k \in \mathbf{Z}_{\geq 0}$ for all k, $H_w(z)$ is entire. The function $I_w(z)$ will be studied in §4 of [JoL 92c], from a Fourier theoretic point of view.

Example 1. We show here how the classical Gauss formula is a special case of Theorem 4.1. Let

$$L = \{n \in \mathbf{Z}_{\geq 0}\}$$

and

$$a(n) = 1 \quad \text{for all } n.$$

The theta function can be written as

$$\theta_z(t) = \sum_{n=0}^{\infty} e^{-(z+n)t} = \frac{e^{-zt}}{1 - e^{-t}}.$$

The principal part is simply

$$P_0 \theta_z(t) = \frac{1}{t},$$

so $p_0 = -1$ and $p_1 = 0$. This allows us to compute the polynomial $S_w(z)$ and obtain the equation

$$S_w(z) = \mathrm{CT}_{s=0}\left[\frac{\Gamma(s)}{w^s}\right] = -\gamma - \log w.$$

Therefore, by combining the Example from §2 and Theorem 4.1, we have

$$D'_L/D_L(z+w) = -\Gamma'/\Gamma(z+w) - \gamma$$
$$= \int_0^\infty \left[\frac{e^{-zt}}{1-e^{-t}} - \frac{1}{t}\right] e^{-wt} dt - \gamma - \log w.$$

Now set $w = 1$ to get

$$-\Gamma'/\Gamma(z+1) = \int_0^\infty \left[\frac{e^{-(z+1)t}}{1-e^{-t}} - \frac{e^{-t}}{t}\right] dt,$$

which is the classical Gauss formula.

Example 2. Let (L, A) be the sequences of eigenvalues and multiplicities associated to the Laplacian that acts on smooth sections of a power of the canonical sheaf over a compact hyperbolic Riemann surface. Then one can combine Example 2 from §3, Theorem 4.1 and the Lerch formula to establish the main theorem of [DP 86] and [Sa 87], without using the Selberg trace formula. In brief, this theorem states that the Selberg zeta function $Z_X(s)$ is expressible, up to universal gamma-like functions, in terms of the regularized product $D_L(s(s-1))$.

§5. Stirling Formula

As in previous sections, we work with sequences (L, A) that satisfy the three convergence conditions **DIR 1**, **DIR 2** and **DIR 3**, and such that the associated theta series satisfies the three asymptotic conditions **AS 1**, **AS 2** and **AS 3**. With this, we consider the asymptotic behavior of certain functions when $\text{Re}(z) \to \infty$, by which we mean that we allow $\text{Re}(z) \to \infty$ in some sector in the right half plane. The point of such a restriction is that in such a sector, $\text{Re}(z)$ and $|z|$ have the same order of magnitude, asymptotically. Specifically, we shall determine the asymptotics of

$$-\text{CT}_{s=0}\xi(s, z) \quad \text{as} \quad x = \text{Re}(z) \to \infty.$$

These asymptotics apply in the spectral case to

$$-\log D_L(z) = \text{CT}_{s=0}\xi(s, z),$$

and, therefore, can be viewed as a generalization of the classical Stirling formula.

To begin, we will study the asymptotics of

$$\xi(s, z) = \int_0^\infty \theta(t) e^{-zt} t^s \frac{dt}{t} \quad \text{for } \text{Re}(z) \to \infty.$$

Fix $\text{Re}(q) > 0$ and, as before, let $P_q \theta$ be as in **AS 2**. Let us write

$$\xi(s, z) = J_1(s, z) + J_2(s, z) + J_3(s, z)$$

where

(1) $$J_1(s, z) = \int_0^1 (\theta(t) - P_q\theta(t)) e^{-zt} t^s \frac{dt}{t},$$

(2) $$J_2(s, z) = \int_1^\infty \theta(t) e^{-zt} t^s \frac{dt}{t},$$

(3) $$J_3(s, z) = \int_0^1 P_q\theta(t) e^{-zt} t^s \frac{dt}{t}.$$

Recall that

$$m(q) = \max \deg B_p \quad \text{for} \quad \text{Re}(p) = \text{Re}(q).$$

The terms in (1) and (2) are handled in the two following lemmas.

Lemma 5.1. *Let* $x = \text{Re}(z)$ *and fix* q *with* $\text{Re}(q) > 0$. *Then*

$$J_1(0, z) = O(x^{-\text{Re}(q)}(\log x)^{m(q)}) \text{ for } x \to \infty.$$

Proof. Recall that

$$\theta(t) - P_q\theta(t) = O(t^{\text{Re}(q)}|\log t|^{m(q)}) \text{ for } t \to 0.$$

Since $\text{Re}(q) > 0$, $J_1(s, z)$ is holomorphic at $s = 0$ and

$$J_1(0, z) = \int_0^1 (\theta(t) - P_q\theta(t))e^{-zt}\frac{dt}{t}.$$

For any complex number s_0, we have the power series expansion

$$\frac{\Gamma(s)}{x^s} = \left(\sum_{k=-1}^{\infty} c_k(s_0)(s-s_0)^k\right) \cdot \left(x^{-s_0} \sum_{l=0}^{\infty} \frac{(-\log x)^l}{l!}(s-s_0)^l\right)$$

(4) $$= x^{-s_0} \sum_{n=-1}^{\infty} \left[\sum_{k+l=n} \frac{c_k(s_0)(-\log x)^l}{l!}\right](s-s_0)^n,$$

from which we have, by **AS 2**,

$$|J_1(0, z)| \ll \int_0^1 e^{-xt} t^{\text{Re}(q)}|\log t|^{m(q)}\frac{dt}{t}$$

$$\leq CT_{s=\text{Re}(q)}\left[(\partial_s)^{m(q)}\frac{\Gamma(s)}{x^s}\right] + \int_1^{\infty} e^{-xt}t^{\text{Re}(q)}(\log t)^{m(q)}\frac{dt}{t}$$

$$= x^{-\text{Re}(q)}\left[\sum_{k+l=m(q)} \frac{c_k(q)(-\log x)^l}{l!} \cdot n!\right] + O(e^{-x}/x)$$

as $x \to \infty$. This completes the proof of the lemma. $\square$

Lemma 5.2. *We have*
$$J_2(0, z) = O(e^{-x}/x) \quad \text{for} \quad x \to \infty.$$

Proof. From **AS 1**, $J_2(s, z)$ is holomorphic at $s = 0$ and
$$J_2(0, z) = \int_1^\infty \theta(t) e^{-zt} \frac{dt}{t},$$
from which the lemma follows, since $\theta(t) = O(e^{ct})$ for some $c > 0$ and so, for x sufficiently large,
$$|J_2(0, z)| \leq K \int_1^\infty e^{-(x-c)t} dt = \frac{K}{x - c} e^{-(x-c)},$$
yielding the stated estimate. □

As for $J_3(s, z)$, recall that
$$J_3(s, z) = \sum_{\text{Re}(p) < \text{Re}(q)} \left[\int_0^1 b_p(t) e^{-zt} t^{s+p} \frac{dt}{t} \right].$$

Lemma 5.3. *We have*
$$J_3(0, z) = \sum_{\text{Re}(p) < \text{Re}(q)} \text{CT}_{s=0} B_p(\partial_s) \left[\frac{\Gamma(s+p)}{z^{s+p}} \right]$$
$$+ O(e^{-x}/x) \quad \text{for } x \to \infty.$$

Proof. Simply write
$$\int_0^1 e^{-zt} b_p(t) t^{s+p} \frac{dt}{t} = B_p(\partial_s) \left[\frac{\Gamma(s+p)}{z^{s+p}} \right] - \int_1^\infty e^{-zt} b_p(t) t^{s+p} \frac{dt}{t}.$$
and as in Lemma 5.2, note that, for any s,
$$\int_1^\infty e^{-zt} b_p(t) t^{s+p} \frac{dt}{t} = O(e^{-x}/x) \quad \text{for } x \to \infty.$$

□

Combining the above three lemmas, we have established the following theorem, which we refer to as the **generalized Stirling's formula**.

Theorem 5.4. *Suppose q is such that $\operatorname{Re}(q) > 0$. Let*

$$\mathbf{B}_q(z) = \sum_{\operatorname{Re}(p) < \operatorname{Re}(q)} \operatorname{CT}_{s=0} B_p(\partial_s) \left[\frac{\Gamma(s+p)}{z^{s+p}} \right] \in \mathbf{C}\langle z \rangle [\log z].$$

Then for $x \to \infty$,

$$\operatorname{CT}_{s=0} \xi(s, z) = \mathbf{B}_q(z) + O(x^{-\operatorname{Re}(q)} (\log x)^{m(q)}).$$

It is important to note that Theorem 5.4 shows that the asymptotics of $\operatorname{CT}_{s=0} \xi(s, z)$ as $\operatorname{Re}(z) \to \infty$ are governed by the asymptotics of $\theta(t)$ as $t \to 0$. Also, in the spectral case, the Lerch formula (Theorem 2.1) applies to give

$$-\log D_L(z) = \operatorname{CT}_{s=0} \xi_L(s, z),$$

and, hence, Theorem 5.4 determines the asymptotics of the regularized product $\log D_L(z)$ as $\operatorname{Re}(z) \to \infty$ and, consequently, the asymptotics of the the characteristic determinant

$$\log \mathbf{D}_L(z) \quad \text{as } \operatorname{Re}(z) \to \infty.$$

Remark 1. In Remark 1 of §4 we defined the polynomials B_p^* by the formula

$$\operatorname{CT}_{s=0} B_p(\dot{\partial}_s) \left[\frac{\Gamma(s+p)}{z^{s+p}} \right] = z^{-p} B_p^*(\log z).$$

Using these polynomials, one can write

$$\mathbf{B}_q(z) = \sum_{\operatorname{Re}(p) < \operatorname{Re}(q)} z^{-p} B_p^*(\log z),$$

so Theorem 5.4 becomes the statement that

$$\operatorname{CT}_{s=0} \xi(s, z) = \sum_{\operatorname{Re}(p) < \operatorname{Re}(q)} z^{-p} B_p^*(\log z) + O(x^{-\operatorname{Re}(q)} (\log x)^{m(q)})$$

as $x = \operatorname{Re}(z) \to \infty$. This restatement emphasizes the point that the asymptotics of $\theta(t)$ as $t \to 0$ determine the asymptotics of $\operatorname{CT}_{s=0} \xi(s, z)$ as $\operatorname{Re}(z) \to \infty$.

To further develop the asymptotics given in Theorem 5.4, one can use the following lemma.

Lemma 5.5. *For any $p \in \mathbf{C}$, we have*

$$\mathrm{CT}_{s=0}\left[\frac{\Gamma(s+p)}{z^{s+p}}\right]$$
$$= \begin{cases} \Gamma(p)z^{-p} & p \notin \mathbf{Z}_{\leq 0} \\ [-c_{-1}(n)\log z + c_0(n)]z^{-p} & p = -n \in \mathbf{Z}_{\leq 0}, \end{cases}$$

where

$$c_{-1}(n) = \frac{(-1)^n}{n!} \quad \text{and} \quad c_0(n) = \frac{(-1)^n}{n!}(-\gamma).$$

One can prove Lemma 5.5 by using the power series expansion (4) and the formula

$$\Gamma(s+1) = s\Gamma(s) \quad \text{and} \quad \Gamma'(1) = -\gamma,$$

which comes from the expansion

$$\Gamma(s) = \frac{1}{s} - \gamma + O(s).$$

Proposition 5.6. *In the special case, we have, for $\mathrm{Re}(q) > 0$,*

$$\mathbf{B}_q(z) = \sum_{n \in \mathbf{Z}_{\geq 0}} \frac{b_{-n}(-1)^n}{n!}(-\gamma - \log z)z^n + \sum_{\substack{p \notin \mathbf{Z}_{\leq 0} \\ \mathrm{Re}(p) < \bar{\mathrm{Re}}(q)}} b_p\Gamma(p)z^{-p};$$

and so

$$\mathrm{CT}_{s=0}\xi(s,z) = \sum_{n \in \mathbf{Z}_{\geq 0}} \frac{b_{-n}(-1)^n}{n!}(-\gamma - \log z)z^n +$$
$$\sum_{\substack{p \notin \mathbf{Z}_{\leq 0} \\ \mathrm{Re}(p) < \bar{\mathrm{Re}}(q)}} b_p\Gamma(p)z^{-p} + O(x^{-\mathrm{Re}(q)}(\log x)^{m(q)})$$

for $\mathrm{Re}(z) \to \infty$.

Example 1. Assume L is such that

$$\theta(t) = \frac{b_{-1}}{t} + b_0 + O(t) \quad \text{for } t \to 0.$$

By taking $q = 1$, Proposition 5.6 then states that

$$\mathrm{CT}_{s=0}\xi(s,z) = b_{-1}z(\log z + \gamma) - b_0(\log z + \gamma) + O(z^{-1})$$
$$= (b_{-1}z - b_0)\log z + \gamma(b_{-1}z - b_0) + O(z^{-1})$$

as $\mathrm{Re}(z) \to \infty$. A particular example of this is when $L = \mathbf{Z}_{\geq 0}$. Recall that the Lerch formula (Theorem 2.1) states

$$\log \Gamma(z) = \frac{1}{2}\log 2\pi + \mathrm{CT}_{s=0}\xi(s,z) - \gamma(z - \frac{1}{2}) - z.$$

By direct computation one sees that

$$\theta(t) = \sum_{n=0}^{\infty} e^{-nt} = \frac{1}{t} + \frac{1}{2} + O(t),$$

so $b_{-1} = 1$ and $b_0 = 1/2$. Therefore,

$$\mathrm{CT}_{s=0}\xi(s,z) = (z - \frac{1}{2})\log z + \gamma(z - \frac{1}{2}) + O(|z|^{-1})$$

as $\mathrm{Re}(z) \to \infty$. From this we conclude that

$$\log \Gamma(z) = \frac{1}{2}\log 2\pi + (z - \frac{1}{2})\log z - z + O(|z|^{-1})$$

as $\mathrm{Re}(z) \to \infty$. This is the classical Stirling formula.

The reader is referred to [Vo 87] for interesting examples of the Stirling formula arising from sequences of eigenvalues associated to differential operators, and to [Sa 87] for the example of the Barnes double gamma function, which appears a factor in the functional equation of the Selberg zeta function associated to a finite volume hyperbolic Riemann surface.

As an application of the same method used to derive Stirling's formula (Theorem 5.4), we derive an asymptotic development of the integral transform

(5) $$\phi(x) = \int_a^{\infty} e^{-xu} \mathrm{CT}_{s=0}\xi(s,u)du,$$

for x real, positive, and $x \to 0$. We fix $a > 0$ and
$$a > -\text{Re}(\lambda_k) \quad \text{for all } k.$$
This function will appear in our work [JoL 92b], but, for now, let us simply view (5) as a special value of an incomplete Laplace-Mellin transform.

We use the decomposition of $\xi(s, u)$ as a sum
$$\xi(s, u) = J_1(s, u) + J_2(s, u) + J_3(s, u)$$
as at the beginning of this section, and deal with the integral transform of each term separately, so
$$\phi(x) = \phi_1(x) + \phi_2(x) + \phi_3(x)$$
where
$$\phi_\nu(x) = \int_a^\infty e^{-xu} \text{CT}_{s=0} J_\nu(s, u) du.$$

The next lemmas lead to Theorem 5.11, which states the asymptotic behavior of $\phi(x)$ as the real variable x approaches zero from the right. We start with the transform of J_1.

Lemma 5.7. *Let $h(t)$ be a bounded measurable function on $[0, 1]$ and assume that for some q with $\text{Re}(q) \geq 2$,*
$$h(t) = O(t^{\text{Re}(q)}) \quad \text{as } t \to 0.$$
Let
$$F(x) = \int_0^1 \frac{h(t)}{x+t} \frac{dt}{t},$$
and let $[q] = [\text{Re}(q)]$. Then F is $C^{[q]-2}$ on $[0, 1]$ and has the Taylor development
$$F(x) = c_0 + \cdots + c_{[q]-3} x^{[q]-3} + O(x^{[q]-2}) \quad \text{as } x \to 0.$$

Proof. The lemma follows directly from Taylor's theorem by differentiating under the integral sign, which is justified by the stated assumptions. □

Remark 2. Lemma 5.7 will be applied to an integral of the form

$$\int_0^1 \frac{h(t)}{x+t} e^{-(x+t)} \frac{dt}{t} = e^{-x} \int_0^1 \frac{h(t) e^{-t}}{x+t} \frac{dt}{t},$$

which is a product of e^{-x} and an integral of exactly the type considered in Lemma 5.7.

Lemma 5.8. *Let $[q] = [\mathrm{Re}(q)]$. Then the function ϕ_1 is of class $C^{[q]-2}$ and has the Taylor expansion*

$$\phi_1(x) = \int_a^\infty e^{-xu} J_1(0, u) du = g_q(x) + O(x^{[q]-3}) \text{ as } x \to 0$$

with the Taylor polynomial g_q of degree $< \mathrm{Re}(q) - 3$.

Proof. From (1) we have, by interchanging the order of integration,

$$\int_a^\infty e^{-xu} J_1(0, u) du = \int_0^1 \frac{\theta(t) - P_q \theta(t)}{x+t} e^{-a(x+t)} \frac{dt}{t}.$$

We now apply Lemma 5.7 to the function

$$h(t) = \frac{\theta(t) - P_q \theta(t)}{t} e^{-a(x+t)},$$

noting that, by **AS 2**,

$$h(t) = O(t^{\mathrm{Re}(q)-1}),$$

as $t \to 0$. With this, the proof of the lemma is complete. $\square$

Lemma 5.9. *The function*

$$\phi_2(z) = \int_a^\infty e^{-zu} J_2(0, u) du$$

which is defined for $\mathrm{Re}(z) > 0$, has a holomorphic extension to include a neighborhood of $z = 0$.

Proof. We have

$$\int_a^\infty e^{-zu} J_2(0, u) du = \int_1^\infty \frac{\theta(t)}{z+t} e^{-a(z+t)} \frac{dt}{t} = e^{-az} \int_1^\infty \frac{\theta(t) e^{-at}}{z+t} \frac{dt}{t}$$

from which the stated result immediately follows. $\square$

Having studied the transforms of J_1 and J_2, we now deal with the transform of J_3.

Lemma 5.10. *Given p and B_p, there is a function h_p, meromorphic in s and entire in z, to be given explicitly below, such that for $\mathrm{Re}(z) > 0$, we have*

$$\int_a^\infty e^{-zu} B_p(\partial_s) \left[\frac{\Gamma(s+p)}{u^{s+p}} \right] du$$

$$= B_p(\partial_s) \left[\frac{\pi z^{s+p-1}}{\sin[\pi(s+p)]} \right] + h_p(s, z),$$

and, in particular,

$$\int_a^\infty e^{-zu} \mathrm{CT}_{s=0} B_p(\partial_s) \left[\frac{\Gamma(s+p)}{u^{s+p}} \right] du$$

$$= \mathrm{CT}_{s=0} B_p(\partial_s) \left[\frac{\pi z^{s+p-1}}{\sin[\pi(s+p)]} \right] + h_p(z),$$

where

$$h_p(z) = \mathrm{CT}_{s=0} h_p(s, z).$$

Proof. It suffices to assume that $-\mathrm{Re}(s + p)$ is a large, non-integral, real number, from which the result follows by analytic continuation in s. For this, note that

$$\int_a^\infty e^{-zu} B_p(\partial_s) \left[\frac{\Gamma(s+p)}{u^{s+p}} \right] du = B_p(\partial_s) \int_a^\infty e^{-zu} \left[\frac{\Gamma(s+p)}{u^{s+p}} \right] du.$$

From this we have

$$\int_a^\infty e^{-zu} u^{1-s-p} \frac{du}{u} = \left[\frac{\Gamma(1-s-p)}{z^{1-s-p}} - \int_0^a e^{-zu} u^{1-s-p} \frac{du}{u} \right]$$

$$= \frac{\pi z^{s+p-1}}{\sin[\pi(s+p)]\Gamma(s+p)} - \sum_{k=0}^\infty \frac{(-z)^k}{k!} \frac{a^{1-s-p+k}}{1-s-p+k}.$$

The lemma follows by putting

(6) $$h_p(s,z) = -\sum_{k=0}^\infty \frac{(-z)^k}{k!} B_p(\partial_s) \left[\frac{\Gamma(s+p) a^{1-s-p+k}}{1-s-p+k} \right],$$

and

$$h_p(z) = -\sum_{k=0}^\infty \frac{(-z)^k}{k!} \mathrm{CT}_{s=0} \left[B_p(\partial_s) \left[\frac{\Gamma(s+p) a^{1-s-p+k}}{1-s-p+k} \right] \right],$$

□

Remark 3. A direct calculation shows that there exists a polynomial $B_p^\#$ of degree at most $\deg B_p + 1$ such that

(7) $$\mathrm{CT}_{s=0} B_p(\partial_s) \left[\frac{\pi z^{s+p-1}}{\sin[\pi(s+p)]} \right] = z^{p-1} B_p^\#(\log z).$$

The possible zero of $\sin[\pi(s+p)]$ at $s=0$ accounts for the possibility of $\deg B_p^\#$ exceeding $\deg B_p$. Further, by combining Remark 1 and Lemma 5.10, we arrive at the formula

$$\int_a^\infty e^{-zu} u^{-p} B_p^*(\log u) du = z^{p-1} B_p^\#(\log z) + h_p(z).$$

As is clear from (6), the function h_p does depend on the choice of a. We note that the power series $h_p(z)$ and the polynomials $B_p^\#$

are given by universal formulas, depending linearly on B_p^*, hence on B_p.

As is shown in the proof of Lemma 5.3, we can write

$$J_3(s,z) = \sum_{\mathrm{Re}(p)<\mathrm{Re}(q)} \left(B_p(\partial_s) \left[\frac{\Gamma(s+p)}{z^{s+p}}\right] - I_p(s,z) \right),$$

where

$$I_p(s,z) = \int_1^\infty e^{-zt} b_p(t) t^{s+p} \frac{dt}{t}.$$

By interchanging order of integration, we can write

$$\int_a^\infty e^{-uz} I_p(s,u)du = \int_1^\infty \frac{e^{-a(z+t)}}{z+t} b_p(t) t^{s+p} \frac{dt}{t}.$$

Therefore, Lemma 5.9 applies to imply the existence of a function f_p, holomorphic in a neighborhood of 0, such that

$$f_p(z) = \int_a^\infty e^{-uz} I_p(s,u)du,$$

so we can write

$$\phi_3(z) = \sum_{\mathrm{Re}(p)<\mathrm{Re}(q)} [z^{p-1} B_p^\#(\log z) + h_p(z) + f_p(z)].$$

With all this, we can combine Lemmas 5.8, 5.9, and 5.10 and obtain the following theorem.

Theorem 5.11. Let $\{g_q\}$, ϕ_2, $\{B_p^\#\}$, $\{f_p\}$ and $\{h_p\}$ be the above sequences of polynomials and entire functions for those p for which $B_p \neq 0$. Then for each q with $\mathrm{Re}(q) > 3$

$$\int_a^\infty e^{-tu} \mathrm{CT}_{s=0} \xi(s,u) du =$$

$$\sum_{\mathrm{Re}(p)<\mathrm{Re}(q)} [t^{p-1} B_p^\#(\log t) + h_p(t) + f_p(t)]$$

$$+ g_q(t) + \phi_2(t) + O(t^{[q]-3})$$

107

as the real variable t approaches zero from the right.

Example 2. In the case $L = \mathbf{Z}_{\geq 0}$ we have that

(8) $$\theta(t) = \sum_{n=0}^{\infty} e^{-nt} = \frac{1}{1-e^{-t}} = \sum_{n=-1}^{\infty} b_n t^n$$

as t approaches zero. In particular, the sequence $\{p\}$ is simply $\mathbf{Z}_{\geq -1}$ and all polynomials B_p have degree zero. Since

$$\mathrm{CT}_{s=0}\left[\frac{\pi t^{s+n-1}}{\sin[\pi(s+n)]}\right] = (-t)^n \frac{\log t}{t},$$

we have, by using the Lerch formula (Theorem 2.1), the equation

(9) $$\int_1^{\infty} e^{-tu} \log \Gamma(u) du = \frac{\log t}{t}\left(\sum_{n=-1}^{\infty} b_n(-t)^n\right) + h(t).$$

Let $q \to \infty$ in Theorem 5.11. Using the absolute convergence of (8), we get (9). In order to deduce (9) from Theorem 5.11, we have used that, upon letting q approach infinity, the power series (8) converges in a neighborhood of the origin. In general, questions concerning the convergence of the above stated power series, which are necessarily questions concerning the growth of the coefficients b_n, must be addressed. For now, let us complete this example by combining the above results to conclude that

$$\int_1^{\infty} e^{-tu} \log \Gamma(u) du = \frac{\log t}{t} \frac{1}{1-e^t} + h(t).$$

This formula verifies calculations that appear in [Cr 19] (see also [JoL 92b]).

Remark 4. The formal power series arising from the asymptotic expansion in Theorems 5.4 and 5.11 (letting $\mathrm{Re}(q) \to \infty$) are interesting beyond their truncations mod $O(t^{[q]-3})$. In important applications, and notably to positive elliptic operators, these power series are convergent, and define entire functions. This is part of the theory of Volterra operators, c.f. [Di 78], Chapter XXIII, (23.6.5.3).

§6. Hankel Formula

The gamma function and the classical zeta function are well known to satisfy a Hankel formula, that is they are representable as an integral over a Hankel contour. The existence of such a formula depends on the integrand having at least an analytic continuation over the Hankel contour, and expecially having an analytic continuation around 0. Indeed, the integrands in these classical Hankel transforms are essentially theta functions. Unfortunately, theta functions cannot always be analytically continued around 0. For instance, $\sum e^{-n^2 t}$ cannot, although $\sum e^{-nt}$ can. Thus the analogues of the classical Hankel transform representations are missing in general.

However, we shall give here one possible Hankel type formula expressing the Hurwitz xi function associated to the sequences (L, A) as a complex integral of the regularized harmonic series. Observe that the Hankel formula which we prove here is different from the classical Hankel representation of the gamma function or the zeta function. We note that such a formula was used by Deninger for the Riemann zeta function (see §3 of [De 92]).

As in previous sections, let (L, A) be sequences whose Dirichlet series satisfy the three convergence conditions **DIR 1**, **DIR 2** and **DIR 3**, and whose associated theta function satisfies the asymptotic conditions **AS 1**, **AS 2** and **AS 3**. We suppose that the sequence $L = \{\lambda_k\}$ is such that

$$\operatorname{Re}(\lambda_k) > 0 \quad \text{for all } k.$$

Since we deal with arbitrary Dirichlet series $\zeta(s) = \sum a_k \lambda_k^{-s}$, it is not the case in general that there is a regularized product whose logarithmic deriviative gives the constant term $\operatorname{CT}_{s=1}\xi(s,z)$; indeed, this exists only in the case when $a_k \in \mathbf{Z}$ for all k. However, important applications will be made to the spectral case, meaning when $a_k \in \mathbf{Z}_{>0}$ for all k, in which case such a regularized product D_L exists. Thus, for this case, we record here once more the

Basic Identity:

$$R(z) = \operatorname{CT}_{s=1}\xi(s, z) = D'_L/D_L(z).$$

All the formulas of this section involving

$$R(z) = \operatorname{CT}_{s=1}\xi(s, z)$$

may then be used with $R(z)$ replaced by $D'_L/D_L(z)$ in the applications to the spectral case.

Recall from **DIR 2**, **DIR 3** and Theorem 1.8 that for any integer $n > \sigma_0$ we have the formula

$$\partial_z^n R(z) = (-1)^n CT_{s=1}\xi(s+n,z) = (-1)^n \Gamma(n) \sum_{k=1}^{\infty} \frac{a_k}{(z+\lambda_k)^n}.$$

To begin, let us establish the following general lemma.

Lemma 6.1. *With assumptions as above, for every sufficiently large positive integer m, there exists a real number T_m with $m \leq T_m \leq m+1$ and a real number σ_2 such that*

$$|R(z)| = O(|z|^{\sigma_2}) \quad \text{for} \quad |z| = T_m, \quad \text{and as} \quad T_m \to \infty.$$

Proof. For every sufficiently large positive integer m, let us write

(1)
$$\begin{aligned}\partial_z^n R(z) = (-1)^n \Gamma(n) \sum_{|\lambda_k| \leq 2m} \frac{a_k}{(z+\lambda_k)^n} \\ + (-1)^n \Gamma(n) \sum_{|\lambda_k| > 2m} \frac{a_k}{(z+\lambda_k)^n}.\end{aligned}$$

If we restrict z to the annulus $m < |z| < m+1$, we have the estimate

$$\sum_{|\lambda_k|>2m} \frac{a_k}{(z+\lambda_k)^n} \leq \sum_{|\lambda_k|>2m} \frac{|a_k|}{(|\lambda_k|-m)^n} \leq 2 \sum_{|\lambda_k|>2m} \frac{|a_k|}{|\lambda_k|^n}$$

which is bounded independent of m provided $n \geq \sigma_0$.

By the convergence condition **DIR 2(b)**, we have that

$$\#\{\lambda_k : m < |\lambda_k| < m+1\} = O(m^{\sigma_1}),$$

so there are $O(m^{\sigma_1+1})$ terms in the first sum. Also, we conclude the existence a sequence $\{T_m\}$ of real numbers, tending to infinity and with $m < T_m < m+1$, such that if $|z| = T_m$, then the distance from z to any $-\lambda_k$ is at least $cm^{-\sigma_1}$ with a suitable small constant c. From **DIR 2(a)** we have that

$$|a_k| = O(|\lambda_k|^{\sigma_0}),$$

so, if $|\lambda_k| < 2m$, we have for those k for which $|\lambda| < 2m$, the bound $|a_k| = Cm^{\sigma_0}$, for a suitable large constant C. With all this, the first sum can be bounded by

$$O(m^{\sigma_0} \cdot m^{\sigma_1+1} \cdot m^{n\sigma_1}).$$

If $|z| = T_m$, we can write this bound as stating

(2) $$\sum_{|\lambda_k| \leq 2m} \frac{a_k}{(z+\lambda_k)^n} = O(|z|^{1+\sigma_0+(n+1)\sigma_1}).$$

This establishes that $\partial_z^n R(z)$ has polynomial growth on the (increasing) circles centered at the origin with radius T_m. From the expression

$$\partial_z^n R(z) = (-1)^n \int_0^\infty \theta(t) e^{-zt} t^n \frac{dt}{t},$$

one has that $\partial_z^n R(z)$ is bounded for $z \in \mathbf{R}_{>0}$ with z sufficiently large. Upon integrating the function $\partial_z^n R(z)$ n times along a path consisting of $\mathbf{R}_{>0}$ and the circle $|z| = T_m$, we conclude that $R(z)$ itself has polynomial growth on the circles $|z| = T_m$. Indeed, from (2) and the fact that $\sigma_0 \geq -\mathrm{Re}(p_0)$, we have

$$|R(z)| = O(|z|^{\sigma_0+(n+1)(\sigma_1+1)}) \quad \text{for} \quad |z| = T_m, \quad \text{and as} \quad T_m \to \infty.$$

This completes the proof of the lemma. □

It should noted that the proof of Lemma 6.1 explicitly constructs the value of σ_2. Indeed, since one can take $n < \sigma_0 + 1$, σ_2 can be written in terms of σ_0 and σ_1.

Having established this preliminary lemma, we can now proceed with our Hankel formula for the regularized harmonic series R. Let δ be a fixed positive number and assume

$$\delta < |\lambda_k| \quad \text{for all } \lambda_k \in L.$$

111

Let C_δ denote the contour consisting of:
- the lower edge of the cut from $-\infty$ to $-\delta$ in the cut plane $\mathbf{U}_L$;
- the circle S_δ, given by $w = \delta e^{i\phi}$ for ϕ ranging from $-\pi$ to π;
- and the upper edge of the cut from $-\delta$ to $-\infty$ in the cut plane $\mathbf{U}_L$.

If $G(z)$ is a meromorphic function in $\mathbf{U}_L$, then

$$\int_{C_\delta} G(w)\,dw = \int_{-\infty}^{-\delta} + \int_{S_\delta} + \int_{-\delta}^{-\infty} G(w)\,dw.$$

Symbolically, let us set

$$\int_{-\infty}^{-\delta} + \int_{-\delta}^{-\infty} = \int_{C_\delta} - \int_{S_\delta}.$$

When taking the sum of the two integrals on the negative real axis, it is of course understood that for the second integral, we deal with the analytic continuation of $G(w)$ over the circle S_δ.

We call the result of the following theorem the **Hankel formula**.

Theorem 6.2. *Let $s \in \mathbf{C}$ be such that $\mathrm{Re}(s) > \sigma_2 + 1$. Then for any z with $\mathrm{Re}(z)$ sufficiently large, we have*

$$\xi(s,z) = \frac{\Gamma(s)}{2\pi i} \int_{C_\delta} R(z-w) w^{-s}\,dw.$$

Proof. Let $T \in \mathbf{R}_{>0}$ be such that $|\lambda_k| \neq T$ for all k, and let $C_{\delta,T}$ denote the contour consisting of:
- the lower edge of the cut from $-T$ to $-\delta$ in $\mathbf{U}_L$;
- the circle S_δ, given by $w = \delta e^{i\phi}$ for ϕ ranging from $-\pi$ to π;
- the upper edge of the cut from $-\delta$ to $-T$ in $\mathbf{U}_L$;
- and the circle S_T, given by $w = Te^{i\phi}$ for ϕ ranging from π to $-\pi$.

Combining the residue theorem with Theorem 3.4(b), we get

$$\frac{1}{2\pi i} \int_{C_{\delta,T}} R(z-w)w^{-s}dw = \sum_{0<|z+\lambda_k|<T} \frac{a_k}{(z+\lambda_k)^s}.$$

Now let $T \to \infty$ along the sequence $\{T_m\}$ that was constructed in Lemma 6.1. With this, the integral over S_T will go to zero as T approaches infinity if s is such that $\operatorname{Re}(s) > \sigma_2 + 1$. For these values of s, the limit of the sum above is the Hurwitz zeta function, and the theorem is proved. □

For the remainder of this section we will study the analytic continuation of the integral given in Theorem 6.2. The proof of Theorem 6.2 shows that the problem in extending the region for which the integral converges arises because of the asymptotic behavior of $R(z-t)$ for $-\operatorname{Re}(t)$ large. In any case, the following lemma takes care of the integral over the circle S_δ.

Lemma 6.3. *For any fixed δ sufficiently small, and any z such that $\operatorname{Re}(z)$ is sufficiently large, the integral*

$$\frac{1}{2\pi i} \int_{S_\delta} R(z-w)w^{-s}dw$$

is holomorphic for all $s \in \mathbf{C}$.

Proof. Since $\operatorname{Re}(z)$ is sufficiently large, the integrand is absolutely bounded on S_δ. □

Lemma 6.3 implies that the analytic continuation of the Hankel formula to any s for which $\operatorname{Re}(s) \leq \sigma_2 + 1$ must take account the asymptotic behavior of $R(z-w)$ for fixed z and as $-\operatorname{Re}(w) \to \infty$. We can use the generalized Stirling's formula, Theorem 5.4, to extend the Hankel formula in much the same way **AS 2** is used to extend the Hurwitz xi function (see, in particular, Lemma 1.3 and Theorem 1.5).

Lemma 6.4. *For fixed $\delta > 0$, for any z with $\operatorname{Re}(z)$ sufficiently large and for any q and p for which $\operatorname{Re}(q) > \operatorname{Re}(p)$, the integral*

$$\frac{1}{2\pi i}\left(\int_{-\infty}^{-\delta} + \int_{-\delta}^{-\infty}\right)[R(z-t) - \mathbf{B}_q(z-t)]\,t^{-s}dt$$

is holomorphic for $\text{Re}(s) = \sigma > -\text{Re}(q) + 1$.

Proof. For any $T > 1$, the integral over the segments from $-T$ to $-\delta$, both the lower cut and the upper cut, is finite for all s since the integrand is absolutely bounded. By Theorem 5.4, we have

$$\left| \frac{1}{2\pi i} \left(\int_{-\infty}^{-T} + \int_{-T}^{-\infty} \right) [R(z-t) - \mathbf{B}_q(z-t)] t^{-s} dt \right|$$

$$\ll \int_{-\infty}^{-T} |t|^{-\sigma - \text{Re}(q)} dt,$$

which converges if $-\sigma - \text{Re}(q) < -1$, as asserted. $\square$

Proposition 6.5. *Let q be such that $\text{Re}(q) > 1$, and let δ be a sufficiently small positive number such that $|\lambda_k| > \delta$ for all k. Then for any z with $\text{Re}(z)$ sufficiently large,*

$$\text{CT}_{s=0} \xi(s, z) =$$

$$\frac{1}{2\pi i} \left(\int_{-\infty}^{-\delta} + \int_{-\delta}^{-\infty} \right) [R(z-t) - \mathbf{B}_q(z-t)] (-\gamma - \log t) \, dt$$

$$+ \text{CT}_{s=0} \frac{\Gamma(s)}{2\pi i} \left(\int_{-\infty}^{-\delta} + \int_{-\delta}^{-\infty} \right) \mathbf{B}_q(z-t) t^{-s} dt$$

$$+ \frac{1}{2\pi i} \int_{S_\delta} R(w - t)(-\gamma - \log w) \, dw.$$

Proof. Write the integral over C_δ as the sum of integrals over S_δ and the line segments between $-\infty$ and $-\delta$. Since

$$\text{CT}_{s=0} \frac{\Gamma(s)}{w^s} = \text{CT}_{s=0} \left(\left[\frac{1}{s} - \gamma + O(s^2) \right] [1 - s \log w + O(s)] \right)$$

(3) $\qquad\qquad\quad = -\gamma - \log w,$

we have, by Lemma 6.3,

$$\text{CT}_{s=0} \frac{\Gamma(s)}{2\pi i} \int_{S_\delta} R(z-w) w^{-s} dw = \frac{1}{2\pi i} \int_{S_\delta} R(z-w)(-\gamma - \log w) \, dw.$$

Lemma 6.4 yields a similar result for the integral of $R - \mathbf{B}_q$. □

Next, we let δ approach zero and show that the integral over S_δ will go to zero.

Lemma 6.6. *For any q and p for which $\text{Re}(q) > \text{Re}(p)$, and for any z such that $\text{Re}(z)$ is sufficiently large, we have*

$$\lim_{\delta \to 0} \int_{S_\delta} R(z-w) \log w \, dw = 0$$

and

$$\lim_{\delta \to 0} \left[\left(\int_{-\infty}^{-\delta} + \int_{-\delta}^{-\infty} \right) [R(z-t) - \mathbf{B}_q(z-t)] \log t \, dt \right]$$

$$= \left(\int_{-\infty}^{0} + \int_{0}^{-\infty} \right) [R(z-t) - \mathbf{B}_q(z-t)] \log t \, dt.$$

Proof. If $\text{Re}(z)$ is sufficiently large, the functions

$$R(z-t) \quad \text{and} \quad R(z-t) - \mathbf{B}_q(z-t)$$

are bounded as t approaches zero. The second term is bounded by $t^{-\text{Re}(q)}$ as $t \to 0$. Therefore, the first integral is bounded by a multiple of

$$\left| \int_{S_\delta} \log t \, dt \right| \ll \delta \log \delta,$$

which approaches zero as δ approaches zero. The same estimate proves the second assertion. □

Combining Lemma 6.4, Proposition 6.5 and Lemma 6.6, we have

Theorem 6.7. *With notation as above,*

$$CT_{s=0}\xi(s,z) =$$

$$\frac{1}{2\pi i}\left(\int_{-\infty}^{0} + \int_{0}^{-\infty}\right)[R(z-t) - \mathbf{B}_q(z-t)](-\gamma - \log t)\, dt$$

$$+ \lim_{\delta \to 0} CT_{s=0} \frac{\Gamma(s)}{2\pi i}\left(\int_{-\infty}^{-\delta} + \int_{-\delta}^{-\infty}\right)\mathbf{B}_q(z-t)t^{-s}dt.$$

One can view the content of Theorem 6.7 as a type of regularized form of the Fundamental Theorem of Calculus. One should note the presence of the term $-\gamma - \log t$ in Theorem 6.7, which is a function that also appeared in our general Stirling formula; see Proposition 5.6 of the previous section.

To conclude, let us note that if instead of considering the Hurwitz xi function we would have studied the Hurwitz zeta function we would have obtained the following result.

Theorem 6.8. *With notation as above,*

$$CT_{s=0}\zeta'(s,z) =$$

$$\frac{1}{2\pi i}\left(\int_{-\infty}^{0} + \int_{0}^{-\infty}\right)[R(z-t) - \mathbf{B}_q(z-t)](-\log t)\, dt$$

$$+ \lim_{\delta \to 0} CT_{s=0} \partial_s \frac{1}{2\pi i}\left(\int_{-\infty}^{-\delta} + \int_{-\delta}^{-\infty}\right)\mathbf{B}_q(z-t)t^{-s}dt.$$

The proof of Theorem 6.8 follows that of Theorem 6.7 with the only change being the use of the formula

$$CT_{s=0}\partial_s w^{-s} = -\log w$$

in place of (3).

§7. Mellin Inversion Formula

So far we have considered sequences (L, A) that satisfy the three convergence conditions **DIR 1**, **DIR 2** and **DIR 3** and whose associated theta function θ satisfies the three asymptotic conditions **AS 1**, **AS 2** and **AS 3**. From these assumptions we then derived properties concerning various transforms. We now want to perform other operations, so we reconsider these axioms *ab ovo*. Especially, we shall consider the inverse Mellin transform which gives θ in terms of ζ. Throughout this section we shall assume

$$\text{Re}(\lambda_k) > 0 \quad \text{for all} \quad k.$$

Recall that Theorem 1.12 proved that the convergence conditions **DIR 1**, **DIR 2** and **DIR 3** imply the asymptotic conditions **AS 1** and **AS 3**. With this, Theorem 1.11 applies to show that for $\text{Re}(s)$ sufficiently large we have

$$\zeta(s) = \sum_{k=1}^{\infty} \frac{a_k}{\lambda_k^s} = \frac{1}{\Gamma(s)} \mathbf{M}\theta(s).$$

As previously stated, the asymptotic condition **AS 2** is quite delicate. In this section, we will show how, by imposing additional assumptions of meromorphy and certain growth conditions on ζ, the asymptotic condition **AS 2** follows. To begin, we need the following lemma which addresses the question of convergence of the partial series

$$\sum_{k=1}^{N-1} \frac{a_k}{\lambda_k^s},$$

to $\zeta(s)$ as we let $N \to \infty$.

Lemma 7.1. *Assume that the sequences (L, A) satisfy the convergence conditions* **DIR 1**, **DIR 2** *and* **DIR 3**, *and let σ_0 be as in* **DIR 2(a)**. *For each $N \geq 1$ let*

$$\psi_N = \sup |\arg(\lambda_k)| \quad \text{for all} \quad k \geq N,$$

so $\psi_N < \frac{\pi}{2}$ for all N sufficiently large.

(a) *If $s = \sigma + it$ is such that*

$$\text{Re}(s) = \sigma > \sigma_0,$$

then for N sufficiently large

$$\left|\zeta(s) - \sum_{k=1}^{N-1} a_k \lambda_k^{-s}\right| \leq e^{|t|\psi_N} \cdot \sum_{k=N}^{\infty} |a_k||\lambda_k|^{-\sigma}.$$

(b) For all s in any fixed compact subset of the half plane

$$\text{Re}(s) = \sigma > \sigma_0,$$

the convergence

$$\lim_{N \to \infty} \left|\zeta(s) - \sum_{k=1}^{N-1} a_k \lambda_k^{-s}\right| = 0,$$

is uniform.

Proof. If we write $\log \lambda_k = \log|\lambda_k| + i \arg(\lambda_k)$, we have the bound

$$\left|\zeta(s) - \sum_{k=1}^{N-1} a_k \lambda_k^{-s}\right| \leq \sum_{k=N}^{\infty} |a_k| \cdot |\lambda_k^{-s}|$$

$$= \sum_{k=N}^{\infty} |a_k| \cdot |e^{-s \log \lambda_k}|$$

$$\leq \sum_{k=N}^{\infty} |a_k| \cdot e^{-\sigma \log |\lambda_k| + |t|\psi_N}.$$

If we define the **absolute zeta** to be

$$\zeta_{\text{abs}}(\sigma) = \sum_{k=1}^{\infty} |a_k||\lambda_k|^{-\sigma}$$

and

$$\zeta_{\text{abs}}^{(N)}(\sigma) = \zeta_{\text{abs}}(\sigma) - \sum_{k=1}^{N-1} |a_k||\lambda_k|^{-\sigma},$$

then we have shown that

$$\left|\zeta(s) - \sum_{k=1}^{N-1} a_k \lambda_k^{-s}\right| \leq e^{|t|\psi_N} \cdot \zeta_{\text{abs}}^{(N)}(\sigma),$$

which establishes part (a). Note that if $\sigma > \sigma_0$, then, by the convergence condition **DIR 2(a)**,

$$\lim_{N \to \infty} \zeta_{\text{abs}}^{(N)}(\sigma) = 0,$$

which shows that the upper bound in (a) approaches zero as N approaches ∞, for fixed s with $\text{Re}(s)$ sufficiently large. As for (b), since ψ_N is bounded, if s lies in a compact subset K of the half plane $\text{Re}(s) > \sigma_0$, then $|t|\psi_N$ is bounded independent of N for all $s \in K$. This completes the proof of the lemma. $\square$

Let φ be a suitable function, which will be appropriately characterized below. For any $\sigma \in \mathbf{R}$, let $\mathcal{L}(\sigma)$ be the vertical line $\text{Re}(s) = \sigma$ in $\mathbf{C}$. Under suitable conditions on φ which guarantee the absolute convergence of the following integral, we define the **vertical transform** $\mathbf{V}_\sigma \varphi$ of φ to be

$$\mathbf{V}_\sigma \varphi(t) = \frac{1}{2\pi i} \int_{\mathcal{L}(\sigma)} \varphi(s) \Gamma(s) t^{-s} ds.$$

From Stirling's formula (see Theorem 5.4 and, specifically, Example 1 of §5), one sees that on vertical lines $\mathcal{L}(\sigma)$, the gamma function has the decaying behavior

(2) $$\Gamma(s) = O_c(e^{-c|s|}) \quad \text{for every } c \text{ with } 0 < c < \pi/2,$$
$$\text{and } |s| \to \infty,$$

where, as indicated, the implied constant depends on c. Previously we studied the Mellin transform and, from Theorem 1.10, we have

$$\zeta(s) = \frac{1}{\Gamma(s)} \mathbf{M}\theta(s) = \sum_{k=1}^{\infty} \frac{a_k}{\lambda_k^s},$$

where
$$\theta(t) = \sum_{k=1}^{\infty} a_k e^{-\lambda_k t}.$$

We shall now study the **inversion formula**

(3) $$\theta(t) = \mathbf{V}_\sigma \zeta(t) = \frac{1}{2\pi i} \int_{\mathcal{L}(\sigma)} \zeta(s)\Gamma(s)t^{-s}ds,$$

which is valid if
$$\sigma > -\mathrm{Re}(\lambda_k) \quad \text{for all } k.$$

The inversion formula (3) is essentially a standard elementary inversion obtained for each individual term from the relation

(4) $$e^{-u} = \frac{1}{2\pi i} \int_{\mathcal{L}(\sigma)} \Gamma(s) t^{-s} ds.$$

The classical proof of (4) comes from an elementary contour integration along a large rectangle going to the left, using the fact that
$$\mathrm{res}_{-n} \Gamma(s) = \frac{(-1)^n}{n!}.$$

The relation (4) is then applied by putting $u = \lambda_k t$ and summing over k. More precisely:

Proposition 7.2. *Assume that ζ satisfies* **DIR 1**, **DIR 2** *and* **DIR 3**, *and let σ_0 be as defined in* **DIR 2(a)**. *Then for every $\delta > 0$, the associated theta series converges absolutely and uniformly for $t \geq \delta > 0$ and*
$$\theta = \mathbf{V}_\sigma \zeta \quad \text{for } \sigma > \sigma_0.$$

Proof. Let $g = \mathbf{V}_\sigma \zeta$ be the vertical transform of ζ with $\sigma > \sigma_0$. By (4) we have

(5) $$g(t) - \sum_{k=1}^{N-1} a_k e^{-\lambda_k t} = \frac{1}{2\pi i} \int_{\mathcal{L}(\sigma)} \left[\zeta(s) - \sum_{k=1}^{N-1} a_k \lambda_k^{-s} \right] \Gamma(s) t^{-s} ds.$$

For N sufficiently large, take ψ_N as in Lemma 7.1, and let us write $s = \sigma + iu$. With this we have the bound

$$\left| g(t) - \sum_{k=1}^{N-1} a_k e^{-\lambda_k t} \right| \leq \frac{\zeta_{abs}^{(N)}(\sigma)}{2\pi} \int_{\mathcal{L}(\sigma)} e^{|u|\psi_N} |\Gamma(\sigma + iu)| t^{-\sigma} du$$

$$= \frac{\zeta_{abs}^{(N)}(\sigma)}{2\pi} t^{-\sigma} \| e^{|u|\psi_N} \Gamma(\sigma + iu) \|_{1,\sigma},$$

where

$$\| e^{|u|\psi_N} \Gamma(\sigma + iu) \|_{1,\sigma}$$

is the L^1-norm of $e^{|u|\psi_N} \Gamma(\sigma + iu)$ on the vertical line $\mathcal{L}(\sigma)$. By the convergence condition **DIR 3** and (2), we conclude there exists a constant C, independent of N, such that

(6) $$\left| g(t) - \sum_{k=1}^{N-1} a_k e^{-\lambda_k t} \right| \leq C \zeta_{abs}^{(N)}(\sigma) t^{-\sigma}.$$

If $t \geq \delta > 0$, (6) can be written as

$$\left| g(t) - \sum_{k=1}^{N-1} a_k e^{-\lambda_k t} \right| \leq C \zeta_{abs}^{(N)}(\sigma) \delta^{-\sigma}.$$

If $\sigma > \sigma_0$, $\zeta_{abs}^{(N)}(\sigma)$ tends to zero when $N \to \infty$. This shows that the theta series

$$\theta_{L,A}(t) = \theta(t) = \sum_{k=1}^{\infty} a_k e^{-\lambda_k t}$$

converges absolutely and uniformly for $t \geq \delta > 0$, as asserted in the statement of the proposition. $\square$

Remark 1. Proposition 7.2, in particular the inequality (6), shows that the three convergence conditions **DIR 1**, **DIR 2** and **DIR 3** implies the asymptotic condition **AS 3**. This provides another proof of the first part of Theorem 1.12.

Since **AS 2** was used to show that $\mathbf{M}\theta(s)$ has a meromorphic continuation, it is natural to expect some type of meromorphy condition on ζ in order to prove that the vertical transform $\mathbf{V}_\sigma \zeta$ satisfies the asymptotic condition **AS 2**. Independently of any σ, we define the **domain of V**, and denote it by $\text{Dom}(\mathbf{V})$, to be the space of functions φ satisfying the following conditions:

V 1. φ is meromorphic and has only a finite number of poles in every right half plane.

V 2. $\varphi\Gamma$ is L^1-integrable on every vertical line where $\varphi\Gamma$ has no pole.

V 3. Let $\sigma_1 < \sigma_2$ be real numbers. There exists a sequence $\{T_m\}$, with $m \in \mathbf{Z}$ and

$$T_m \to \infty \text{ if } m \to \infty \text{ and } T_m \to -\infty \text{ if } m \to -\infty,$$

such that uniformly for $\sigma \in [\sigma_1, \sigma_2]$, we have

$$(\varphi\Gamma)(\sigma + iT_m) \to 0 \text{ as } |m| \to \infty.$$

Very often one has the simpler condition:

V 3'. $\varphi(s)\Gamma(s) \to 0$ for $|s| \to \infty$ and s lying within any vertical strip of finite width.

Even though the condition **V 3'** is satisfied in many cases, it is necessary to have **V 3** as stated. The reason for condition **V 3** is that we shall let a rectangle of integration tend to infinity, and we need that the integral of $\varphi\Gamma$ on the top and bottom of the rectangle tends to 0. Throughout, it is understood that when we take the vertical transform, we select σ such that $\varphi\Gamma$ has no pole on $\mathcal{L}(\sigma)$. The vertical transform $\mathbf{V}_\sigma\varphi$ of φ depends on the choice of σ, of course, and we shall determine this dependence in a moment.

Remark 2. In the more standard cases, such as zeta functions of number fields or modular forms, or easier kinds of Selberg zeta functions, there is no difficulty in verifying the three conditions **V 1**, **V 2**, **V 3**, or, usually, **V 3'**. In fact, functions φ of this kind have usually at most polynomial growth in vertical strips, so their product with the gamma function decreases exponentially in vertical strips. In certain other interesting cases, it may be more difficult to prove this polynomial growth, and there are cases where it remains to be determined exactly what is the order of growth in vertical strips. The proof of polynomial growth depends on functional equations and Euler products in the classical cases.

We shall determine conditions under which we get **AS 2** for $\mathbf{V}_\sigma \varphi$. Immediately from the restricted growth of $\varphi \Gamma$ on vertical lines **V 1**, the asymptotic behavior of $\varphi \Gamma$ on horizontal lines **V 3**, and Cauchy's formula, we have:

Lemma 7.3. *Let $\varphi \in \mathrm{Dom}(\mathbf{V})$, and let $\sigma' < \sigma$ be such that $\varphi \Gamma$ has no pole on $\mathcal{L}(\sigma')$ and $\mathcal{L}(\sigma)$. Let $\mathcal{R}(\sigma', \sigma)$ be the strip (the infinite rectangle) defined by the inequalities*

$$\sigma' < \mathrm{Re}(s) < \sigma,$$

and let $\{-p\}$ be the sequence of poles of $\varphi \Gamma$. Then

$$\mathbf{V}_\sigma \varphi(t) = \mathbf{V}_{\sigma'} \varphi(t) + \sum_{-p \in \mathcal{R}(\sigma', \sigma)} \mathrm{res}_{-p}[\varphi(s) \Gamma(s) t^{-s}].$$

To continue, let us analyze the sum

$$\sum_{-p \in \mathcal{R}(\sigma', \sigma)} \mathrm{res}_{-p}[\varphi(s) \Gamma(s) t^{-s}]$$

which was obtained in Lemma 7.3. For this, let us write

(7) $$t^{-s} = t^p \cdot t^{-(s+p)} = t^p \cdot \sum_{n=0}^{\infty} \frac{(-\log t)^n}{n!} (s+p)^n.$$

Let $d_p = -\mathrm{ord}_{-p}[\varphi \Gamma]$ and consider the Laurent expansion at $s = -p$:

(8) $$\varphi(s) \Gamma(s) = \sum_{k=-d_p}^{\infty} c_k (s+p)^k,$$

so

$$\mathrm{res}_{-p}[\varphi(s) \Gamma(s) t^{-s}] = \sum_{k+n=-1} c_k \frac{(-\log t)^n}{n!} \cdot t^p.$$

Then from (7) and (8) we see immediately that there exists a polynomial B_p of degree d_p such that

(9) $$\operatorname{res}_{-p}[\varphi(s)\Gamma(s)t^{-s}] = B_p(\log t)t^p.$$

As in §1, given a sequence $\{p\}$ of complex numbers, ordered by increasing real parts that tend to infinity, and given a sequence of polynomials $\{B_p\}$ for every p in the sequence, we define

$$P_q(t) = \sum_{\operatorname{Re}(p) < \operatorname{Re}(q)} B_p(\log t)t^p.$$

As before, we put

$$m(q) = \max \deg B_p \text{ for } \operatorname{Re}(p) = \operatorname{Re}(q).$$

With this, we can combine Lemma 7.3 and (9) to obtain:

Theorem 7.4. *Let $\varphi \in \operatorname{Dom}(\mathbf{V})$, and let $\{-p\}$ be the sequence of poles of $\varphi\Gamma$. Let σ be a positive, sufficiently large, real number such that neither φ nor Γ have poles for $\operatorname{Re}(s) \geq \sigma$. Then the vertical transform $\mathbf{V}_\sigma \varphi$ of φ satisfies the asymptotic condition* **AS 2**, *or, briefly stated,*

$$\mathbf{V}_\sigma \varphi(t) \sim \sum_p B_p(\log t)t^p \text{ as } t \to 0.$$

Proof. If we let $s = \sigma' + it$ with σ' a large negative number, then we have the estimate

$$|\mathbf{V}_{\sigma'}\varphi(t)| \leq \frac{1}{2\pi} \int_{\mathcal{L}(\sigma')} |\varphi(s)\Gamma(s)t^{-s}|\, dt \leq \frac{1}{2\pi} t^{-\sigma'} \|\varphi\Gamma\|_{1,\sigma'},$$

where $\|\varphi\Gamma\|_{1,\sigma'}$ is the L^1-norm of $\varphi\Gamma$ on the line $\mathcal{L}(\sigma')$. By **V 2**, this integral is finite. Now simply combine Lemma 7.3 and (9), and let $t \to 0$ to conclude the proof. □

With all this, we can now summarize the connection between the convergence conditions that apply to sequences (L, A) and the asymptotic conditions that apply to the associated theta series.

Theorem 7.5. *Let (L, A) be sequences that satisfy the three convergence conditions* **DIR 1**, **DIR 2** *and* **DIR 3**, *and assume that the associated zeta function*

$$\zeta_{L,A}(s) = \zeta(s) = \sum_{k=1}^{\infty} a_k \lambda_k^{-s}$$

is in $\mathrm{Dom}(\mathbf{V})$. *Then the associated theta series*

$$\theta_{L,A}(t) = \theta(t) = \sum_{k=1}^{\infty} a_k e^{-\lambda_k t}$$

satisfies the asymptotic conditions **AS 1**, **AS 2** *and* **AS 3**. *The asymptotic expansion of* **AS 2** *is given by (9) and Theorem 7.4 with $\varphi = \zeta$.*

Proof. Theorem 1.12 shows that if (L, A) satisifies the three convergence conditions, then the associated theta series satisfies the asymptotic conditions **AS 1** and **AS 3**. Theorem 7.4 shows that the meromorphy assumption on ζ, namely that ζ is in $\mathrm{Dom}(\mathbf{V})$, implies **AS 2**. □

To conclude this section, we will use Theorem 7.4 to show how, given sequences whose theta series satisfies the asymptotic conditions, one can construct a new sequence with the same property. The following theorem follows immediately from the definitions and the holomorphy of the exponential function.

Theorem 7.6. *Let $z \in \mathbf{C}$ be fixed and assume that (L, A) is such that the associated theta series satisifies the three asymptotic conditions* **AS 1**, **AS 2** *and* **AS 3**. *Then the sequence $(L + z, A)$ satisfies the three asymptotic conditions* **AS 1**, **AS 2** *and* **AS 3**.

Proof. The theta function associated to $(L + z, A)$ is simply $e^{-zt}\theta_{L,A}(t)$, hence the proof follows.

Theorem 7.7. *Let $r > 0$ and assume that (L, A) is such that the associated theta series satisifies the three asymptotic conditions* **AS 1**, **AS 2** *and* **AS 3**. *Then the theta series associated to*

$$(\{\lambda_k^{1/r}\}, A) = (L^{1/r}, A)$$

satisfies satisfies the three asymptotic conditions **AS 1**, **AS 2** and **AS 3**. If the exponents $\{p\}$ in the asymptotic expansion of $\theta_{L,A}$ near $t = 0$ are real, then so are the exponents in the asymptotic expansion of $\theta_{L^{1/r},A}$ near $t = 0$.

Proof. Since $\zeta(s)$ is meromorphic as a function of s, so is $\zeta(rs)$, and we can apply Theorem 7.4. The reality statement is immediate from the proof of Lemma 7.3.

Let us consider two zeta functions ζ_1 and ζ_2 corresponding to the sequences (L_1, A_1) and (L_2, A_2), respectively. Let (L_3, A_3) be the **tensor product**, which we define so that L_3 is the family of all products

$$L_3 = \{\lambda_k \lambda'_j\} \text{ with } \lambda_k \in L_1 \text{ and } \lambda'_j \in L_2;$$

while A_3 is the family of all products $\{a_k a'_j\}$, so then the zeta function ζ_3 is simply the product

$$\zeta_3 = \zeta_1 \zeta_2.$$

When written in full, the zeta function ζ_3 is

$$\zeta_3(s) = \sum_{k,j}^{\infty} \frac{a_k a'_j}{(\lambda_k \lambda'_j)^s},$$

and the associated theta series θ_3 reads

$$\theta_3(t) = [\theta_1 \otimes \theta_2](t) = \sum_{k,j}(a_k a'_j)e^{-(\lambda_k \lambda'_j)t}.$$

To study ζ_3 and θ_3, we shall apply Lemma 7.1. For this, we define the **truncated zeta function** to be

$$\zeta^{(N)}(s) = \sum_{k=N}^{\infty} a_k \lambda_k^{-s} = \zeta(s) - \sum_{k=1}^{N-1} a_k \lambda_k^{-s}.$$

Theorem 7.8. Let (L_1, A_1) and (L_2, A_2) be sequences such that the associated theta series satisify the three asymptotic conditions **AS 1**, **AS 2** and **AS 3**. Assume there exists an $\epsilon > 0$ such that for all j and k, we have

$$-\frac{\pi}{2} + \epsilon \leq \arg(\lambda_k) + \arg(\lambda'_j) \leq \frac{\pi}{2} - \epsilon,$$

or, in other words, the tensor product series satisfies **DIR 3**. Further, assume there exists some N such that the truncated zeta functions

$$\zeta_1^{(N)}, \zeta_2^{(N)}, \text{ and the product } \zeta_1^{(N)}\zeta_2^{(N)}$$

are in Dom(**V**). Then the theta series associated to the tensor product (L_3, A_3) satisfies satisfies the three asymptotic conditions **AS 1**, **AS 2** and **AS 3**. If the exponents in the asymptotic expansion **AS 2** are real for θ_1 and θ_2, then they are also real for $\theta_3 = \theta_1 \otimes \theta_2$.

Proof. Since θ_1 and θ_2 satisify **AS 1**, it is immediate that θ_3 satisfies **AS 1**. Also, it is immediate that the tensor product (L_3, A_3) satisifes the convergence conditions **DIR 1** and **DIR 2** since the product of two absolutely convergent Dirichlet series is absolutely convergent. Therefore, by Corollary 1.10, we know that the zeta functions ζ_1 and ζ_2 are meromorphic in **C**. By assuming that the tensor product series satisifies **DIR 3**, Proposition 7.2 applies to show that $\theta_3 = \theta_1 \otimes \theta_2$ satisifes **AS 3**. Finally, by assuming that the truncated zeta functions are in Dom(**V**), we may apply Theorem 7.4 to conclude that θ_3 satisfies **AS 2**.

In the case of real exponents for the asymptotic expansions of θ_1 and θ_2, we know, again by Corollary 1.10, that the poles of ζ_1 and ζ_2 are of the form $-(p+n)$ with real p and $n \in \mathbf{Z}$. Hence, by the proof of Lemma 7.3, the exponents of the asymptotic expansion for θ_3, which are the poles of ζ_3, are also real, thus concluding proof of the theorem. □

Example 1. Let L be the sequence of eigenvalues associated to a Laplacian that acts on C^∞ functions on a compact hyperbolic Riemann surface X. The parametrix construction of the heat kernel shows that L satisfies the three asymptotic conditions **AS 1**, **AS 2** and **AS 3**. From Theorem 7.6 we have that $L - 1/4$ satisfies the three asymptotic conditions, and Theorem 7.7 implies that the sequence

$$\sqrt{L - 1/4} = \{\sqrt{\lambda_k - 1/4}\} = \{r_k\}$$

also satisfies the three asymptotic conditions. The sequence

$$1/2 + \sqrt{-1} \cdot \sqrt{L - 1/4} = \{1/2 + ir_k\}$$

is precisely the set of zeros $\{\rho\}$ with $\text{Im}(\rho) > 0$ for the Selberg zeta function associated to X. Note that the general Cramér theorem proved in [JoL 92b] also proves that the sequence $\sqrt{L - 1/4}$ satisfies the three asymptotic conditions **AS 1**, **AS 2** and **AS 3**.

Essentially the same argument, applied in reverse, holds for general zeta functions, such as those associated to the theta function as in Cramér's theorem ([JoL 92b]).

Example 2. The Dedekind zeta function of a number field

$$\zeta(s) = \sum_{\mathfrak{a}} \mathbf{N}\mathfrak{a}^{-s} = \sum_{k=1}^{\infty} a_k k^{-s}$$

satisfies all three conditions **DIR 1**, **DIR 2**, and **DIR 3**, and lies in $\text{Dom}(\mathbf{V})$. Here a_k is the number of ideals $\mathfrak{a}$ with $\mathbf{N}\mathfrak{a} = k$. Therefore, by Theorem 7.5, the associated theta function

$$\theta(t) = \sum_{k=1}^{\infty} a_k e^{-kt}$$

satisfies the three asymptotic conditions, especially **AS 2**. Note that this theta function is different from the theta function used in the classical (Hecke) proof of the functional equation. A similar remark of course holds for the L-series.

TABLE OF NOTATION

Because of an accumulation of conditions and notation, and their use in the current series of papers, we tabulate here the main objects and conditions that we shall consider.

We let $L = \{\lambda_k\}$ and $A = \{a_k\}$ be sequences of complex numbers. To these sequences we associate various objects.

A **Dirichlet series** or **zeta function**:

$$\zeta_{A,L}(s) = \zeta(s) = \sum_{k=1}^{\infty} a_k \lambda_k^{-s},$$

and, more generally, for each positive integer N, the **truncated Dirichlet series**

$$\zeta_{A,L}^{(N)}(s) = \zeta^{(N)}(s) = \sum_{k=N}^{\infty} a_k \lambda_k^{-s}.$$

The sequences L and A (or Dirichlet series) may be subject to the following conditions:

DIR 1. For every positive real number c, there is only a finite number of k such that $\text{Re}(\lambda_k) \leq c$.

DIR 2.
   (a) The Dirichlet series
   $$\sum_k \frac{a_k}{\lambda_k^{\sigma}}$$
   converges absolutely for some real σ. Equivalently, we can say that there exists some $\sigma_0 \in \mathbf{R}_{>0}$ such that
   $$|a_k| = O(|\lambda_k|^{\sigma_0}) \text{ for } k \to \infty.$$
   (b) The Dirichlet series
   $$\sum_k \frac{1}{\lambda_k^{\sigma}}$$
   converges absolutely for some real σ. Specifically, let σ_1 be a real number for which
   $$\sum_k \frac{1}{|\lambda_k|^{\sigma_1}} < \infty.$$

We define m_0 to be the smallest *integer* ≥ 1 such that

$$\sum_{1}^{\infty} \frac{|a_k|}{|\lambda_k|^{m_0}} < \infty.$$

DIR 3. There is a fixed $\epsilon > 0$ such that for all k sufficiently large, we have

$$-\frac{\pi}{2} + \epsilon \leq \arg(\lambda_k) \leq \frac{\pi}{2} - \epsilon.$$

Equivalently, there exists positive constants C_1 and C_2 such that for all k,

$$C_1|\lambda_k| \leq \operatorname{Re}(\lambda_k) \leq C_2|\lambda_k|.$$

A **theta series** or **theta function**:

$$\theta_{A,L}(t) = \theta(t) = a_0 + \sum_{k=1}^{\infty} a_k e^{-\lambda_k t},$$

a **reduced theta series**

$$\theta_{L,A}^{(1)}(t) = \theta^{(1)}(t) = \sum_{k=1}^{\infty} a_k e^{-\lambda_k t},$$

and, more generally, for each positive integer N, the **truncated theta series**

$$\theta_{L,A}^{(N)}(t) = \theta^{(N)}(t) = \sum_{k=N}^{\infty} a_k e^{-\lambda_k t}.$$

The **asymptotic exponential polynomials** for integers $N \geq 1$:

$$Q_N(t) = a_0 + \sum_{k=1}^{N-1} a_k e^{-\lambda_k t}.$$

We are also given a sequence of complex numbers

$$\{p\} = \{p_0, \ldots, p_j, \ldots\}$$

with

$$\operatorname{Re}(p_0) \leq \operatorname{Re}(p_1) \leq \cdots \leq \operatorname{Re}(p_j) \leq \ldots$$

increasing to infinity. To every p in this sequence, we associate a polynomial B_p and we set

$$b_p(t) = B_p(\log t).$$

We then define:

The **asymptotic polynomials at 0**:

$$P_q(t) = \sum_{\operatorname{Re}(p) < \operatorname{Re}(q)} b_p(t) t^p.$$

We define

$$m(q) = \max \deg B_p \quad \text{for} \quad \operatorname{Re}(p) = \operatorname{Re}(q)$$

and

$$n(q) = \max \deg B_p \quad \text{for} \quad \operatorname{Re}(p) < \operatorname{Re}(q).$$

Let $\mathbf{C}\langle T \rangle$ be the algebra of polynomials in T^p with arbitrary complex powers $p \in \mathbf{C}$. Then, with this notation, $P_q(t) \in \mathbf{C}[\log t]\langle t \rangle$.

A function f on $(0, \infty) = \mathbf{R}_{>0}$ may be subject to **asymptotic conditions**:

AS 1. Given a positive number C and $t_0 > 0$, there exists N and $K > 0$ such that

$$|f(t) - Q_N(t)| \leq K e^{-Ct} \text{ for } t \geq t_0.$$

AS 2. For every q, we have

$$f(t) - P_q(t) = O(t^{\operatorname{Re}(q)} |\log t|^{m(q)}) \text{ for } t \to 0.$$

We will write **AS 2** as

$$f(t) \sim \sum_p b_p(t) t^p.$$

AS 3. Given $\delta > 0$, there exists an $\alpha > 0$ and a constant $C > 0$ such that for all N and $0 < t \leq \delta$ we have

$$|\theta(t) - Q_N(t)| \leq C/t^\alpha.$$

Given a polynomial B_p a direct calculation shows that there exist polynomials B_p^*, $\widetilde{B}_p$ and $B_p^\#$ of degree $\leq \deg B_p + 1$ such that:

$$\mathrm{CT}_{s=0} B_p(\partial_s) \left[\frac{\Gamma(s+p)}{z^{s+p}}\right] = z^{-p} B_p^*(\log z)$$

$$\mathrm{CT}_{s=1} B_p(\partial_s) \left[\frac{\Gamma(s+p)}{z^{s+p}}\right] = z^{-p-1} \widetilde{B}_p(\log z)$$

$$\mathrm{CT}_{s=0} B_p(\partial_s) \left[\frac{\pi z^{s+p-1}}{\sin[\pi(s+p)]}\right] = z^{p-1} B_p^\#(\log z)$$

The possible pole of $\Gamma(s+p)$ at $s = 0$ or $s = 1$ and the possible zero of $\sin[\pi(s+p)]$ at $s = 0$ accounts for the possibility of $\deg B_p^*$, $\deg \widetilde{B}_p$, or $\deg B_p^\#$ exceeding $\deg B_p$.

Part II

A Parseval Formula for Functions with a Singular Asymptotic Expansion at the Origin

Introduction

We shall determine the Fourier transform of a fairly general type of function φ which away from the origin has derivatives that are of bounded variation on $\mathbf{R}$ and are in $L^1(\mathbf{R})$, but at the origin the function has a principal part which is a generalized polynomial in the variable x and $\log x$, namely

$$\varphi(x) = \sum_p B_p(\log x) x^p + O\left(|\log x|^m\right),$$

where $\{p\}$ ranges over a finite number of complex numbers with $\operatorname{Re}(p) < 0$, for all p, B_p is a polynomial, and m is some positive integer. Thus the Fourier transform is determined as a distribution, and more generally as a functional on a large space of test functions also to be described explicitly. Alternatively, we may say that we are proving the Parseval formula for such a function φ and its Fourier transform.

Aside from the Parseval formula having intrinsic interest in pure Fourier analysis, it arises in a natural way in analytic number theory, in the theory of differential and pseudo differential operators, and more generally in the theory of regularized products as developed in [JoL 92a].

In the so-called "explicit formulas" of number theory, one proves essentially that the sum of a suitable function taken over the primes is equal to the sum of the Fourier-Mellin transform taken over the zeros of the zeta function. Classically, only very special cases were given (see Ingham [In 32]), and Weil was the first to observe that the formula was valid on a rather large space of test functions, and could be expressed as an equality of functionals. The sum over the primes includes a term at infinity, which amounts to an integral of the test function against the logarithmic derivative of the gamma function, taken over a vertical line $\operatorname{Re}(s) = a$ [We 52]. Weil proved what amounted to a Parseval formula, by determining what amounted to the Fourier transform of the logarithmic derivative of the gamma function on a vertical line in an explicit form. Weil's functional was reproduced with some additional details (making use of some general results concerning general Schwarz distributions) in [La 70]. However, the form in which Weil (and [La 70]) left the functional at infinity still required what appeared as too complicated arguments to identify it with the classical forms in the classical special cases. Barner reformulated the Weil functional in a more practical form [Ba 81], [Ba 90] and also extended the domain of validity of the formula. Our Parseval formula includes as a special case the formulas of Weil and Barner for the gamma function.

In spectral theory or in the theory of regularized series and products, there arises a regularized harmonic series R and a regularized product (or regularized determinant) D. As a corollary of the Parseval formula, we determine the Fourier transform of R, and of the logarithmic derivative $D'/D(a + it)$ on vertical lines of the form $a + it$, where D is the regularized product from §3 of [JoL 92a]. This determination is applied in our general version of explicit formulas [JoL 93]. Several other applications will also be given in subsequent papers, including for instance functional equations for general theta functions.

As to the proof, we shall first give a special case which covers the classical case of number theory and Barner's formulation. In this special case, the Parseval formula involves only the Dirac functional applied to the test function, whereas the general case concerning arbitrary regularized determinants involves higher derivatives of the test function. These higher derivatives come from the polar part in the asymptotic expansion of the theta function at 0. The special case occurs when this polar part consists only of $1/x$.

Our proof in the special case is more direct than Barner's or Weil's, and our formulation of the result already exhibits some general principles which were not immediately apparent in previous proofs. In particular, we avoid what now appear as detours by formulating lemmas in pure Fourier analysis showing how the singularity behaves under Fourier transform. These lemmas are used in the special and general case. We are indebted to Peter Jones for a proof of one of these lemmas, which has independent interest in general Fourier analysis.

§1. A Theorem on Fourier Integrals

This section is preliminary, and proves some lemmas on Fourier integrals which will be used in the next section.

We recall the **Fourier transform**

$$f^\wedge(t) = \frac{1}{\sqrt{2\pi}} \int_{-\infty}^{\infty} f(x) e^{-itx} dx.$$

We shall be concerned with Fourier inversion. For $f \in L^1(\mathbf{R})$ and $A > 0$ we define

$$f_A(x) = \frac{1}{\pi} \int_{-\infty}^{\infty} f(y) \frac{\sin A(x-y)}{x-y} dy = \frac{1}{\sqrt{2\pi}} \int_{-A}^{A} f^\wedge(t) e^{itx} dt.$$

The middle expression with the sine comes from the last expression and the definition of $f^\wedge$, after an application of Fubini's theorem and the evaluation of a simple integral of elementary calculus. Let

$$f^-(x) = f(-x).$$

We are interested in seeing how f_A converges to f, that is we want the inversion formula $f^{\wedge\wedge} = f^-$ in the form

$$f(x) = \lim_{A \to \infty} f_A(x).$$

We shall give conditions under which the inversion formula is true.

We let the **Schwartz space** $\text{Sch}(\mathbf{R})$ be the vector space of functions which are infinitely differentiable, and such that the function and all its derivatives tend rapidly to 0 at infinity. That f **tends rapidly to 0 at infinity** means that for all polynomials P the function Pf is bounded. Then the Schwartz space is self dual, that is

$$\text{Sch}(\mathbf{R})^\wedge = \text{Sch}(\mathbf{R}).$$

An elementary result of analysis asserts that the formula $f^{\wedge\wedge} = f^-$ is true for f in the Schwartz space. We shall assume such an elementary result, and extend it to functions which are less smooth, namely we shall deal with functions of bounded variation. All the background material needed is contained in [La 93]. We let $BV(\mathbf{R})$

denote the space of complex valued functions of bounded variation on **R**, i.e. the space of functions of bounded variation on each finite interval $[a, b]$, and such that the total variations are uniformly bounded for all $[a, b]$, in other words there exists $B > 0$ such that

$$V(f, a, b) \leq B \text{ for all } [a, b].$$

We let

$$V_{\mathbf{R}}(f) = \sup_{[a,b]} V(f, a, b).$$

Remark. *If $f \in \mathrm{BV}(\mathbf{R}) \cap L^1(\mathbf{R})$ then $f(x) \to 0$ as $x \to \pm\infty$.*

Indeed, if $|f(x)| \geq c > 0$ for infinitely many x tending to infinity, since $f \in L^1(\mathbf{R})$ there are infinitely many y such that $|f(y)| \leq c/2$ (say), and so the function could not be of bounded variation.

We let $d\mu_f$ be the Riemann-Stieltjes measure associated with f, and sometimes abbreviate $d\mu_f$ by df.

The function f_A exhibits different behavior near 0 and at infinity. Its behavior will be described in part in the lemmas below. We shall need the function

$$S(x) = \int_0^x \frac{\sin t}{t} dt,$$

so S is continuous and bounded on **R**. In fact, S is bounded by the area under the first arch of $(\sin t)/t$ (between 0 and π), as follows at once by the alternating nature of the integrand.

Lemma 1.1. *Let $f \in \mathrm{BV}(\mathbf{R}) \cap L^1(\mathbf{R})$. Then for $A > 0$ the function f_A is bounded, and in fact there is a uniform bound, independent of A:*

$$\|f_A\|_\infty \leq \frac{1}{\pi} \|S\|_\infty V_{\mathbf{R}}(f),$$

where $\| \ \|$ is the sup norm.

Proof. Integrating by parts yields

$$f_A(x) = \int_{-\infty}^{\infty} f(x - t) \frac{d}{dt} S(At) dt$$

$$= -\int_{-\infty}^{\infty} S(At) df(t - x).$$

The desired bound follows by the standard absolute estimates. We have used here that over a finite interval $[a, b]$, the two terms $f(b)S(Ab)$ and $f(a)S(Aa)$ tend to 0 as $a \to -\infty, b \to \infty$ because S is bounded, and f tends to 0. This concludes the proof. $\square$

We recall a classical theorem from Fourier analysis giving natural conditions under which Fourier inversion holds, especially at a discontinuity. For this purpose, we shall say that a function f is **normalized at a point** x if

$$f(x) = \frac{1}{2}[f(x+) + f(x-)].$$

Thus the right and left limits of f exist at x, and the value of f at x is the midpoint. We say that f is **normalized** if f is normalized at every $x \in \mathbf{R}$.

Theorem 1.2. *Let $f \in \mathrm{BV}(\mathbf{R}) \cap L^1(\mathbf{R})$, and suppose f is normalized. Then f_A is bounded independently of A and*

$$\lim_{A \to \infty} f_A(x) = f(x) \text{ for all } x \in \mathbf{R}.$$

For a proof, see Titchmarsh [Ti 48].

The remainder of this section is devoted to examining the uniformity of the convergence in Theorem 1.2. To begin, we mention a very special case.

Lemma 1.3. *There exists a function $\alpha \in \mathrm{Sch}(\mathbf{R})$ such that $\alpha^\wedge$ has compact support, and $\alpha(0) \neq 0$. For such a function, we have*

$$\alpha_A = \alpha$$

for all A sufficiently large.

Proof. Let $\beta \in C_c^\infty(\mathbf{R})$ be an even function ≥ 0 with $\beta(0) > 0$. Let $\alpha = \beta^\wedge$. Then $\beta = \alpha^\wedge$ has compact support, and the direct evaluation of the Fourier integral together with the definition of α_A shows that the other conditions are satisfied. $\square$

The following quantitative formulation of the Riemann-Lebesgue lemma is proved by Barner in [Ba 90], Satz 82 in §21, to which the reader is referred for a proof.

Lemma 1.4. *Assume:*

(a) $g \in BV(\mathbf{R})$.

(b) $g(x) = O(|x|^\epsilon)$ *for some* $\epsilon > 0$ *as* $x \to 0$.

Then the improper integral that follows exists for $A > 0$ and satisfies the bound

$$\int_0^\infty g(y) e^{iAy} \frac{dy}{y} = O_g(A^{-\frac{\epsilon}{1+\epsilon}}).$$

Also, we need the following elementary lemma.

Lemma 1.5. *For all $0 < a < b$ and $A > 0$ we have*

$$\left| \int_a^b \frac{\sin At}{t} \, dt \right| \leq \frac{3}{Aa}.$$

Proof. This follows from the change of variables $u = At$, the alternating nature of the integrand, and an elementary area estimate. □

The main result of this section is the following uniform version of Theorem 1.2.

Theorem 1.6. *Let $g \in BV(\mathbf{R}) \cap L^1(\mathbf{R})$ and assume*

$$g(x) = O(|x|^\epsilon) \text{ for } |x| \to 0.$$

Let $\delta = \min(\frac{1}{8}, \frac{\epsilon}{16})$. Then for all $A > 1$,

$$g_A(x) - g_A(0) = O_g(|x|^\delta) \text{ for } |x| \to 0,$$

the estimate on the right being independent of A.

Theorem 1.6 will be proved using a series of lemmas. Before continuing, let us state the following corollary of Theorem 1.6 that further refines the uniformity of the pointwise convergence result stated in Theorem 1.2.

Corollary 1.7. *Assume g has M derivatives and all the functions $g, g^{(1)}, \ldots, g^{(M)}$ are in $BV(\mathbf{R}) \cap L^1(\mathbf{R})$. Assume that*

$$g(x) = O(|x|^{M+\epsilon}) \quad \text{for } |x| \to 0.$$

Then:

(a) *The function g_A has M derivatives $(g_A)^{(1)}, \ldots, (g_A)^{(M)}$ and*

$$(g_A)^{(k)} = \left(g^{(k)}\right)_A \quad \text{for } k = 1, \ldots M,$$

so, without ambiguity, we can write the derivatives of g_A as $g_A^{(1)}, \ldots, g_A^{(M)}$.

(b) *We have*

$$g_A(x) - \sum_{k=0}^{M} g_A^{(k)}(0) \frac{x^k}{k!} = O(|x|^{M+\delta}) \quad \text{for } |x| \to 0,$$

the estimate on the right being independent of A.

Proof. Let h be a differentiable function on $\mathbf{R}$, with derivative h', such that $h, h' \in BV(\mathbf{R}) \cap L^1(\mathbf{R})$. Since $d/du((\sin Au)/u)$ is bounded, one can interchange derivative and integral (see page 357 of [La 83]). These calculations yield that

$$\frac{d}{dx} h_A(x) = h_A^{(1)}(x) = \frac{d}{dx} \int_{-\infty}^{\infty} h(y) \frac{\sin A(y-x)}{(y-x)} dy$$

$$= \int_{-\infty}^{\infty} h(y) \frac{d}{dx} \left[\frac{\sin A(y-x)}{(y-x)} \right] dy$$

$$= \int_{-\infty}^{\infty} h(y) \left(-\frac{d}{dy}\right) \left[\frac{\sin A(y-x)}{(y-x)} \right] dy$$

$$= \int_{-\infty}^{\infty} \frac{d}{dy} h(y) \left[\frac{\sin A(y-x)}{(y-x)} \right] dy$$

$$= \left[\frac{d}{dx} h\right]_A (x).$$

The integration by parts step is valid since $(\sin Au)/u$ is bounded and $h(x)$ approaches zero as x approaches infinity. This proves (a) by induction, letting $h = g^{(k)}$. As to (b), if we apply Theorem 1.6 to the function $g_A^{(M)}(x)$ we get

$$g_A^{(M)}(x) - g_A^{(M)}(0) = O_g(|x|^\delta).$$

By repeatedly integrating this equation from 0 to x and applying (a), we get (b), thus proving the corollary. □

The remainder of this section is devoted to the proof of Theorem 1.6. To do so, we will write

$$g_A(x) = \int_{-1}^{1} g(y) \frac{\sin A(x-y)}{x-y} dy + \left(\int_{-\infty}^{-1} + \int_{1}^{\infty} \right) g(y) \frac{\sin A(x-y)}{x-y} dy,$$

and investigate the finite and infinite intervals separately. For notational simplicity, we will consider the integrals over the intervals $[0, 1]$ and $[1, \infty)$, with the analysis over the intervals $(-\infty, -1]$ and $[-1, 0]$ being identical to that over the corresponding positive intervals. We begin with the analysis over the finite intervals.

Proposition 1.8. *Let $g \in \mathrm{BV}(\mathbf{R}) \cap L^1(\mathbf{R})$. Assume in addition that there exists $\varepsilon > 0$ such that*

$$g(x) = O(|x|^\varepsilon) \text{ for } x \to 0.$$

Then there is $\delta > 0$ such that for all $A \geq 1$ we have

$$\int_0^1 g(y) \left[\frac{\sin Ay}{y} - \frac{\sin A(x-y)}{x-y} \right] dy = O_g(|x|^\delta) \text{ for } x \to 0,$$

the estimate on the right being independent of A.

Proof. We owe the proof of this proposition to Peter Jones, who showed that one can take $\delta = \min(\frac{1}{8}, \frac{\varepsilon}{16})$, as stated in Theorem 1.6. We need to split the integral over various intervals, depending on A and x. We first settle the easiest case.

Case I. Suppose $|x| < A^{-4}$ and, say, $x > 0$. Then

$$\left| \int_0^1 g(y) \left[\frac{\sin Ay}{y} - \frac{\sin A(x-y)}{x-y} \right] dy \right| \ll \|g\|_1 x^{1/2}.$$

Proof. Since

$$\left| \frac{d}{dy} \left(\frac{\sin Ay}{y} \right) \right| \ll A^2,$$

we use the Mean Value Theorem and the hypothesis $x < A^{-4}$ to get the bound

$$\left| \frac{\sin Ay}{y} - \frac{\sin A(x-y)}{x-y} \right| \ll A^2 x \leq x^{1/2},$$

and then we estimate the desired integral in the coarsest way with the sup norm to conclude the proof of the present case. $\square$

Case II. Suppose $|x| \geq A^{-4}$, so $A^{-1} \leq |x|^{1/4}$.

In this case, we will bound each separate integral without using the difference of sines. Lemma 1.4 takes care of the term with $(\sin Ay)/y$, giving a bound of $x^{\epsilon/8}$. The next lemma takes care of the term with $(\sin A(x-y))/(x-y)$.

Lemma 1.9. *With the implied constant in $\ll$ depending on $\|g\|_1$, $\|g\|_\infty$, and $V_{\mathbf{R}}(\dot{g})$, we have for $x \geq A^{-4}$:*

$$\left| \int_0^1 g(y) \frac{\sin A(x-y)}{x-y} dy \right| \ll x^{1/8} + x^{\epsilon/16},$$

the estimate on the right independent of A.

Proof. We split the integral

$$\int_0^1 g(y) \frac{\sin A(x-y)}{x-y} dy$$

$$= \int_{|x-y| \leq A^{-1}} + \int_{A^{-1} \leq |x-y| \leq A^{-1} x^{-1/8}} + \int_{A^{-1} x^{-1/8} \leq |x-y|}$$

$$= I_1 + I_2 + I_3.$$

For the first integral I_1, we change variable putting $u = x - y$. Then the limits of integration for u are $|u| \leq A^{-1}$, and we get a bound

(3) $$|I_1| \leq \|g\|_1 \, AA^{-1} \ll (x + A^{-1})^\varepsilon \ll x^{\varepsilon/4}.$$

For the second integral I_2, again with $u = x - y$, we find that

$$|I_2| \leq \|g\|_\infty \int_{A^{-1}}^{A^{-1}x^{-1/8}} \frac{du}{u} \ll \|g\|_\infty \log 1/x$$

where the interval over which we take the sup norm of g is

$$A^{-1} \leq |x - y| < A^{-1}x^{-1/8}.$$

Using the growth estimate $g(x) = O(|x|^\varepsilon)$ for x near zero and the assumption $A^{-1} \leq x^{1/4}$, we get

(4) $$\begin{aligned}|I_2| &\ll (x + A^{-1}x^{-1/8})^\varepsilon \log(1/x) \\ &\ll (x + x^{1/8})^\varepsilon \log(1/x) \ll x^{\varepsilon/16}.\end{aligned}$$

For the third integral I_3, since the set of discontinuities of a function of bounded variation is countable, we can select x_0 such that g is continuous at x_0 and

$$x + A^{-1}x^{-1/8} \leq x_0 \leq x + 2A^{-1}x^{-1/8}.$$

For simplicity, we just look at $y \geq x_0$. The other piece for I_3 is done in the same way. We then decompose the integral over $y \geq x_0$ into a sum:

$$\begin{aligned}\int_{y \geq x_0} g(y) \frac{\sin A(x-y)}{x-y} dy &= \int_{y \geq x_0} g(x_0) \frac{\sin A(x-y)}{x-y} dy \\ &\quad + \int_{y \geq x_0} (g(y) - g(x_0)) \frac{\sin A(x-y)}{x-y} dy \\ &= J_1 + J_2,\end{aligned}$$

and we estimate J_1, J_2 successively. For the integral J_1, using Lemma 1.5, we find:

$$|J_1| \ll |g(x_0)| \frac{1}{A(x_0 - x)}$$
$$\ll (x + 2A^{-1}x^{-1/8})^\varepsilon A^{-1} A x^{1/8}$$
(5) $$\ll x^{1/8}.$$

For the integral J_2, let χ_y be the characteristic function of $[0, y)$, that is put

$$\chi(y, t) = \chi_y(t) = \begin{cases} 1 & \text{if } t < y \\ 0 & \text{if } t \geq y. \end{cases}$$

If g is continuous at y, then

$$g(y) - g(x_0) = \int_{x_0}^{1} \chi(y, t) d\mu_g(t).$$

This is a convenient expression to plug into an application of Fubini's theorem which gives

$$J_2 = \int_{x_0}^{1} \int_{x_0}^{1} \chi(y, t) \frac{\sin A(x - y)}{x - y} d\mu_g(t) dy$$

$$= \int_{x_0}^{1} \left[\int_{x_0}^{1} \chi(y, t) \frac{\sin A(y - x)}{y - x} dy \right] d\mu_g(t)$$

$$= \int_{x_0}^{1} \left[\int_{t}^{1} \frac{\sin A(y - x)}{y - x} dy \right] d\mu_g(t)$$

$$= \int_{x_0}^{1} \left[\int_{t-x}^{1-x} \frac{\sin Au}{u} du \right] d\mu_g(t).$$

Hence by Lemma 1.5, we find

$$|J_2| \leq \int_{x_0}^{1} \frac{3}{A(t - x)} |d\mu_g(t)| \ll A^{-1}(x_0 - x)^{-1} V_{\mathbf{R}}(g)$$
(6) $$\ll A^{-1} A x^{1/8} = x^{1/8},$$

which concludes the estimate of the second integral J_2, and therefore of the integral I_3. Thus (3), (4), (5), and (6) conclude the proof of Lemma 1.9. □

Having taken care of all cases, we have completed the proof of Proposition 1.8.

To finish the proof of Theorem 1.6, we need the following proposition.

Proposition 1.10. *Let $g \in BV(\mathbf{R}) \cap L^1(\mathbf{R})$. Then for all $A \geq 1$ we have*

$$\int_1^\infty g(y) \left[\frac{\sin Ay}{y} - \frac{\sin A(x-y)}{x-y} \right] dy = O_g(|x|^{1/4}) \text{ for } x \to 0,$$

the estimate on the right being independent of A.

The proof of Proposition 1.10, which is much easier than that of Proposition 1.8, will be given through the following lemmas.

Lemma 1.11. *Assume $g \in BV(\mathbf{R}) \cap L^1(\mathbf{R})$. Then*

$$\left| \int_1^\infty g(y) \frac{\sin Ay}{y} dy \right| \leq \frac{3}{A} V_\mathbf{R}(g).$$

Proof. Extend g to $[0, \infty)$ by defining $g(t) = 0$ if $0 \leq t < 1$. For fixed $t \in (0,1)$ and $b > 1$, consider the integral

$$\int_t^b g(y) \frac{\sin Ay}{y} dy.$$

Let

$$S_A(x) = -\int_x^\infty \frac{\sin Ay}{y} dt$$

Using integration by parts we have

$$\int_t^b g(y) \frac{\sin Ay}{y} dy = g(y) S_A(y) \Big|_t^b - \int_t^b S_A(y) dg(y).$$

The point evaluations of $g(y)S_A(y)$ will be zero as b approaches infinity, since $g(t) = 0$, $g(b)$ approaches zero, and $S_A(y)$ is bounded. Also, by Lemma 1.5

$$\left| \int_t^b S_A(y) dg(y) \right| \leq \sup_{[t,b]} |S_A| \, V_{\mathbf{R}}(g) \leq \frac{3}{At} V_{\mathbf{R}}(g).$$

Therefore, we have shown, after letting t approach 1 and b approach ∞ that

$$\left| \int_1^\infty g(y) \frac{\sin Ay}{y} dy \right| \leq \frac{3}{A} V_{\mathbf{R}}(g),$$

as asserted. □

We now consider the integral in Proposition 1.10 in two separate cases, when $x \leq 1/A^4$ and when $x > 1/A^4$.

Case I. *Assume $x \leq 1/A^4$. Then there is a universal constant C such that*

$$\left| \int_1^\infty g(y) \left[\frac{\sin Ay}{y} - \frac{\sin A(x-y)}{x-y} \right] dy \right| \leq (C \|g\|_1) x^{1/2},$$

the estimate on the right independent of A.

Proof. By the Mean Value Theorem we have

$$\left| \frac{\sin A(y-x)}{y-x} - \frac{\sin Ay}{y} \right| \leq CA^2 x.$$

So, we can bound the integral in question by

$$CA^2 x \|g\|_1 \leq (C \|g\|_1) x^{1/2}. \quad \square$$

Case II. Assume $x > 1/A^4$ and, without loss of generality, we can also assume that $x \leq 1/2$. Then

$$\left| \int_1^\infty g(y) \left[\frac{\sin Ay}{y} - \frac{\sin A(x-y)}{x-y} \right] dy \right| \leq C_g x^{1/4},$$

where the constant C_g is independent of A, and depends only on g.

Proof. Let us estimate the integrals separately. In fact, write the integrals as

$$\int_1^\infty g(y) \frac{\sin A(y-x)}{y-x} dy - \int_1^\infty g(y) \frac{\sin Ay}{y} dy$$

$$= \int_{1-x}^\infty g(u+x) \frac{\sin Au}{u} du - \int_1^\infty g(y) \frac{\sin Ay}{y} dy$$

$$= \int_{1-x}^1 g(u+x) \frac{\sin Au}{u} du + \int_1^\infty g(u+x) \frac{\sin Au}{u} du$$

$$- \int_1^\infty g(y) \frac{\sin Ay}{y} dy.$$

The first integral is bounded by $\|g\|_\infty x$ since $(\sin Au)/u$ is bounded on $[1/2, 1]$. Lemma 1.11 applies to bound the second integral, independent of x, as well as the third integral. The bound achieved is

$$\|g\|_\infty x + \frac{6}{A} V_{\mathbf{R}}(g).$$

Since $1/A < x^{1/4}$, Case II is proved.

Combining the two cases above, we have completed our proof of Proposition 1.10. With all this, the proof of Theorem 1.6 is completed by combining Proposition 1.8 and Proposition 1.10.

§2. A Parseval formula

We recall the definition of the hermitian product

$$\langle f, g \rangle = \int_{-\infty}^{\infty} f(x)\bar{g}(x)dx.$$

For f, g in the Schwartz space, we assume the elementary **Parseval Formula**

$$\langle f, g \rangle = \langle f^\wedge, g^\wedge \rangle.$$

The point of this section is to prove this formula under less restrictive conditions. We shall extend conditions of Barner [Ba 90], for which the formula is true. The basic facts from real analysis (functions of bounded variation and Stieltjes integral, Fourier transforms under smooth conditions) used here are contained in [La 83].

We call the following conditions from [Ba 90] the **basic conditions** on a function f:

Condition 1. $f \in BV(\mathbf{R}) \cap L^1(\mathbf{R})$.

Condition 2. There exists $\varepsilon > 0$ such that

$$f(x) = f(0) + O(|x|^\varepsilon) \quad \text{for } x \to 0.$$

Condition 3. f is normalized, as defined in §1.

In addition to functions satisfying the basic conditions, we shall deal with a special pair of functions, arising as follows. We suppose given:

- A Borel measure μ on $\mathbf{R}^+$ such that $d\mu(x) = \psi(x)dx$, where ψ is some bounded, (Borel) measurable function.

- A measurable function φ on $\mathbf{R}^+$ such that:
  (a) The function $\varphi_0(x) = \varphi(x) - 1/x$ is bounded as x approaches zero.
  (b) Both functions $1/x$ and $\varphi(x)$ are in $L^1(|\mu|)$ outside a neighborhood of zero.

Thus we impose two asymptotic conditions on φ, one condition near zero and one condition at infinity. We call (μ, φ) a **special**

pair. We define the functional

$$W_{\mu,\varphi}(\alpha) = \int_0^\infty \left(\varphi(x)\alpha(x) - \frac{\alpha(0)}{x} \right) d\mu(x).$$

Its "Fourier transform" as a distribution is the function $W_{\mu,\varphi}^\wedge$ such that

$$W_{\mu,\varphi}^\wedge(t) = \frac{1}{\sqrt{2\pi}} \int_0^\infty \left(\varphi(x)e^{-itx} - \frac{1}{x} \right) d\mu(x)$$

$$= \frac{1}{\sqrt{2\pi}} W_{\mu,\varphi}(\bar\chi_t) \quad \text{where} \quad \chi_t(x) = e^{itx}.$$

The following theorem generalizes the Barner-Weil formula [Ba 90].

Theorem 2.1. *Let f satisfy the three basic conditions, and let (μ,φ) be a special pair. Then*

$$\lim_{A\to\infty} \int_{-A}^A f^\wedge(t) W_{\mu,\varphi}^\wedge(t) dt = W_{\mu,\varphi}(f^-)$$

$$= \int_0^\infty \left[\varphi(x) f(-x) - \frac{f(0)}{x} \right] d\mu(x).$$

Proof. At a certain point in the proof, we shall need to distinguish two cases, but we proceed as far as we can go without such distinction, according to a rather standard pattern of proof. We have:

$$\int_{-A}^A f^\wedge(t) W_{\mu,\varphi}^\wedge(t) dt$$

$$= \frac{1}{2\pi} \int_{-A}^A dt \int_{-\infty}^\infty f(y) e^{-ity} dy \int_0^\infty \left[\varphi(x) e^{-itx} - \frac{1}{x} \right] d\mu(x)$$

$$= \frac{1}{2\pi} \int_{-A}^A dt \iint_{\mathbf{R}\times\mathbf{R}^+} f(y) \left[\varphi(x) e^{-it(x+y)} - \frac{e^{-ity}}{x} \right] dy\, d\mu(x).$$

By the assumptions on (μ, φ) and f, we can interchange the integrals which are absolutely convergent. We then perform the inner integration with respect to t, and the expression becomes

$$= \iint_{\mathbf{R}\times\mathbf{R}+} \frac{f(y)}{\pi} \left[\varphi(x) \frac{\sin A(x+y)}{x+y} - \frac{\sin Ay}{xy}\right] dy d\mu(x)$$

$$= \int_{\mathbf{R}+} \int_{\mathbf{R}} \frac{f(-y)}{\pi} \left[\varphi(x) \frac{\sin A(x-y)}{x-y} - \frac{\sin Ay}{xy}\right] dy d\mu(x)$$

$$= \int_{\mathbf{R}+} \left[\varphi(x) f_A(-x) - \frac{f_A(0)}{x}\right] d\mu(x).$$

At this point, we are finished with the proof in the case $f = \alpha$ is a function satisfying the conditions of Lemma 1.3, namely $\alpha_A = \alpha$ for sufficiently large A. For the general case, we write the above integral as a sum

$$= \int_{\mathbf{R}+} (\varphi(x) - \frac{1}{x}) f_A(-x) d\mu(x) + \int_{\mathbf{R}+} \frac{1}{x}(f_A(-x) - f_A(0)) d\mu(x).$$

Note that $\varphi(x) - 1/x$ is in $L^1(|\mu|)$ by our assumptions on (μ, φ).

Our final step is to prove that we can take the limit as $A \to \infty$ under the integral sign in both terms for an arbitrary f satisfying the basic conditions. In the first integral, one can apply the dominated convergence theorem by the boundedness of f_A (see Theorem 1.2) and the assumptions on $\varphi(x) - 1/x$. As to the second integral, we split the integral

$$\int_{\mathbf{R}+} = \int_0^1 + \int_1^\infty.$$

Again, we apply the dominated convergence theory to the integral over $[1, \infty)$, by Theorem 1.2, the boundedness of f_A, and the assumption that $1/x$ is in $L^1(\mu)$. The more difficult part is the integral over the inteval $[0, 1]$.

Let α be as in Lemma 1.3 and such that $\alpha(0) = f(0)$, which can be achieved after multiplying α by a constant. Let $g = f - \alpha$. The formula of Theorem 2.1 is linear in f, and it is immediately verified that α and hence g satisfies the basic conditions and

$$g(x) = O(|x|^\epsilon) \text{ for } x \to 0.$$

Having proved the formula for α, we are reduced to proving it for g. Recall that Theorem 1.6 states that for a positive δ, which depends on ϵ,

$$g_A(x) - g_A(0) = O(|x|^\delta) \text{ for } x \to 0$$

uniformly in A. Therefore, the dominated convergence theorem again applies since $x^{-1+\delta}$ is integrable over $[0,1]$, and we get, since $g(0) = 0$,

$$\lim_{A \to \infty} \int_0^1 \frac{1}{x}(g_A(-x) - g_A(0))d\mu(x) = \int_0^1 \frac{1}{x}g(-x)d\mu(x).$$

After taking the limit as $A \to \infty$ under the integrals we see that the final expression above becomes

$$\int_{\mathbf{R}+} (\varphi(x) - \frac{1}{x})f(-x)d\mu(x) + \int_{\mathbf{R}+} \frac{1}{x}(f(-x) - f(0))d\mu(x),$$

which proves the theorem. □

§3. The General Parseval Formula

Following the ideas in [JoL 92a], we now prove a general Parseval formula associated to measurable functions with arbitrary principal part, thus generalizing the results of §2 in which the function $\varphi(x)$ was required to have principal part equal to $1/x$.

Suppose we are given:

- A Borel measure μ on $\mathbf{R}^+$ such that $d\mu(x) = \psi(x)dx$, where ψ is some bounded (Borel) measurable function.

- A measurable function φ on $\mathbf{R}^+$ having the following properties. There is a function $P_0(x) \in \mathbf{C}[\log x]\langle x\rangle$, which we shall write as

$$P_0(x) = \sum_{\mathrm{Re}(p)<0} b_p(x)x^p \quad \text{with} \quad b_p(x) = B_p(\log x) \in \mathbf{C}[\log x]$$

such that:

(a) There is some integer $m > 0$ such that

$$\varphi(x) - P_0(x) = O(|\log x|^m) \quad \text{for } x \to 0.$$

(b) Let M be the largest integer $< -\mathrm{Re}(p_0)$, so that

$$-1 \leq M + \mathrm{Re}(p_0) < 0.$$

Then both functions $x^M P_0(x)$ and $\varphi(x)$ are in $L^1(|\mu|)$ outside a neighborhood of zero.

From condition (a) and the power series expansion of e^{itx} one obtains the existence of functions $u_k(x)$ such that

$$\varphi(x)e^{itx} - \sum_{k=0}^{M} u_k(x)(it)^k = O(|\log x|^m) \quad \text{as } x \to 0.$$

The functions $u_k(x)$ come from the expression

$$\sum_{k+\mathrm{Re}(p)<0} \frac{b_p(x)x^{p+k}}{k!}(it)^k = \sum_{k=0}^{M} u_k(x)(it)^k.$$

153

As before, the above requirements impose two asymptotic conditions on φ, one condition near zero and one condition at infinity, and we call such a pair (μ, φ) a **special pair**. We define the functional

$$W_{\mu,\varphi}(\alpha) = \int_0^\infty \left(\varphi(x)\alpha(x) - \sum_{k=0}^M u_k(x)\alpha^{(k)}(0) \right) d\mu(x).$$

Its "Fourier transform" as a distribution is the function $W^\wedge_{\mu,\varphi}(t)$ such that

$$W^\wedge_{\mu,\varphi}(t) = \frac{1}{\sqrt{2\pi}} \int_0^\infty \left(\varphi(x)e^{-itx} - \sum_{k=0}^M u_k(x)(-it)^k \right) d\mu(x)$$

$$= \frac{1}{\sqrt{2\pi}} W_{\mu,\varphi}(\bar\chi_t) \quad \text{where} \quad \chi_t(x) = e^{itx}.$$

The following theorem generalizes Theorem 2.1.

Theorem 3.1. *Assume f and its first M derivatives satisfy the three basic conditions, and let (μ, φ) be a special pair. Then*

$$\lim_{A\to\infty} \int_{-A}^A f^\wedge(t) W^\wedge_{\mu,\varphi}(t) dt = W_{\mu,\varphi}(f^-)$$

$$= \int_0^\infty \left[\varphi(x) f(-x) - \sum_{k=0}^M u_k(x)(-1)^k f^{(k)}(0) \right] d\mu(x).$$

Proof. The proof is essentially identical to the proof of Theorem

2.1. For completeness, let us present the details. We have:

$$\int_{-A}^{A} f^{\wedge}(t) W_{\mu,\varphi}^{\wedge}(t) dt$$

$$= \int_{-A}^{A} \frac{dt}{2\pi} \int_{-\infty}^{\infty} f(y) e^{-ity} dy \int_{0}^{\infty} \left[\varphi(x) e^{-itx} - \sum_{k=0}^{M} u_k(x)(-it)^k \right] d\mu(x)$$

$$= \int_{-A}^{A} \frac{dt}{2\pi} \iint_{\mathbf{R} \times \mathbf{R}^+} f(y) \left[\varphi(x) e^{-itx} - \sum_{k=0}^{M} u_k(x)(-it)^k \right] e^{-ity} dy d\mu(x).$$

As in the proof of Theorem 2.1, the assumptions on (μ, φ) and f allow us to interchange the integrals which are absolutely convergent. By integrating with respect to t, this expression becomes

$$\iint_{\mathbf{R} \times \mathbf{R}^+} \frac{f(y)}{\pi} \left[\varphi(x) \frac{\sin A(x+y)}{x+y} - \sum_{k=0}^{M} u_k(x) \left(\frac{d}{dy}\right)^k \left[\frac{\sin Ay}{y}\right] \right] dy d\mu(x).$$

Continuing, we have:

$$\iint_{\mathbf{R} \times \mathbf{R}^+} \frac{f^-(y)}{\pi} \left[\varphi(x) \frac{\sin A(x-y)}{x-y} - \sum_{k=0}^{M} u_k(x) \left(\frac{-d}{dy}\right)^k \left[\frac{\sin Ay}{y}\right] \right] dy d\mu(x)$$

$$= \int_{\mathbf{R}^+} \left[\varphi(x) f_A(-x) - \sum_{k=0}^{M} u_k(x)(-1)^k f_A^{(k)}(0) \right] d\mu(x).$$

In the above steps we have used the differentiation formula

$$\frac{1}{2} \int_{-A}^{A} (-it)^k e^{-ity} dt = \left(\frac{d}{dy}\right)^k \left[\frac{\sin Ay}{y}\right]$$

and the integration by parts formula

$$\int_{-\infty}^{\infty} \left[\left(-\frac{d}{dy}\right)^k \frac{\sin Ay}{y} \right] \frac{f(-y)}{\pi} dy = (-1)^k f_A^{(k)}(0),$$

155

which is valid by Lemma 1.1 and the arguments given in the proof of Corollary 1.7.

At this point, we are finished with the proof in the case $f = \alpha$ is a function satisfying the conditions of Lemma 1.3, namely $\alpha_A = \alpha$ for sufficiently large A. For the general case, we write the integral as the sum

$$\int_{\mathbf{R}^+} f_A(-x)(\varphi(x) - P_0(x))d\mu(x)$$

$$+ \int_1^\infty \left[f_A(-x)P_0(x) - \sum_{k=0}^M u_k(x)(-1)^k f_A^{(k)}(0) \right] d\mu(x)$$

$$+ \int_0^1 \left[f_A(-x)P_0(x) - \sum_{k=0}^M u_k(x)(-1)^k f_A^{(k)}(0) \right] d\mu(x).$$

The proof of Theorem 3.1 now finishes as did the proof of Theorem 2.1. By an appropriate extension of Lemma 1.3, choose an α for which $\alpha_A = \alpha$ for sufficiently large A and the numbers $\alpha(0), \ldots, \alpha^{(M)}(0)$ have been chosen to agree with the first M derivatives of f at zero, and set $g = f - \alpha$. The above integrals are linear in the function f, so proving the theorem for g will imply the theorem for f, so we work with g. By Theorem 1.2, the boundedness of g_A (as stated in Lemma 1.1), and assumption (b) above, one can apply the dominated convergence theorem to the first two integrals above. Using Corollary 1.7 and the definition of the function $u_k(x)$, we can write the third integral as

$$\int_0^1 \left[g_A(-x)P_0(x) - \sum_{k=0}^M u_k(x)(-1)^k g_A^{(k)}(0) \right] d\mu(x)$$

$$= \int_0^1 P_0(x) \left[g_A(-x) - \sum_{k=0}^M \frac{g_A^{(k)}(0)}{k!}(-x)^k \right] d\mu(x).$$

By Corollary 1.7, the integrand is bounded by $Cx^{M+\text{Re}(p_0)+\delta}$ and

$$M + \text{Re}(p_0) + \delta \geq -1 + \delta,$$

so $Cx^{M+\text{Re}(p_0)+\delta}$ is integrable. The dominated convergence theorem applies, and the theorem is proved. $\square$

§4. The Parseval Formula for $I_w(a+it)$

To conclude our investigation of the regularized harmonic series, let us show how Theorem 3.1 applies to prove a Parseval formula associated to the regularized harmonic series encountered in §3 of [JoL 92a]. We will assume the notation defined in §3.

Recall as in §4 of [JoL 92a] that the classical Gauss formula states that for $\mathrm{Re}(z) > 0$,

$$-\Gamma'/\Gamma(z+1) = \int_0^\infty \left[\frac{e^{-zx}}{1-e^{-x}} - \frac{1}{x}\right] e^{-x} dx.$$

If we let $z = a + it$ with $a > -1$, then we get

(1) $$-\Gamma'/\Gamma(a+1+it) = \int_0^\infty \left[\varphi_a(x) e^{-itx} - \frac{1}{x}\right] d\mu(x)$$

where

$$\varphi_a(x) = \frac{e^{-ax}}{1-e^{-x}} \quad \text{and} \quad d\mu(x) = e^{-x} dx.$$

One can view (1) as a type of regularized Fourier transform representation of the gamma function. Finally, by the change of variables t to t/b then x to bx for $b > 0$, (1) becomes

$$-\Gamma'/\Gamma(a+1+i\frac{t}{b}) = \int_0^\infty \left[\frac{be^{-abx}}{1-e^{-bx}} e^{-itx} - \frac{1}{x}\right] e^{-bx} dx.$$

Next we will present a generalization of (1) making use of results in §5 of [JoL 92a]. That is, associated to any Dirichlet series ζ satisfying **DIR 1** and **DIR 2**, as defined in §1 of [JoL 92a], and whose associated theta function $\theta(t)$ satisfies **AS 1**, **AS 2** and **AS 3**, we will realize the regularized harmonic series $I_w(z)$, whose definition will be recalled below, as a regularized Fourier transform. Theorem 2.1 applies to the classical Parseval formula involving the gamma function, which is used in the Barner-Weil explicit formula. In this section we will use our regularized Fourier transform to present a general Parseval formula.

Recall that the principal part of the theta function is

$$P_0\theta(x) = \sum_{\operatorname{Re}(p)<0} b_p(x)x^p,$$

so
$$\theta(x) - P_0\theta(x) = O(|\log x|^m) \quad \text{as } x \to 0.$$

As in [JoL 92a], let $\theta_z(x) = e^{-zx}\theta(x)$. By expanding e^{-zx} in a power series, we see that the principal part of $\theta_z(x)$ is

(2) $$P_0\theta_z(x) = P_0\left[e^{-zx}\theta(x)\right] = \sum_{\operatorname{Re}(p)+k<0} \frac{b_p(x)x^{p+k}}{k!}(-z)^k.$$

From §4 of [JoL 92a] we recall the following result.

Theorem 4.1. *For any fixed complex w with*

$$\operatorname{Re}(w) > \max_k\{-\operatorname{Re}(\lambda_k)\} \quad \text{and} \quad \operatorname{Re}(w) > 0,$$

the integral

$$I_w(z) = \int_0^\infty [\theta_z(x) - P_0\theta_z(x)]\, e^{-wx} dx$$

is convergent for $\operatorname{Re}(z) > 0$. Further, $I_w(z)$ has a meromorphic continuation to all $z \in \mathbf{C}$ with simple poles at $-\lambda_k + w$ with residue a_k.

Remark. In the spectral case, when $\zeta(s) = \sum a_k \lambda_k^{-s}$ with $a_k \in \mathbf{Z}_{\geq 0}$, Theorem 4.1 of [JoL 92a] states that

$$D'_L/D_L(z+w) = I_w(z) + S_w(z)$$

where $S_w(z)$ is a polynomial in z of degree $< -\operatorname{Re}(p_0)$, with coefficients whose dependence on L is through b_p for $\operatorname{Re}(p) < 0$, and whose dependence on w is through elements in $\mathbf{C}[\log w]\langle w\rangle$. Also, in §4 of [JoL 92a] it is shown that Theorem 4.1 yields the classical Gauss formula in the case $L = \mathbf{Z}_{\geq 0}$.

To continue, let us work with the principal part of the theta function. If we restrict the variable z in (2) to a vertical line by letting $z = a + it$ we get

$$
\begin{aligned}
P_0 \theta_z(x) &= \sum_{\mathrm{Re}(p)+k<0} \frac{b_p(x) x^{p+k}}{k!} (-a - it)^k \\
(3) &= \sum_{k < -\mathrm{Re}(p_0)} c_k(a, x)(-it)^k,
\end{aligned}
$$

where the coefficients

$$c_k(a, x) = c_k(a, x, \zeta)$$

depend on the variables a, x and on ζ through the coefficients of t^p for $\mathrm{Re}(p) < 0$ (see **AS 2**). With this, the integral in Theorem 4.1 can be written as
(4)
$$I_w(z) = \int_0^\infty \left[\theta(x) e^{-x(a+it)} - \sum_{k < -\mathrm{Re}(p_0)} c_k(a, x)(-it)^k \right] e^{-wx} dx.$$

Thus, we obtain:

Corollary 4.2. For any $w \in \mathbf{C}$ with $\mathrm{Re}(w) > \max_k\{-\mathrm{Re}(\lambda_k)\}$ and $\mathrm{Re}(w) > 0$, and any $a \in \mathbf{R}^+$ define

$$d\mu_w(x) = e^{-wx} dx \quad \text{and} \quad \theta_a(x) = \theta(x) e^{-ax}.$$

Then

$$I_w(a + it) = \int_0^\infty \left[\theta_a(x) e^{-itx} - \sum_{k < -\mathrm{Re}(p_0)} c_k(a, x)(-it)^k \right] d\mu_w(x).$$

This is our desired generalization of (1) and will be referred to as the **regularized Fourier transform** representation of a regularized harmonic series. Then Theorem 3.1 yields:

Theorem 4.3. *Assume f and its first M derivatives satisfy the three basic conditions. For any $w \in \mathbf{C}$ with*

$$\mathrm{Re}(w) > \max_k\{-\mathrm{Re}(\lambda_k)\} \quad \text{and} \quad \mathrm{Re}(w) > 0,$$

and any $a \in \mathbf{R}^+$, define

$$d\mu_w(x) = e^{-wx}dx \quad \text{and} \quad \theta_a(x) = \theta(x)e^{-ax}.$$

Then

$$\lim_{A \to \infty} \frac{1}{\sqrt{2\pi}} \int_{-A}^{A} f^{\wedge}(t) I_w(a+it) dt$$

$$= \int_0^\infty \left[\theta_a(x)f(-x) - \sum_{k < -\mathrm{Re}(p_0)} c_k(a,x) f^{(k)}(0) \right] d\mu_w(x).$$

In the spectral case when $L = \mathbf{Z}_{\geq 0}$ Theorem 4.3 is the classical Barner-Weil formula.

BIBLIOGRAPHY

[Ba 81] BARNER, K.: On Weil's explicit formula. *J. reine angew. Math.* **323,** 139-152 (1981).

[Ba 90] BARNER, K.: Einführung in die Analytische Zahlentheorie. Preprint (1990).

[BeKn 86] BELAVIN, A. A., and KNIZHNIK, V. G.: Complex geometry and the theory of quantum string. *Sov. Phy. JETP* **2,** 214-228 (1986).

[BrS 85] BRÜNING, J., and SEELEY, R.: Regular singular asymptotics. *Adv. Math.* **58,** 133-148 (1985).

[CaV 90] CARTIER, P., and VOROS, A.: Une nouvelle interprétation de la formule des traces de Selberg. pp. 1-68, volume 87 of *Progress in Mathematics*, Boston: Birkhauser (1990).

[Cr 19] CRAMÉR, H.: Studien über die Nullstellen der Riemannschen Zetafunktion. *Math. Z.* **4,** 104-130 (1919).

[De 92] DENINGER, C.: Local L-factors of motives and regularized products. *Invent. Math.* **107,** 135-150 (1992).

[DP 86] D'HOKER, E., and PHONG, D.: On determinants of Laplacians on Riemann surfaces. *Comm. Math. Phys.* **105,** 537-545 (1986).

[Di 78] DIEUDONNÉ, J.: *Éléments d' Analyse, Vol. VII,* Paris: Gauthier-Villars (1978), reprinted as *Treatise on Analysis Volume 10-VII* San Diego: Academic Press (1988).

[DG 75] DUISTERMAAT, J. J. and GUILLEMIN, V. W.: The spectrum of positive elliptic operators and periodic bicharacteristics. *Invent. Math.* **29,** 39-79 (1975).

[Fr 86] FRIED, D.: Analytic torsion and closed geodesics on hyperbolic manifolds. *Invent. Math.* **84,** 523-540 (1986).

[G 77] GANGOLLI, R.: Zeta functions of Selberg's type for compact space forms of symmetric space of rank one. *Ill. J. Math.* **21,** 1-42 (1977).

[Gr 86] GRUBB, G.: *Functional Calculus of Pseudo-Differential Boundary Problems.* Progress in Mathematics **65** Boston: Birkhauser (1986).

[Ha 49] HARDY, G. H.: *Divergent Series.* Oxford: Oxford University Press (1949).

[In 32] INGHAM, A.: *The Distribution of Prime Numbers,* Cambridge University Press, Cambridge, (1932).

[JoL 92a] JORGENSON, J., and LANG, S.: Some complex analytic properties of regularized products and series. This volume.

[JoL 92b] JORGENSON, J., and LANG, S.: On Cramér's theorem for general Euler products with functional equations. To appear in *Math. Annalen.*

[JoL 92c] JORGENSON, J., and LANG, S.: A Parseval formula for functions with a singular asymptotic expansion at the origin. This volume.

[JoL 92d] JORGENSON, J., and LANG, S.: Artin formalism and heat kernels. To appear in *J. reine angew. Math.*

[JoL 93] JORGENSON, J., and LANG, S.: Explicit formulas and regularized products. Yale University Preprint (1993).

[Ku 88] KUROKAWA, N.: Parabolic components of zeta functions. *Proc. Japan Acad., Ser A* **64,** 21-24 (1988).

[La 70] LANG, S.: *Algebraic Number Theory,* Menlo Park: Addison-Wesley (1970), reprinted as Graduate Texts in Mathematics **110,** New York: Springer-Verlag (1986).

[La 83] LANG, S.: *Real Analysis,* Menlo Park: Addison-Wesley (1983).

[La 85] LANG, S.: *Complex Analysis,* Graduate Texts in Mathematics **103,** New York: Springer-Verlag (1985).

[La 93] LANG, S.: *Real and Functional Analysis, Third Edition*, New York: Springer-Verlag (1993).

[McS 67] McKEAN, H. P., and SINGER, I. M.: Curvature and eigenvalues of the Laplacian. *J. Differ. Geom.* **1**, 43-70 (1967).

[MP 49] MINAKSHISUNDARAM, S., and PLEIJEL, A.: Some properties of the eigenfunctions of the Laplace operator on Riemannian manifolds. *Can. J. Math.* **1**, 242-256 (1949).

[RS 73] RAY, D., and SINGER, I.: Analytic torsion for complex manifolds. *Ann. Math.* **98**, 154-177 (1973).

[Sa 87] SARNAK, P.: Determinants of Laplacians. *Comm. Math. Phys.* **110**, 113-120 (1987).

[Su 85] SUNADA, T.: Riemannian coverings and isospectral manifolds. *Ann. Math.* **121**, 169-186 (1985).

[TaZ 91] TAKHTAJAN, L. A., and ZOGRAF, P. G.: A local index theorem for families of $\bar{\partial}$-operators on punctured Riemann surfaces and a new Kähler metric on their moduli spaces. *Comm. Math. Phys.* **137**, 399-426 (1991).

[Ti 48] TITCHMARSH, E. C.: *Introduction to the Theory of Fourier-Integrals, 2nd Edition* Oxford University Press, Oxford (1948).

[Ve 81] VENKOV, A. B., and ZOGRAF, P. G.: On analogues of the Artin factorization formulas in the spectral theory of automorphic functions connected with induced representations of Fuschsian groups, *Soviet Math. Dokl.*, **21**, 94-96 (1981)

[Ve 83] VENKOV, A. B., and ZOGRAF, P. G.: On analogues of the Artin factorization formulas in the spectral theory of automorphic functions connected with induced representations of Fuschsian groups, *Math. USSR- Izv.* **21**, 435-443 (1983).

[Vo 87] VOROS, A.: Spectral functions, special functions, and the Selberg zeta function. *Comm. Math. Phys* **110**, 439-465 (1987).

[We 52] WEIL, A.: Sur les "formules explicites" de la théorie des nombres premiers, *Comm. Lund* (vol. dédié à Marcel Riesz), 252-265 (1952).

[We 72] WEIL, A.: Sur les formules explicites de la théorie des nombres, *Izv. Mat. Nauk (Ser. Mat.)* **36,** 3-18 (1972).

Artin formalism and heat kernels

By *Jay Jorgenson**) and *Serge Lang* at New Haven

Introduction

The Artin formalism for L-series of number fields ([Ar 23] and [Ar 30]) is known to be satisfied in topology, in a context of the characteristic polynomial in linear algebra, when one has an endomorphism acting functorially on some representation spaces associated with a finite Galois covering [La 56]. In the present paper, we show that Artin's formalism is satisfied in the infinite-dimensional analogue of a one-parameter group e^{-tA}, where A is a positive self-adjoint operator. The kernel representing the operator e^{-tA} is called the associated heat kernel and could be called the characteristic kernel, since it plays the role of the characteristic polynomial in the finite-dimensional case. For a brief, elegant introduction to heat kernels, see lecture 3 of [Fa 92], and for a more extensive discussion, [Ch 84] and [BGV 92].

Quite generally, given a sequence of complex numbers $\Lambda = \{\lambda_k\}$ one can define several generalized symmetric functions associated to Λ. Of particular interest are the following spectral functions whose *formal* definitions we now recall.

The *theta function* $\theta_\Lambda(t)$:

$$\theta_\Lambda(t) = \sum_{\lambda_k} e^{-\lambda_k t}.$$

The *Laplace-Mellin transform* $\xi_\Lambda(s, z)$:

$$\xi_\Lambda(s, z) = \int_0^\infty \theta_\Lambda(t) e^{-zt} t^s \frac{dt}{t}.$$

The *Hurwitz zeta function* $\zeta_\Lambda(s, z)$:

$$\zeta_\Lambda(s, z) = \sum_{\lambda_k} \frac{1}{(z + \lambda_k)^s}.$$

---

*) The first author acknowledges support from NSF grant DMS-89-05661.

The *spectral zeta function* $\zeta_\Lambda(s)$:

$$\zeta_\Lambda(s) = \sum_{\lambda_k \neq 0} \lambda_k^{-s}.$$

The *spectral xi function* $\xi_\Lambda(s)$:

$$\xi_\Lambda(s) = \Gamma(s)\zeta_\Lambda(s).$$

The *regularized determinant* of the sequence $\Lambda + z$:

$$\mathbf{D}_\Lambda(z) = \exp\left(-\zeta'_\Lambda(0, z)\right).$$

The derivative is taken with respect to the variable s.

The *regularized determinant* of the sequence Λ itself:

$$\mathbf{D}_\Lambda(0) = \exp\left(-\zeta'_\Lambda(0)\right).$$

Such a sequence Λ appears as the eigenvalues of a Laplacian acting on C^∞ sections of a metrized vector sheaf $\mathscr{E}$ over a compact Riemannian manifold X. In this case one can express the above spectral functions in terms of the trace of the heat kernel associated to the Laplacian.

In this paper we shall consider finite coverings of X or direct sum decompositions of $\mathscr{E}$ and show that the trace of the heat kernel and the above functions satisfy Artin's formalism. These functions will sometimes be viewed as lying in an additive group of functions and sometimes lying in a multiplicative group of functions, so the Artin formalism may be written in additive or multiplicative notation, as the case may be. We define L-functions associated to the heat kernel, proving Artin's formalism for these functions, then associating the above spectral functions to the trace of the heat kernel. The proof of Artin's formalism in the context of heat kernels requires only the existence and uniqueness of the heat kernel, coming from the fact that it is a global object, meaning a distinguished section of a certain sheaf, but uniquely defined through local conditions. Any other object fulfilling such conditions will also satisfy Artin's formalism. In addition, any object which is a homomorphic image of the heat kernel (see the definition following Propostion 2.1) satisfies Artin's formalism.

The Artin formalism in the case of L-series of number fields is reproduced on page 233 of [La 86]. The Artin formalism has appeared before in some contexts of spectral zeta functions in dealing with prime geodesics as analogues of primes in number fields. Sarnak uses special cases on Riemann surfaces to investigate analogues of Tcheboratarev's theorem especially concerning the distribution of such geodesics in homology classes [Sa 81] (see also [KaS 87] and [PS 88]). Sunada used a special case of this formalism to construct isospectral manifolds that are not isometric, in a way analogous to the method used to construct non-isomorphic number fields that have the same zeta function ([Su 85]).

The paper is organized as follows. After presenting necessary notation and preliminary material in §1, we will prove in §2 that the trace of the heat kernel satisfies Artin's formalism (the properties that a function must fulfill in order to satisfy Artin's formalism will be stated in §2). In §3 we study the Hurwitz xi function and the characteristic determinant, and show that these functions satisfy Artin's formalism. In the case of compact hyperbolic Riemann surfaces, the characteristic determinant is simply related to the Selberg zeta function, thus we show that the Selberg zeta function satisfies Artin's formalism (see, for example, [Ve 79], [VeZ 81] and [VeZ 83] for a proof using the trace formula, and [DP 86] or [Sa 87] for a proof of the connection between the Selberg zeta function and the characteristic determinant, again using the trace formula). In §4 we will use Artin's formalism for the characteristic determinant in order to present a generalization of Kronecker's second limit formula and Jacobi's derivative formula, which are known results from the theory of the Riemann theta function. Finally, in §5, we will show that Artin's formalism can be applied to prove the classical Riemann theta relations and the functional equation of the Riemann theta function.

Before continuing, let us give an example of a classical result involving the Riemann theta function together with the spectral interpretation. Recall that the Riemann theta function with characteristics α and β in the variables $z \in \mathbf{C}$ and τ in the upper half plane $\mathfrak{h}$ is defined by the series

$$\theta\begin{bmatrix}\alpha\\\beta\end{bmatrix}(z,\tau) = \sum_{k=-\infty}^{\infty} \exp\left(\pi i (n+\alpha)^2 \tau + 2\pi i (n+\alpha)(z+\beta)\right).$$

Let $q_\tau = \exp(2\pi i \tau)$ and define

$$\Delta(\tau) = q_\tau \prod_{k=1}^{\infty} (1 - q_\tau^n)^{24}.$$

A (weak) form of the Jacobi derivative formula states that

$$\left| \theta\begin{bmatrix}0\\0\end{bmatrix}(0,\tau) \theta\begin{bmatrix}1/2\\0\end{bmatrix}(0,\tau) \theta\begin{bmatrix}0\\1/2\end{bmatrix}(0,\tau) \right|^8 = 2^8 |\Delta(\tau)|.$$

A spectral interpretation of this formula is the following. Let X denote the elliptic curve realized as $\mathbf{C}$ modulo the lattice generated by 1 and τ. The Jacobi derivative formula becomes the statement that the regularized determinant associated to the Laplacian, which acts on the trivial sheaf $\mathcal{O}$ relative to a translation invariant metric, satisfies Artin's formalism when considering X covering itself through the isogeny multiplication by two. The determinant is viewed as lying in a multiplicative group of functions.

The techniques we use to establish Artin's formalism are quite elementary, involving only well-known representation theory of finite groups and the local characterization of a globally unique object, namely the heat kernel. However, the number of classical formulas that follow as a consequence of our work is striking. For the sake of brevity, we have chosen to present corollaries of the formalism in a few, previously studied cases, namely Selberg's zeta function on hyperbolic Riemann surfaces and Riemann's theta relations on complex abelian varieties. Also, we will describe how one can use the Artin formalism to obtain the classical Gauss multiplication formula for the gamma function. The development of other identities involving special functions via the Artin formalism will be left for future investigation.

§1. Notation and background material

In this section we recall basic material on spectral theory, topology of vector sheaves and (finite group) representation theory that will be used. For more detailed discussions of these topics, the reader is referred to the following texts: Chapter I of [At 67] for the foundations of group actions on vector sheaves; [BGV 92], Chapters 1 and 2, Chapters I and VI of [Ch 84] and Lecture 3 of [Fa 92] for spectral theory; and [La 93], Chapter XVIII for representation theory.

Let Y denote a compact Riemannian manifold of real dimension n, and let $\mathscr{E}$ denote a metrized vector sheaf on Y of rank $\operatorname{rk}(\mathscr{E})$. The inner product on $\mathscr{E}$ will be denoted by $\langle \cdot, \cdot \rangle_{\mathscr{E}}$, and the volume element on Y will be denoted by vol_Y. Assume that $\mathscr{E}$ is equipped with a connection, and let $\Delta_{\mathscr{E}}$ denote the associated positive Laplacian that acts on smooth sections of $\mathscr{E}$. A section

$$y \mapsto \psi(y; Y, \mathscr{E}) = \psi(y)$$

is an eigensection of $\Delta_{\mathscr{E}}$ with associated eigenvalue $\lambda(Y, \mathscr{E})$ if

(1.1) $$\Delta_{\mathscr{E}} \psi - \lambda(Y, \mathscr{E}) \psi = 0.$$

Let $\lambda_0(Y, \mathscr{E})$ denote the zero eigenvalue, counted with multiplicity $m_0(Y, \mathscr{E})$. Let us denote the sequence of *non-zero* eigenvalues by the sequence

$$0 < \lambda_1(Y, \mathscr{E}) \leq \lambda_2(Y, \mathscr{E}) \leq \ldots,$$

where each non-zero eigenvalue in the sequence is repeated according to its multiplicity. It is classical that the set of orthonormal eigensections $\{\psi_j\}$ forms a complete basis of the Hilbert space of L^2 sections of $\mathscr{E}$.

The sheaf $\mathscr{E} \boxtimes \bar{\mathscr{E}}$ over $Y \times Y$ is the sheaf whose fiber over (y_1, y_2) is $\mathscr{E}_{y_1} \otimes \bar{\mathscr{E}}_{y_2}$ (see page 74 of [BGV 92]). The *heat kernel* $K(Y, \mathscr{E})(y_1, y_2, t)$ is the integral kernel on Y which inverts the parabolic operator

(1.2) $$\mathbf{L}_{\mathscr{E}} = \Delta_{\mathscr{E}} + \frac{\partial}{\partial t},$$

acting on the space of smooth sections of $\mathscr{E}$ with values of t in $\mathbf{R}$. More precisely, the heat kernel is the section of $\mathscr{E} \boxtimes \bar{\mathscr{E}} \times \mathbf{R}^+$ uniquely characterized by the following property. Let s be a smooth section of $\mathscr{E}$ and let

$$(K * s)(y_1, t) = \int_Y \langle K(Y, \mathscr{E})(y_1, y_2, t), s(y_2) \rangle_{\mathscr{E}} \cdot \operatorname{vol}_Y(y_2).$$

Then

$$\mathbf{L}_{\mathscr{E}}(K * s) = 0 \quad \text{and} \quad \lim_{t \to 0} (K * s)(y_1, t) = s(y_1).$$

For all y_1 and y_2 on Y and positive t, one can formally express the heat kernel as

$$(1.3) \qquad K_{\mathscr{E}}(y_1, y_2, t) = \sum_{j=0}^{\infty} e^{-\lambda_j(Y,\mathscr{E})t} \psi_j(y_1; Y, \mathscr{E}) \otimes \bar{\psi}_j(y_2; Y, \mathscr{E}).$$

If we combine (1.1) and (1.2) the equality asserted in (1.3) is formally true. Indeed, the series in (1.3) converges uniformly and absolutely, thus verifying the equation (see page 91 of [BGV 92] or page 140 of [Ch 84]). For any positive integer N, there are smooth sections

$$b_{-n/2}(y; Y, \mathscr{E}), b_{-(n-1)/2}(y; Y, \mathscr{E}), \ldots, b_{(N-1)/2}(y; Y, \mathscr{E})$$

of $\mathscr{E}$ such that if we define

$$P_N K(Y, \mathscr{E})(y, t) = \sum_{k=-n}^{2N-1} b_{k/2}(y; Y, \mathscr{E}) \otimes \bar{b}_{k/2}(y; Y, \mathscr{E}) t^{k/2},$$

then

$$(1.4) \qquad K(Y, \mathscr{E})(y, y, t) - P_N K(Y, \mathscr{E})(y, t) = O(t^{N/2}) \quad \text{as } t \to 0.$$

If $y_1 \neq y_2$, then there is a constant c such that

$$(1.5) \qquad K(Y, \mathscr{E})(y_1, y_2, t) = O(e^{-c/t}) \quad \text{as } t \to 0$$

(see page 154 of [Ch 84] or page 82 of [BGV 92]).

Taking the scalar product on $\mathscr{E}$ we obtain the function

$$\langle K(Y, \mathscr{E})(y, y, t) \rangle_{\mathscr{E}}$$

of y on Y (see page 95 of [La 87]). Upon integrating this over Y with respect to vol_Y, we obtain the *trace of the heat kernel* $\mathrm{tr}\, K(Y, \mathscr{E})(t)$, which is defined to be

$$\mathrm{tr}\, K(Y, \mathscr{E})(t) = \int_Y \langle K(Y, \mathscr{E})(y, y, t) \rangle_{\mathscr{E}} \cdot \mathrm{vol}_Y(y).$$

From §2.6 of [BGV 92], we recall the classical Weyl's law which states that the trace of the heat kernel satisfies the asymptotic formula

$$\mathrm{tr}\, K(Y, \mathscr{E})(t) = \frac{\mathrm{vol}_Y(Y)\,\mathrm{rk}(\mathscr{E})}{(4\pi t)^{n/2}} + O(t^{-n/2+1}) \quad \text{as } t \to 0.$$

We also recall:

Proposition 1.1. *Let $\Lambda = \Lambda(\mathscr{E})$ denote the sequence of eigenvalues of the Laplacian $\Delta_{\mathscr{E}}$ which acts on smooth sections of the metrized vector sheaf $\mathscr{E}$ over the Riemannian manifold Y. Then the theta function $\theta_{\Lambda(\mathscr{E})}(t)$ is the trace of the heat kernel,*

$$\theta_{\Lambda(\mathscr{E})}(t) = \mathrm{tr}\, K(Y, \mathscr{E})(t),$$

or
$$\sum_{j=0}^{\infty} e^{-\lambda_j(Y,\mathscr{E})t} = \int_Y \langle K(Y,\mathscr{E})(y,y,t)\rangle_{\mathscr{E}} \cdot \mathrm{vol}_Y(y).$$

Let $m_0(Y,\mathscr{E})$ denote the dimension of the zero eigenspace of $\Delta_{\mathscr{E}}$. From the above proposition it is immediate that

$$m_0(Y,\mathscr{E}) = \lim_{t\to\infty} \mathrm{tr}\, K(Y,\mathscr{E})(t).$$

Let us define

$$P_N \,\mathrm{tr}\, K(Y,\mathscr{E})(t) = \int_Y \langle P_N K(Y,\mathscr{E})(y,t)\rangle_{\mathscr{E}} \mathrm{vol}_Y(y),$$

and

$$\beta_{k/2}(Y,\mathscr{E}) = \int_Y \langle b_{k/2}(y;Y,\mathscr{E}), b_{k/2}(y;Y,\mathscr{E})\rangle_{\mathscr{E}} \mathrm{vol}_Y(y).$$

From Proposition 1.1, one can express the other spectral functions presented in the introduction in terms of the trace of the heat kernel.

Corollary 1.2. *The spectral functions associated to $\Lambda(\mathscr{E})$ are expressible in terms of the trace of the heat kernel as follows.*

The Laplace-Mellin transform $\xi_{\Lambda(\mathscr{E})}(s,z)$:

$$\xi_{\Lambda(\mathscr{E})}(s,z) = \int_0^{\infty} [\mathrm{tr}\, K(Y,\mathscr{E})(t)] e^{-zt} t^s \frac{dt}{t}.$$

The Hurwitz zeta function $\zeta_{\Lambda(\mathscr{E})}(z,z)$:

$$\zeta_{\Lambda(\mathscr{E})}(s,z) = \frac{1}{\Gamma(s)} \xi_{\Lambda(\mathscr{E})}(s,z).$$

The spectral zeta function $\zeta_{\Lambda(\mathscr{E})}(s)$:

$$\zeta_{\Lambda(\mathscr{E})}(s) = \frac{1}{\Gamma(s)} \int_0^{\infty} (\mathrm{tr}\, K(Y,\mathscr{E})(t) - m_0(Y,\mathscr{E})) t^s \frac{dt}{t}.$$

The spectral xi function $\xi_{\Lambda(\mathscr{E})}(s)$:

$$\xi_{\Lambda(\mathscr{E})}(s) = \Gamma(s)\zeta_{\Lambda(\mathscr{E})}(s).$$

The regularized determinant of the sequence $\Lambda(\mathscr{E}) + z$:

$$\mathbf{D}_\Lambda(z) = \exp(-\zeta'_{\Lambda(\mathscr{E})}(0,z)).$$

The regularized determinant of the sequence $\Lambda(\mathscr{E})$ itself:

$$\mathbf{D}_{\Lambda(\mathscr{E})}(0) = \exp\left(-\zeta'_{\Lambda(\mathscr{E})}(0,0)\right) = \exp\left(-\zeta'_{\Lambda(\mathscr{E})}(0)\right).$$

The Laplace-Mellin transform, which we also call the Hurwitz xi function, and the Hurwitz zeta function are meromorphic in the region

$$\mathbf{C} \times \{\mathbf{C} \setminus \mathbf{R}_{\leq 0}\}.$$

The spectral zeta function and the spectral xi function are meromorphic. The regularized determinant $\mathbf{D}_{\Lambda(\mathscr{E})}(z)$ is holomorphic, and $\mathbf{D}_{\Lambda(\mathscr{E})}(0)$ is a positive real number.

Proof. All assertions follow from the small time asymptotics of the heat kernel (1.4), the long time asymptotics of the heat kernel and the definition of the gamma function. We refer to [JoL 92] for further details in a more general situation.

Let us next recall necessary material from (finite group) representation and the topology of vector sheaves. Assume that Y admits a fixed point free action by a finite group G, and assume that the metric on Y is such that G acts as isometries. We call such a manifold a G-manifold. We let

$$p = p_X^Y : Y \to X = G \setminus Y$$

be the natural quotient with induced metric. By a G-sheaf $\mathscr{E}$ over a G-manifold we mean a metrized sheaf $\mathscr{E}$ such that the total space of the corresponding vector bundle over Y, with metrics from Y and $\mathscr{E}$, is a G-manifold and for any point y on Y and $g \in G$, the associated map

$$g_{\mathscr{E}} : \mathscr{E}_y \to \mathscr{E}_{gy}$$

is a metric isomorphism (see §1.6 of [At 67]). Let

$$q_X^Y : \mathscr{E} \to \mathscr{E}_{X,Y} = G \setminus \mathscr{E}$$

denote the quotient sheaf of $\mathscr{E}$ on X, with its quotient metric.

Definition 1.3. Let $\mathscr{E}$ be a G-sheaf on a G-manifold Y. For any $g \in G$, define the *g-shifted trace of the heat kernel* to be

$$\operatorname{tr} K(Y, \mathscr{E})(g, t) = \int_Y \langle K(Y, \mathscr{E})(y, gy, t) \rangle_{\mathscr{E}} \cdot \operatorname{vol}_Y(y).$$

Since $\mathscr{E}$ is a G-sheaf on the G-manifold Y, the g-shifted heat trace is a well defined function of g and t. Note that if g is not the identity element in G, (1.5) implies the existence of a constant $c > 0$ such that as t approaches zero,

$$\operatorname{tr} K(Y, \mathscr{E})(g, t) = O(e^{-c/t}).$$

Let π denote a finite-dimensional unitary representation of the group G of dimension d_π with representation space V_π. Assume that V_π is given its natural L^2 metric. If $\mathscr{E}$ is a

G-sheaf on Y, then so is $\mathscr{E} \otimes V_\pi$. There is a unique action, denoted by $g \mapsto \pi_\mathscr{E}(g)$, of G on section $\mathscr{E} \otimes V_\pi$ as follows. Suppose s is a decomposable section of $\mathscr{E} \otimes V_\pi$, with $s = s_\mathscr{E} \otimes v$, where $s_\mathscr{E}$ is a section of $\mathscr{E}$ and $v \in V_\pi$. Then

$$(1.6) \qquad (\pi_\mathscr{E}(g)s)(y) = g s_\mathscr{E}(y) \otimes \pi_\mathscr{E}(g) v.$$

For simplicity of notation, we write $\pi(g)$ instead of $\pi_\mathscr{E}(g)$.

We will use the notation $\mathscr{E}(\pi)_{X,Y}$ to denote the quotient sheaf of $\mathscr{E} \otimes_\mathbf{C} V_\pi$ on X, symbolically written as

$$\mathscr{E}(\pi)_{X,Y} = (G, \pi) \backslash (\mathscr{E} \otimes V_\pi).$$

We are using the subscript (X, Y) to indicate that we started with a sheaf over Y and have constructed a sheaf over X. Note that the sheaf $\mathscr{E}(\pi)_{X,Y}$ has rank

$$\mathrm{rk}(\mathscr{E}(\pi)_{X,Y}) = \mathrm{rk}(\mathscr{E}) d_\pi,$$

and

$$\mathrm{vol}(Y) = \mathrm{vol}(X)[G:1].$$

Given a G-sheaf $\mathscr{E}$ on Y and a finite dimensional unitary representation π of G with representation space V_π, the (G, π) action (1.6) applies to the heat kernel $K(Y, \mathscr{E} \otimes V_\pi)$ to produce a section of the sheaf $\mathscr{E}(\pi)_{X,Y} \boxtimes \bar{\mathscr{E}}(\pi)_{X,Y}$ over X. Given a section s of the G-sheaf $\mathscr{E}$ over Y, we will use $\mathrm{Tr}_G(s)$ to denote the G-average of s (see [At 67], page 37). If s is a G-invariant section of $\mathscr{E}$, then $\mathrm{Tr}_G(s) = s$. We will also use q_X^Y denote the map which projects G invariant sections of $\mathscr{E}$ over Y to sections of $\mathscr{E}_{X,Y}$.

Let us now consider the case when G has a non-trivial normal subgroup H, so there is an intermediate manifold $Z = H \backslash Y$ between Y and X that covers X with covering group G/H. Let $\pi_{G/H}$ be a representation of the factor group G/H with representation space $V_{\pi_{G/H}}$. Let

$$\mathrm{can}: G \to G/H$$

denote the canonical morphism, and let

$$\pi_G = \pi_{G/H} \circ \mathrm{can}$$

denote the corresponding representation of G. We shall denote (left) coset representatives of H in G by λ. The representation π_G of G is called the inflation of the representation of $\pi_{G/H}$.

Let H denote a subgroup of G of index $[G:H]$, not necessarily normal, and let π_H denote a finite dimensional representation of H with representation space W_{π_H} of complex dimension d_{π_H}. The induced representation

$$\pi_H^G = \mathrm{ind}_H^G(\pi_H)$$

associated to π_H has dimension $[G:H] d_{\pi_H}$. Let $\mathbf{C}\langle G/H \rangle$ denote a complex dimensional vector space of complex dimension $[G:H]$, with coordinates parameterized by left coset representatives of H in G. The induced representation π_H^G of π_H acts on the representation space

$$V_{\pi_H^G} = W_{\pi_H} \otimes_{\mathbf{C}} \mathbf{C}\langle G/H\rangle.$$

We shall use χ_{π_H} to denote the character of π_H and $\chi_{\pi_H^G}$ to denote the character of the induced representation π_H^G.

§ 2. Heat kernels and Artin formalism

Let A be an abelian group. To each triple $(Y/X, \mathscr{E}, \pi)$ consisting of a Galois covering Y/X with covering group G, a G-sheaf $\mathscr{E}$ on Y and a finite dimensional unitary representation π of G with representation space V_π, we suppose associated an element $L(Y/X, \mathscr{E}, \pi)$ in A; and, to each pair $(Y, \mathscr{E})$, we suppose associated an element $\varphi(Y, \mathscr{E})$. We say that this association satisfies the *Artin formalism* if the following conditions are satisfied.

AF 1. Normalization.

$$L(Y/X, \mathscr{E}, \pi) = \varphi(X, \mathscr{E}(\pi)).$$

In particular, if 1 denotes the trivial representation, we have

$$L(X/X, \mathscr{E}, 1) = \varphi(X, \mathscr{E}).$$

AF 2. Additivity. For any finite-dimensional unitary representations π_1 and π_2 of G,

$$L(Y/X, \mathscr{E}, \pi_1 \oplus \pi_2) = L(Y/X, \mathscr{E}, \pi_1) + L(Y/X, \mathscr{E}, \pi_2).$$

AF 3. Inflation. Let H denote a normal subgroup of G, $Z = H\backslash Y$, and $\pi_{G/H}$ a finite-dimensional unitary representation of the quotient group G/H. If π_G denotes the inflation of the representation $\pi_{G/H}$, then

$$L(Z/X, (H\backslash\mathscr{E}), \pi_{G/H}) = L(Y/X, \mathscr{E}, \pi_G).$$

AF 4. Induction. Let H denote a subgroup of G, $Z = H\backslash Y$, and π_H a finite-dimensional unitary representation of H. If $\mathrm{ind}_H^G(\pi_H)$ denotes the induced representation of π_H to G, then

$$L(Y/Z, \mathscr{E}, \pi_H) = L(Y/X, \mathscr{E}, \mathrm{ind}_H^G(\pi_H)).$$

Proposition 2.1. *Let $\alpha: A \to A'$ be a homomorphism of abelian groups. If the association $[L, \varphi]$ satisfies Artin's formalism, then so does $[\alpha \circ L, \alpha \circ \varphi]$.*

In the situation of Proposition 2.1, we shall say that $[\alpha \circ L, \alpha \circ \varphi]$ is a *homomorphic image of* $[L, \varphi]$.

The main result of this section is the following theorem.

Theorem 2.2. *Let A be the abelian group of C^∞ functions on $\mathbf{R}^+$ and define*

$$L(Y/X, \mathscr{E}, \pi)(t) = [Y : X]^{-1} \sum_{g \in G} \chi_\pi(g) \operatorname{tr} K(Y, \mathscr{E} \otimes V_\pi)(g, t)$$

and

$$\varphi(Y, \mathscr{E})(t) = \theta_{\Lambda(\mathscr{E})}(t).$$

Then this association satisfies the Artin formalism.

The remainder of this section is devoted to the proof of Theorem 2.2, which simply comes down to the existence and uniqueness of the heat kernel and standard formulas from the representation theory of finite groups. Each part of Theorem 2.2 will be proved by expressing the appropriate heat kernel in two ways, then computing the trace. We start with direct sums.

Proposition 2.3. *Let $\mathscr{E}$ and $\mathscr{F}$ denote vector sheaves on Y. Then*

$$K(Y, (\mathscr{E} \oplus \mathscr{F}))(y_1, y_2, t) = K(Y, \mathscr{E})(y_1, y_2, t) \oplus K(Y, \mathscr{F})(y_1, y_2, t).$$

Assume that $\mathscr{E}$ is a G-sheaf on Y and that π_1 and π_2 are two finite-dimensional unitary representations of G. Then

$$K(X, \mathscr{E}(\pi_1 \oplus \pi_2)_{X,Y})(x_1, x_2, t) = K(X, \mathscr{E}(\pi_1)_{X,Y})(x_1, x_2, t) \oplus K(X, \mathscr{E}(\pi_2)_{X,Y})(x_1, x_2, t).$$

Proof. The proposition is essentially obvious, because the Laplacian, sections, heat kernels, and passing to the G-quotient decompose as direct sums on the direct sums of vector sheaves.

Of particular interest is when π is the (left) regular representation of G, which we shall denote by π^G_{reg}. In that case, Proposition 2.3 states the following.

Corollary 2.4. *Let $\hat{G}$ denote the set of isomorphism classes of finite-dimensional irreducible representations of G. Then*

$$K(X, \mathscr{E}(\pi^G_{\text{reg}})_{X,Y}) = \bigoplus_{\pi \in \hat{G}} K(X, \mathscr{E}(\pi)_{X,Y} \otimes V_\pi).$$

Proposition 2.3 asserts an equality of distinguished sections, namely heat kernels, of vector sheaves. By taking the trace of the heat kernels in Proposition 2.3 and Corollary 2.4, we obtain the following theorem.

Theorem 2.5. *In Proposition 2.3 and Corollary 2.4, we have, as functions of $t \in \mathbf{R}^+$, the equalities of traces of the heat kernels*:

$$\operatorname{tr} K(Y, (\mathscr{E} \oplus \mathscr{F}))(t) = \operatorname{tr} K(Y, \mathscr{E})(t) + \operatorname{tr} K(Y, \mathscr{F})(t),$$

$$\operatorname{tr} K(X, \mathscr{E}(\pi_1 \oplus \pi_2)_{X,Y})(t) = \operatorname{tr} K(X, \mathscr{E}(\pi_1)_{X,Y})(t) + \operatorname{tr} K(X, \mathscr{E}(\pi_2)_{X,Y})(t),$$

$$\operatorname{tr} K(X, \mathscr{E}(\pi^G_{\text{reg}})_{X,Y})(t) = \sum_{\pi \in \hat{G}} \operatorname{tr} K(X, \mathscr{E}(\pi)_{X,Y} \otimes V_\pi)(t).$$

As with Proposition 2.3, the proof of Theorem 2.5 is immediate from the definition of the direct sum metric on $\mathscr{E} \oplus \mathscr{F}$ and trace of the heat kernel. Theorem 2.5 verifies **AF 2**.

If we let t approach zero in the third equation in Theorem 2.5 and take the coefficient of $t^{-n/2}$ via Weyl's law, we arrive at the relation

$$\text{vol}(X)\,\text{rk}(\mathscr{E})\,d_{\pi^G_{\text{reg}}} = \text{vol}(X)\,\text{rk}(\mathscr{E}) \sum_{\pi \in \hat{G}} \dim(V_\pi)\,\text{mult}(\pi^G_{\text{reg}};\pi),$$

where $\text{mult}(\pi^G_{\text{reg}};\pi)$ denotes the multiplicity of π in π^G_{reg}. Using the fact that

$$d_\pi = \dim(V_\pi) = \text{mult}(\pi^G_{\text{reg}};\pi)$$

and

$$d_{\pi^G_{\text{reg}}} = [G:1],$$

we obtain the relation

$$[G:1] = \sum_{\pi \in \hat{G}} d_\pi^2,$$

which is a well-known formula from the theory of finite group representations and is implicit in Corollary 2.4.

The next step in proving Theorem 2.2 is to verify a stronger form of **AF 1** which states a non-trivial relation concerning the heat kernel itself, rather than its trace, when passing to the quotient by G. Throughout we shall use the notation

$$x_1 = p(y_1), \quad x_2 = p(y_2)$$

to denote the image of points y_1 and y_2 from Y to X.

Proposition 2.6. *Let $\mathscr{E}$ denote a G-sheaf on the G-manifold Y, with quotient manifold $X = G\backslash Y$ and quotient sheaf $G\backslash\mathscr{E}$. Then we have the equality of heat kernels*

$$K(X,(G\backslash\mathscr{E})_{X,Y})(x_1,x_2,t) = q_X^Y \circ \left(\sum_{g \in G} K(Y,\mathscr{E})(y_1, g y_2, t) \right).$$

Symbolically, this can be written as

$$K(X,(G\backslash\mathscr{E})_{X,Y}) = q \circ \text{Tr}_G K(Y,\mathscr{E}).$$

If $\mathscr{E}$ denotes a G-sheaf on Y and π a finite-dimensional unitary representation of G with representation space V_π, then

$$K(X, \mathscr{E}(\pi)_{X,Y})(x_1,x_2,t) = q_X^Y \circ \left(\sum_{g \in G} \pi(g)[K(Y,(\mathscr{E} \otimes V_\pi))](y_1, g y_2, t) \right).$$

Proof. The proof of Proposition 2.6 follows that of Proposition 2.3. By this we mean that one first observes that the sum in question is indeed a section of the sheaf

$$(G\backslash\mathscr{E}) \boxtimes \overline{(G\backslash\mathscr{E})}.$$

Then we note that the action by the Laplacian $\Delta_{X,G\backslash\mathscr{E}}$ and integration commute with the operation Tr_G. By existence and uniqueness of the heat kernel, the proof is complete. □

Formulas similar to that in Proposition 2.6 appear in very special cases in the literature, such as [Ch 84], page 155, equation (50) for the case of the trivial sheaf, or on Riemann surfaces in [He 76], pages 333 and 360 for powers of the canonical sheaf.

Corollary 2.7. *In Proposition 2.6, assume that π is one-dimensional with character χ_π. Then*

$$K(X, \mathscr{E}(\pi)_{X,Y})(x_1, x_2, t) = q_X^Y \circ \left(\sum_{g \in G} \chi_\pi(g) K(Y, \mathscr{E})(y_1, g y_2, t) \right).$$

Upon computing the trace of the expressions in Proposition 2.6, we get:

Theorem 2.8. *In Proposition 2.6, we have, as functions of $t \in \mathbf{R}^+$, the equality of traces of heat kernels*

$$[Y:X] \operatorname{tr} K(X, (G \backslash \mathscr{E})_{X,Y})(t) = \sum_{g \in G} \operatorname{tr} K(Y, \mathscr{E})(g, t).$$

Let χ_π denote the character of the representation π. Then

$$[Y:X] \operatorname{tr} K(X, \mathscr{E}(\pi)_{X,Y})(t) = \sum_{g \in G} \chi_\pi(g) \operatorname{tr} K(Y, \mathscr{E})(g, t).$$

The proof of Theorem 2.8 is a direct consequence of Proposition 2.6 and the definition of the trace of the heat kernel. Note that Theorem 2.8 proves **AF 1**. Also, one should observe the compatibility of Theorem 2.8 with Theorem 2.5 in the case π is reducible since

$$\chi_{\pi_1 \oplus \pi_2} = \chi_{\pi_1} + \chi_{\pi_2}.$$

In Theorem 2.8 we can take a sum over π in $\hat{G}$ and use the orthogonality of group characters to get:

Theorem 2.9. *In Proposition 2.6, we have for any $g \in G$ the equality of heat trace*

$$\operatorname{tr} K(Y, \mathscr{E})(g, t) = \sum_{\pi \in \hat{G}} d_\pi \chi_\pi(g^{-1}) \operatorname{tr} K(X, \mathscr{E}(\pi)_{X,Y})(t).$$

In particular, by taking g to be the identity element in G, we arrive at the equality of heat traces

$$\operatorname{tr} K(Y, \mathscr{E})(t) = \sum_{\pi \in \hat{G}} d_\pi \operatorname{tr} K(X, \mathscr{E}(\pi)_{X,Y})(t).$$

Let us now consider the case when G has a normal subgroup H.

Proposition 2.10. *Let H be a normal subgroup of G, and $Z = H \backslash Y$ be the intermediate covering manifold of X with Galois covering group G/H. Let $\mathscr{E}$ denote a G-sheaf on Y. Then we have the equalities of heat kernels*

$$K(X,(G\backslash\mathscr{E})_{X,Y})(x_1,x_2,t) = q_X^Y \circ \left(\sum_{g \in G} K(Y,\mathscr{E})(y_1, gy_2, t) \right)$$

$$= q_X^Y \circ \left(\sum_{\lambda \in (G/H)} \sum_{h \in H} K(Y,\mathscr{E})(y_1, \lambda h y_2, t) \right)$$

$$= q_X^Z \circ \left(\sum_{\lambda \in (G/H)} K(Z,(H\backslash\mathscr{E})_{Z,Y})(z_1, \lambda z_2, t) \right)$$

Symbolically, this can be written as

$$K(X,(G\backslash\mathscr{E})_{X,Y}) = q_X^Y \circ \mathrm{Tr}_G K(Y,\mathscr{E})$$
$$= q_X^Z \circ \mathrm{Tr}_{G/H} \circ q_{Z,Y} \circ \mathrm{Tr}_H K(Y,\mathscr{E})$$
$$= q_X^Z \circ \mathrm{Tr}_{G/H} K(Z,(H\backslash\mathscr{E})_{Z,Y}).$$

Let $\pi_{G/H}$ be a finite-dimensional unitary representation of the quotient group G/H, and let π_G denote the inflation of $\pi_{G/H}$ to G. Then

$$K(X,\mathscr{E}(\pi_G)_{X,Y})(x_1,x_2,t) = q_X^Y \circ \left(\sum_{g \in G} \pi_G(g)[K(Y,(\mathscr{E} \otimes V_{\pi_G}))](y_1, gy_2, t) \right)$$

$$= q_X^Y \circ \left(\sum_{\lambda \in (G/H)} \sum_{h \in H} \pi_G(\lambda h)[K(Y,(\mathscr{E} \otimes V_{\pi_G}))](y_1, \lambda h y_2, t) \right)$$

$$= q_X^Z \circ \left(\sum_{\lambda \in (G/H)} \pi_{G/H}(\lambda)[K(Z,((H\backslash\mathscr{E}) \otimes V_{\pi_G H})_{Z,Y})](z_1, \lambda z_2, t) \right).$$

Proposition 2.10 is, in a sense, a reformulation of Proposition 2.6, which shows how the heat kernel descends through a finite cover. As a result, we have a corresponding set of resulting formulas, which we will list and refer the reader to the above discussion for proofs.

Corollary 2.11. *In Proposition 2.10, assume that $\pi_{G/H}$ is a one dimensional representation with character $\chi_{\pi_G H}$. Then*

$$K(X,\mathscr{E}(\pi)_{X,Y})(x_1,x_2,t) = q_X^Y \circ \left(\sum_{g \in G} \chi_{\pi_G H}(g) K(Y,(\mathscr{E} \otimes V_{\pi_G}))(y_1, gy_2, t) \right)$$

$$= q_X^Y \circ \left(\sum_{\lambda \in (G/H)} \sum_{h \in H} \chi_{\pi_G H}(\lambda h) K(Y,(\mathscr{E} \otimes V_{\pi_G}))(y_1, \lambda h y_2, t) \right)$$

$$= q_X^Z \circ \left(\sum_{\lambda \in (G/H)} \chi_{\pi_G H}(\lambda) K(Z,((H\backslash\mathscr{E}) \otimes V_{\pi_H})_{Z,Y})(z_1, \lambda z_2, t) \right).$$

As in Theorem 2.5, we have the following trace computations that follow from the formulas in Proposition 2.10.

Theorem 2.12. *In Proposition* 2.10 *and Corollary* 2.11, *we have, as functions of* $t \in \mathbf{R}^+$, *the equality of heat traces*

$$[Y:X]\operatorname{tr} K(X,(G\backslash\mathscr{E})_{X,Y})(t) = [Y:Z] \sum_{\lambda \in G/H} \operatorname{tr} K(Z,(H\backslash\mathscr{E})_{Z,Y})(\lambda, t)$$
$$= \sum_{g \in G} \operatorname{tr} K(Y,\mathscr{E})(g,t).$$

Let $\chi_{\pi_{G,H}}$ denote the character of the representation $\pi_{G/H}$. Then,

$$[Y:X]\operatorname{tr} K(X,\mathscr{E}(\pi_{G/H})_{X,Y})(t) = [Y:Z] \sum_{\lambda \in G/H} \chi_{\pi_{G,H}}(\lambda) \operatorname{tr} K(Z,(H\backslash\mathscr{E})_{Z,Y})(\lambda, t)$$
$$= \sum_{g \in G} \chi_{\pi_G}(g) \operatorname{tr} K(Y,\mathscr{E})(g,t).$$

The second assertion of Theorem 2.12 verifies **AF3**. Note that the second equality is again a consequence of Theorem 2.8 since

$$\operatorname{tr} K(Z,(H\backslash\mathscr{E})_{Z,Y})(\lambda, t) = \sum_{h \in H} \operatorname{tr} K(Y,\mathscr{E})(\lambda h, t).$$

As in Theorem 2.9, we can consider a sum over $(G/H)\hat{}$ and arrive at the following result.

Theorem 2.13. *In Theorem* 2.12, *we have for any* $\lambda \in G/H$,

$$[Y:Z] \sum_{\pi_{G,H} \in (G/H)\hat{}} d_{\pi_{G,H}} \chi_{\pi_{G,H}}(\lambda^{-1}) \operatorname{tr} K(X,\mathscr{E}(\pi_{G/H})_{X,Y})(t)$$
$$= [Y:Z] \operatorname{tr} K(Z,(H\backslash\mathscr{E})_{Z,Y})(\lambda, t)$$
$$= \sum_{\pi_{G,H} \in (G/H)\hat{}} \sum_{g \in G} \chi_{\pi_G}(\lambda^{-1}g) \operatorname{tr} K(Y,\mathscr{E})(g,t).$$

In particular, by taking λ to be the identity in G/H, we arrive at the following equalities of heat traces:

$$[Y:Z] \sum_{\pi_{G,H} \in (G/H)\hat{}} d_{\pi_{G,H}} \operatorname{tr} K(X,\mathscr{E}(\pi_{G/H})_{X,Y})(t)$$
$$= [Y:Z] \operatorname{tr} K(Z,(H\backslash\mathscr{E})_{Z,Y})(t)$$
$$= \sum_{\pi_{G,H} \in (G/H)\hat{}} \sum_{g \in G} \chi_{\pi_G}(g) \operatorname{tr} K(Y,\mathscr{E})(g,t).$$

It remains to prove **AF4**. As above, this follows from first expressing the appropriate heat kernel as the average of heat kernels from a covering manifold, then taking traces. Since the proofs are essentially identical to proofs given above, we will simply list the corresponding formulas.

Proposition 2.14. *Let H denote a subgroup of G, $Z = H \backslash Y$, and $\mathscr{E}$ a G-sheaf on Y. Let π_H be a finite-dimensional unitary representation of H, and $\mathrm{ind}_H^G(\pi_H) = \pi_H^G$ denote the induced representation of π_H to G. Let*

$$\pi_H^G = \bigoplus_{j=1}^{r} \pi_j$$

be a decomposition of the induced representation π_H^G into irreducible representations, counted with multiplicity. Then

$$\mathscr{E}(\pi_H^G)_{X,Y} = \bigoplus_{j=1}^{r} \mathscr{E}(\pi_j)_{X,Y},$$

and

$$K(X, \mathscr{E}(\pi_H^G)_{X,Y}) = \bigoplus_{j=1}^{r} K(X, \mathscr{E}(\pi_j)_{X,Y})(x_1, x_2, t)$$

$$= q_X^Y \circ \left(\sum_{g \in G} \pi_H^G(g) [K(Y, \mathscr{E})](y_1, g y_2, t) \right).$$

Taking trace of the heat kernel expressions given in Proposition 2.14, we arrive at the following result.

Theorem 2.15. *In Proposition 2.14, let χ_{π_H} denote the character of π_H on H and let $\chi_{\pi_H^G}$ denote the character of the induced representation. Then we have, as functions of $t \in \mathbf{R}^+$, the equalities of heat traces*

$$[Y:X] \operatorname{tr} K(X, \mathscr{E}(\pi_H^G)_{X,Y})(t) = [Y:X] \sum_{j=1}^{r} \operatorname{tr} K(X, \mathscr{E}(\pi_j)_{X,Y})(t)$$

$$= \sum_{g \in G} \chi_{\pi_H^G}(g) \operatorname{tr} K(Y, \mathscr{E})(g, t).$$

By Frobenius reciprocity, we have, as functions of $t \in \mathbf{R}^+$, the equalities of heat traces

$$\sum_{g \in G} \chi_{\pi_H^G}(g) \operatorname{tr} K(Y, \mathscr{E})(g, t) = \sum_{\lambda h \in G} \chi_{\pi_H^G}(\lambda h) \operatorname{tr} K(Y, \mathscr{E})(\lambda h, t)$$

$$= \sum_{h \in H} [G:H] \chi_{\pi_H}(h) \operatorname{tr} K(Y, \mathscr{E})(h, t)$$

$$= [Z:X] \operatorname{tr} K(Z, \mathscr{E}(\pi_H)_{Z,Y})(t).$$

The equations in Theorem 2.15 prove **AF 4**, thus completing the proof of Theorem 2.2. As in Theorem 2.9 and Theorem 2.13, we can take the equations in Theorem 2.15, sum over the space of irreducible representations of H, and arrive at the following result.

Theorem 2.16. *Let $\hat{H}$ denote the set of irreducible representations of H, and d_π denote the dimension of $\pi \in \hat{H}$. Then*

$$\sum_{\pi \in \hat{H}} \sum_{j=1}^{r} d_\pi \operatorname{tr} K(X, \mathscr{E}(\pi_j)_{X,Y})(t) = \operatorname{tr} K(Z, (H \backslash \mathscr{E})(\pi_H)_{Z,Y})(t).$$

In the simplest case when π_H is the trivial representation, then Theorem 2.17 reduces to Theorem 2.9 because

$$\bigoplus_{j=1}^{r} \mathscr{E}(\pi_j)_{X,Y} \cong \bigoplus_{\pi \in \hat{H}} (\mathscr{E}(\pi) \otimes V_\pi)_{X,Y}.$$

In Theorem 2.16, if we let t approach zero and take the coefficient of $t^{-n/2}$ in Weyl's law, we arrive at the equation

$$\sum_{\pi \in \hat{H}} \sum_{j=1}^{r} \left(\text{vol}(X) \, \text{rk}\left(\mathscr{E}(\pi_j)\right) \text{mult}\left(\pi_H^G; \pi_j\right) \right) = \text{rk}(\mathscr{E}) \, \text{vol}(Z),$$

which simplifies to the standard formula

$$\sum_{\pi \in \hat{H}} \sum_{j=1}^{r} d_{\pi_j} \text{mult}(\pi_H^G; \pi_j) = [G:H].$$

§ 3. Hurwitz xi function and regularized determinants

In §1 we introduced several basic spectral functions related to the trace of the heat kernel. As a consequence of Theorem 2.2, these spectral functions satisfy Artin's formalism. In this section we will study the Hurwitz xi function and the regularized determinant, and we will examine their spectral properties.

As before, let Y denote a Riemannian manifold with metrized vector sheaf $\mathscr{E}$. Let $K(Y, \mathscr{E})$ denote the heat kernel associated to the metrized sheaf $\mathscr{E}$ and $\text{tr} K(Y, \mathscr{E})(t)$ denote the trace of the heat kernel. Referring to notation in §1, let

$$P_0 \text{tr} K(Y, \mathscr{E})(t) = \sum_{j=-n}^{-1} \beta_{j/2}(Y, \mathscr{E}) t^{j/2}.$$

The function $P_0 \text{tr} K(Y, \mathscr{E})(t)$ is called the *principal part of the heat trace* and is the unique polynomial in the variable $t^{-1/2}$ without constant term such that

(3.1) $$\text{tr} K(Y, \mathscr{E})(t) - P_0 \text{tr} K(Y, \mathscr{E})(t) = O(1) \quad \text{as } t \to 0.$$

Let us *formally* define the function

(3.2) $$I(Y, \mathscr{E})(s, z) = \int_0^\infty [\text{tr} K(Y, \mathscr{E})(t) - P_0 \text{tr} K(Y, \mathscr{E})(t)] e^{-zt} t^s \frac{dt}{t},$$

which (formally) can be written as

$$\xi_{\Lambda(\mathscr{E})}(s, z) = I(Y, \mathscr{E})(s, z) + \sum_{j=-n}^{-1} \frac{\beta_{j/2}(Y, \mathscr{E}) \Gamma(s + j/2)}{z^{s+j/2}}.$$

We have the following result.

Theorem 3.1. *The function $I(Y, \mathscr{E})(s, z)$ has a meromorphic continuation to the region*

$$\mathbf{C} \times \{\mathbf{C} \setminus \mathbf{R}_{\leq 0}\}.$$

Proof. As with Corollary 1.2, the reader is referred to §3 of [JoL 92] for a complete proof which, for the sake of brevity, will not be repeated here. □

For a meromorphic function $G(s)$, we let

$\mathrm{CT}_{s=s_0} G(s) =$ constant term in the Laurent expansion of $G(s)$ at $s = s_0$.

The following result, which we cite from [JoL 92], considers the Laurent expansion of the Hurwitz xi function near $s = 1$.

Theorem 3.2. *Near $s = 1$, the Hurwitz xi function $\xi_{\Lambda(\mathscr{E})}(s, z)$ has the expansion*

$$\xi_{\Lambda(\mathscr{E})}(s, z) = \frac{R_{-1}(1; z)}{s - 1} + R_0(1; z) + O(s - 1)$$

where $R_{-1}(1; z)$ is a polynomial of degree $< n/2$, and

$$R_0(1; z) = \mathrm{CT}_{s=1} \xi_{\Lambda(\mathscr{E})}(s, z)$$

is a meromorphic function in z for all $z \in \mathbf{C}$ whose singularities are simple poles at $z = -\lambda_k$ with residue equal to the multiplicity of λ_k in the sequence $\Lambda(\mathscr{E})$. Furthermore,

$$\mathrm{CT}_{s=1} \xi_{\Lambda(\mathscr{E})}(s, z) = -\partial_z \mathrm{CT}_{s=0} \xi_{\Lambda(\mathscr{E})}(s, z)$$

and

$$R_{-1}(1; z) = \sum_{-j/2 + k = -1} \beta_{j/2}(Y, \mathscr{E}) \frac{(-z)^k}{k!}.$$

Theorem 3.1 tells us that it makes sense to consider the Laurent expansion of the Hurwitz xi function near $s = 1$, as we do in Theorem 3.2, and, as a function of s, the Hurwitz xi function has a simple pole at $s = 1$. From the relation

$$\xi_{\Lambda(\mathscr{E})}(s + 1, z) = -\partial_z \xi_{\Lambda(\mathscr{E})}(s, z),$$

we conclude that the Hurwitz zeta function

$$\zeta_{\Lambda(\mathscr{E})}(s, z) = \Gamma(s)^{-1} \xi_{\Lambda(\mathscr{E})}(s, z)$$

is holomorphic at $s = 0$. Further, by Theorem 3.2 and some elementary complex analysis, we have that

$$-\partial_z \zeta'_{\Lambda(\mathscr{E})}(0, z) = R_0(1; z) + \gamma R_{-1}(1; z)$$

is the logarithmic derivative of some holomorphic function, which we call the regularized determinant.

Definition 3.3. The *regularized determinant* $\mathbf{D}(Y,\mathscr{E})(z)$ of the sequence $\Lambda(\mathscr{E}) + z$ is the unique holomorphic function of z defined by

$$\log \mathbf{D}(Y,\mathscr{E})(z) = -\zeta'_{\Lambda(\mathscr{E})}(0,z).$$

Implicit in Definition 3.3 is the claim that the regularized determinant can be holomorphically extended to the entire complex plane, which is more than is asserted by Theorem 3.1. The proof of this claim is part of the general Lerch formula that is proved in §2 of [JoL 92] and will not be repeated here. Also, let us note that, in the notation of [JoL 92], the *regularized product* of the sequence $\Lambda(\mathscr{E}) + z$ is the meromorphic function $D_{\Lambda(\mathscr{E})}(z)$ defined by

$$\log D_{\Lambda(\mathscr{E})}(z) = -\mathrm{CT}_{s=0}\, \zeta_{\Lambda(\mathscr{E})}(s,z),$$

so

$$D'_{\Lambda(\mathscr{E})}/D_{\Lambda(\mathscr{E})}(z) = R_0(1;z) = \mathbf{D}'_{\Lambda(\mathscr{E})}/\mathbf{D}_{\Lambda(\mathscr{E})}(z) - \gamma R_{-1}(1;z).$$

To continue, let us use the function $I(Y,\mathscr{E})(s,z)$ to derive an integral representation for the regularized determinant.

Theorem 3.4. *The logarithmic derivative of the regularized determinant has the integral representation*

$$(\mathbf{D}'/\mathbf{D})(Y,\mathscr{E})(z) = I(Y,\mathscr{E})(1,z) + \gamma R'_{-1}(1;z)$$

$$+ \sum_{j=-n}^{-1} \beta_{j/2}(Y,\mathscr{E})\, \mathrm{CT}_{s=1}\left[\frac{\Gamma(s+j/2)}{z^{s+j/2}}\right]$$

where $I(Y,\mathscr{E})$ was defined in (3.2).

The proof of Theorem 3.4 is immediate from the bound (3.1) and the definition of the regularized determinant. The reader is referred to §3 and §4 of [JoL 92] for futher details.

Let A be the abelian group generated by all traces of heat kernels. Each one of the functions in the next theorem is obtained by starting from the association of §2, applying a suitable homomorphism to A, and then applying Proposition 2.1.

Theorem 3.5. a) *The principal part of the trace of the heat kernel satisfies Artin's formalism.*

b) *The residue function R_{-1} satisfies Artin's formalism.*

c) *The Hurwitz xi function satisfies Artin's formalism.*

d) *The integral I satisfies Artin's formalism.*

e) *The regularized determinant $\mathbf{D}$ satisfies Artin's formalism.*

Proof. For instance, for the regularized determinant, we first apply the Laplace-Mellin transform, which is a homomorphism of the abelian group A into another abelian group of functions of two variables (s, z). The maps

$$f(s, z) \mapsto f'(0, z) \mapsto \exp(-f'(0, z))$$

define a homomorphism from the abelian group of functions of two variables (s, z), holomorphic in s at $s = 0$, into a multiplicative group of functions of z. Thus we define

$$\mathbf{D}(Y/X, \mathscr{E}, \pi)$$

to be the homomorphic image of $L(Y/X, \mathscr{E}, \pi)$ under the above composite homomorphism, where $L(Y/X, \mathscr{E}, \pi)$ is the function defined in Theorem 2.2. Alternatively, in light of **AF1**, we could also take $\mathbf{D}(Y/X, \mathscr{E}, \pi)$ to be the image of $\theta(X, \mathscr{E}(\pi))$ under these homomorphisms which gives

$$\mathbf{D}(Y/X, \mathscr{E}, \pi) = \mathbf{D}(X, \mathscr{E}(\pi)).$$

This proves the last case. The other cases are treated similarly.

Let us consider the special case when Y is a compact Riemann surface of genus $g(Y) \geq 2$. In this case we have

$$P_0 \operatorname{tr} K(Y, \mathscr{E})(t) = \beta_{-1}(Y, \mathscr{E}) t^{-1} \quad \text{with} \quad \beta_{-1}(Y, \mathscr{E}) = \frac{\operatorname{rk}(\mathscr{E}) \operatorname{vol}(Y)}{4\pi}$$

and $R_{-1}(1; z) = \beta_{-1}(Y, \mathscr{E})$. Thus, we obtain the formula

$$(\mathbf{D}'/\mathbf{D})(Y, \mathscr{E})(z) = I(Y, \mathscr{E})(1, z) + \beta_{-1}(Y, \mathscr{E}) \operatorname{CT}_{s=1} \left[\frac{\Gamma(s-1)}{z^{s-1}} \right] + \gamma \beta_{-1}(Y, \mathscr{E})$$

$$= I(Y, \mathscr{E})(1, z) - \beta_{-1}(Y, \mathscr{E}) \log z.$$

Let π denote a finite-dimensional unitary representation of the fundamental group $\pi_1(Y, \mathbb{Z})$, and $\mathscr{E}$ be a sheaf of the form

$$\mathscr{E} = \mathscr{K}^{j/2} \otimes \mathcal{O}(\pi),$$

where $\mathscr{K}$ is the canonical sheaf on Y and j is an integer. In this setup, Selberg defined a zeta function $Z(Y, \mathscr{E})(s)$ via an Euler product over the primitive hyperbolic conjugacy classes in the fundamental group $\pi_1(Y, \mathbb{Z})$ (see [Sel 56] as well as [He 76]). Equivalently, one can use a formula similar to that in Theorem 3.4 above to study $Z(Y, \mathscr{E})(s)$. Specifically, the Selberg zeta function can be realized as follows.

Let $K_{\mathfrak{h}, j}(\varrho, t)$ denote the heat kernel associated to the Laplacian coming from $\mathscr{K}^{j/2}$ on the hyperbolic upper half plane $\mathfrak{h}$ (since the hyperbolic metric on $\mathfrak{h}$ is homogeneous, the heat kernel is a function of t and the distance ϱ between two points). The Selberg zeta function is given as

(3.3)
$$(Z'/Z)(Y, \mathscr{E})(s)$$

$$= (2s - 1) \int_0^\infty [\operatorname{tr} K(Y, \mathscr{E})(t) - \operatorname{vol}(Y) \operatorname{rk}(\mathscr{E}) K_{\mathfrak{h}, j}(0, t)] e^{-s(s-1)t} dt$$

(see [He 76], page 67 or [Mc 72], page 239). By combining (3.3) with Theorem 3.4, we arrive at the following result.

Theorem 3.6. *For each integer $j \geq 0$, there is a meromorphic function $G_{j/2}(s)$, depending solely on the integer j, having the following property. Let Y denote a compact Riemann surface of genus $g(Y) \geq 2$ with canonical sheaf $\mathscr{K}$, and assume all metrics are hyperbolic. Let π be a finite dimensional representation of the fundamental group $\pi_1(Y, \mathbf{Z})$, and let*

$$\mathscr{E} = \mathscr{K}^{j/2} \otimes \mathscr{O}(\pi).$$

Then

$$\mathbf{D}(Y, \mathscr{E})(s(s-1)) = (G_{j/2}(s))^{(g(Y)-1)\mathrm{rk}(\mathscr{E})} Z(Y, \mathscr{E})(s).$$

Proof. From above, we have

(3.4)
$$(\mathbf{D}'/\mathbf{D})(Y, \mathscr{E})(s(s-1)) - (Z'/Z)(Y, \mathscr{E})(s)$$
$$= (2s-1) \int_0^\infty [\mathrm{vol}(Y) \mathrm{rk}(\mathscr{E}) K_{\mathfrak{h},j}(0,t) - P_0 \mathrm{tr} K(Y, \mathscr{E})(t)] e^{-s(s-1)t} dt$$
$$- \beta_{-1}(Y, \mathscr{E})(2s-1) \log[s(s-1)]$$

where, in this case, we have

$$P_0 \mathrm{tr} K(Y, \mathscr{E})(t) = \beta_{-1}(Y, \mathscr{E}) t^{-1}.$$

It is shown in [He 76] that (3.3) has a meromorphic extension to the entire s-plane, and its singularities are simple poles whose residues are integer multiples of $(g(Y)-1) \mathrm{rk}(\mathscr{E})$ (see [He 76], Theorem 4.11 and the discussion beginning on page 438). This result, together with Theorem 3.4, proves the theorem with

(3.5) $$(g(Y)-1) \mathrm{rk}(\mathscr{E}) G'_{j/2}/G_{j/2}(s) = -\beta_{-1}(Y, \mathscr{E})(2s-1) \log[s(s-1)]$$
$$+ (2s-1) \int_0^\infty [\mathrm{vol}(Y) \mathrm{rk}(\mathscr{E}) K_{\mathfrak{h},j}(0,t) - \beta_{-1}(Y, \mathscr{E}) t^{-1}] e^{-s(s-1)t} dt. \quad \square$$

Since
$$\mathrm{vol}(Y) = 4\pi(g(Y)-1),$$

notice that the factor $\mathrm{vol}(Y) \mathrm{rk}(\mathscr{E})$ appears in each term of (3.5), which means we can write (3.5) as

$$G'_{j/2}/G_{j/2}(s) = (2s-1) \left[\int_0^\infty \left[4\pi K_{\mathfrak{h},j}(0,t) - \frac{1}{t} \right] e^{-s(s-1)t} dt - \log[s(s-1)] \right]$$
$$= (2s-1) \int_0^\infty \left[4\pi K_{\mathfrak{h},j}(0,t) e^{-s(s-1)t} - \frac{e^{-t}}{t} \right] dt.$$

Direct calculations show that

$$G_0(s) = \left(e^{a-s(s-1)} \frac{\Gamma_2(s+1/2)}{\Gamma(s)} (2\pi)^s \right)^2$$

and
$$G_{1/2}(s) = \left(e^{a-s(s-1)}(\Gamma_2(s+1/2))^2(2\pi)^s\right)^2,$$

where $\Gamma(s)$ is the gamma function, $\Gamma_2(s)$ is the Barnes double gamma function, and

$$a = -\frac{1}{4} - \frac{1}{2}\log(2\pi) + 2\zeta'_\mathbf{Q}(-1)$$

with $\zeta_\mathbf{Q}(s)$ denoting the Riemann zeta function (see [CV 90], [DP 86], [Ma 92], [Sa 87], [Vo 87], and [Wi 92]). The case when Y is the compact quotient of any real rank one symmetric space is considered in [Ga 77] and [Wi 92].

Corollary 3.7. *Define Z as in the discussion leading to Theorem 3.6. Then Z satisfies Artin's formalism.*

Corollary 3.7 follows from Theorem 3.5, Theorem 3.6 and the fact that the exponents of the universal functions $G_{j/2}(s)$ are simply the (algebraic) Euler characteristic of $\mathscr{E}$, which, as a topological invariant, is known to satisfy Artin's formalism. Corollary 3.7 was studied and proved through the use of the Selberg trace formula in the papers of Venkov and Zograf [Ve 79], [VeZ 81], and [VeZ 83].

To conclude this section, let us briefly explain how the Artin formalism when applied to a particular Hurwitz zeta function yields the classical multiplication formula for the gamma function. Consider the case where

$$Y = \mathbf{R}/2\pi\mathbf{Z},$$

and the operator in question is the elliptic operator d/dx, which acts on the subspace of C^∞ functions on Y whose negative Fourier coefficients are all zero. The manifold Y admits a fixed point free action by the group $G = \mathbf{Z}/N\mathbf{Z}$, for any $N \in \mathbf{Z}_{>0}$. The Hurwitz zeta function associated to this operator is

$$\zeta_\mathbf{Q}(s, Nz) = \sum_{n=0}^{\infty} \frac{1}{(Nz+n)^s}.$$

By the Artin formalism, we have the equality

$$\zeta_\mathbf{Q}(s, Nz) = \sum_{j=0}^{N-1} N^{-s} \zeta_\mathbf{Q}(s, s+j/N),$$

where the factor of N^{-s} appears because of the area of the quotient manifold $G \backslash Y$ is $1/N$ times the area of Y. The classical Lerch formula (see §2 of [Jo 92]) states

$$\mathbf{D}(z) = \frac{\sqrt{2\pi}}{\Gamma(z)} = \exp\left(-\zeta'_\mathbf{Q}(0, z)\right).$$

This, together with the formula

$$\zeta_\mathbf{Q}(0, z) = -z + \frac{1}{2},$$

yields the identity

$$N^{Nz-1/2}\mathbf{D}(Nz) = \prod_{j=0}^{N-1} \mathbf{D}(z+j/N),$$

which is Gauss's multiplication formula or distribution relation for the gamma function.

Note that in the example, the operator is elliptic but not hermitian or positive. The Artin formalism applies in more general cases than we have given, e.g. to certain pseudo differential operators instead of the positive Laplacian. In this article we limited ourselves to a relatively concrete case whose terminology is standard, to avoid using much more space recalling other terminology which is not so standard. In this way we place emphasis on the Artin formalism itself, but we do want to make explicit the possibility of wider applications as already mentioned in the introduction.

§ 4. Generalized Jacobi and Kronecker formulae

In this section, we show that the classical Jacobi derivative formula and Kronecker's second limit formula are special cases of Artin's formalism when applied to the spectral zeta function associated to line sheaves on tori. These results admit generalizations which amount to Artin's formalism for abelian coverings, especially those arising from multiplication by an integer on a compact real torus. In addition, we establish an identity involving theta functions, derivatives of theta functions and Prym differentials that we derive from spectral theory and, in particular, formulas involving analytic torsion which we cite from [Jo91] and [Jo92].

Recall that if Y is a compact Riemannian manifold with metrized vector sheaf $\mathscr{E}$, the spectral zeta function is defined, for Re(s) sufficiently large, by

$$\zeta(Y,\mathscr{E})(s) = \frac{1}{\Gamma(s)} \int_0^\infty [\operatorname{tr} K(Y,\mathscr{E})(t) - m_0(Y,\mathscr{E})] t^s \frac{dt}{t} = \sum_{n=1}^\infty \lambda_n(Y,\mathscr{E})^{-s}.$$

Proposition 4.1. *Let G be a finite abelian group, and let Y be a G-manifold with G-sheaf $\mathscr{E}$ and quotient manifold X. Then we have the equality of spectral zeta functions*

$$\sum_{\chi \in \hat{G}} \zeta(X,\mathscr{E}(\chi)_{X,Y})(s) = \zeta(Y,\mathscr{E})(s).$$

In particular, by considering the special value at $s = 0$, we obtain

$$\sum_{\chi \in \hat{G}} \zeta(X,\mathscr{E}(\chi)_{X,Y})'(0) = \zeta(Y,\mathscr{E})'(0).$$

Proof. The proof follows directly from Theorem 2.9 and the definition of the spectral zeta function $\zeta(Y,\mathscr{E})(s)$. □

We call the second assertion in Proposition 4.1 the *generalized Jacobi identity* for the Riemannian manifold X. Further simplification of Proposition 4.1 takes place when X is a real compact torus, in which case we need the following lemma.

Lemma 4.2. *Let X denote a real compact torus of real dimension n with trivial sheaf $\mathcal{O}$ and a translation invariant metric. Then*

$$\zeta(X, \mathcal{O})(0) = -1.$$

Proof. Realize X as $\mathbf{R}^n$ modulo the lattice generated by the columns of the matrix B, and let $\tilde{B}$ denote the matrix of the dual lattice. The trace of the heat kernel is simply

$$\operatorname{tr}(X, \mathcal{O})(t) = \sum_{m \in \mathbf{Z}^n} e^{-{}^t m \, \tilde{B} m c t},$$

for some positive constant c that depends on the scale of the metric. By the Poisson summation formula (see page 267 of [La 87] or Proposition 5.2 below), there exist positive constants c_1 and c_2 such that

(4.1) $$\operatorname{tr}(X, \mathcal{O})(t) = c_1 t^{-n/2} + O(e^{-c_2/t}) \quad \text{as } t \to 0.$$

Recall that we can write the spectral zeta function as

(4.2) $$\zeta(X, \mathcal{O})(s) = \frac{1}{\Gamma(s)} \int_0^1 (\operatorname{tr}(X, \mathcal{O})(t) - 1) t^s \frac{dt}{t} + \frac{1}{\Gamma(s)} \int_1^\infty (\operatorname{tr}(X, \mathcal{O})(t) - 1) t^s \frac{dt}{t}.$$

The second integral in (4.2) is holomorphic at $s = 0$, and $\Gamma(s)$ has a first order pole at $s = 0$ with residue one. With this, we can substitute (4.1) into (4.2) and arrive at the equation

$$\zeta(Y, \mathcal{E})(s) = \frac{c_1 s}{s - n/2} - 1 + s \cdot (\text{holo. at } s = 0),$$

from which the lemma follows. □

Theorem 4.3. *Let X denote a real compact torus of real dimension n, with a translation invariant metric. Let k_X be multiplication by k and let X_k denote the kernel of k_X. Then*

$$\sum_{\chi \neq \mathrm{Id}} \zeta(X, \mathcal{O}(\chi)_{X,Y})'(0) = -\log k^n,$$

where the sum is taken over the non-trivial characters of $G \cong X_k$.

Proof. When computing the terms $\zeta(X, \mathcal{O})'(0)$ that appear on both sides of the equation in Proposition 4.1, one must carefully note that one term is relative to the given translation invariant metric, and the other is metric obtained by scaling the given metric by k^{-1}. If one scales the given metric by the constant k^{-1}, the eigenvalues scale by the constant k^n. From the definition of the spectral zeta function, we obtain

$$\zeta_{k^{-1} \cdot g}(Y, \mathcal{E})(s) = k^{-ns} \zeta_g(Y, \mathcal{E})(s),$$

hence,

(4.3) $\qquad \zeta_{k-1,g}(Y,\mathscr{E})'(0) = -\log k^n \cdot \zeta_g(Y,\mathscr{E})(0) + \zeta_g(Y,\mathscr{E})'(0)$.

If we combine (4.3) with Lemma 4.2, the proof is complete. □

For the remainder of this section, we will study Theorem 4.3 in the case X is an algebraic curve. For this, let us recall basic results from the theory of theta functions and complex algebraic geometry associated to algebraic curves. Further details are given in many of the references cited in the bibliography.

Let X denote a non-singular algebraic curve of genus g defined over $\mathbf{C}$, and let $\mathscr{K}$ denote the canonical sheaf on X. Let P be a point of X and

$$\psi_P : X \to J(X)$$

be the Abel-Jacobi map of X into its Jacobian which sends P to the origin. Let

$$W_{g-1,P} = W_{g-1}$$

denote the divisor of $J(X)$, which is equal to the sum of $\psi_P(X)$ taken $g-1$ times. Consider the ensuing principal theta polarization corresponding to the hermitian form $\mathbf{H}_X$ (see Chapter VI of [La 82] or Chapter 13 of [La 83]). The principal theta polarization gives rise to the Riemann theta function and its theta divisor, which we denote by Θ, as a divisor on the Jacobian $J(X)$ (see page 101 of [La 82] or page 334 of [La 83]). There is a well-defined choice of a sheaf $\mathscr{S}$ of degree $g-1$ such that $\mathscr{S}^2$ is isomorphic to the canonical sheaf $\mathscr{K}$ on X. The sheaf $\mathscr{S}$ is characterized as follows. Up to linear equivalence, there is a unique divisor D of degree $g-1$ such that

$$\Theta = W_{g-1,P} - \psi_P(D).$$

Let $\mathscr{S}$ be the unique, up to isomorphism, line sheaf such that $\mathscr{S} \cong \mathcal{O}(D)$. Using Riemann's vanishing theorem, one can show that $\mathscr{S}^2 \cong \mathscr{K}$.

A classical description of the above set-up is as follows. Choose a canonical basis of $H_1(X,\mathbf{Z})$, which we denote by $A_1, B_1, \ldots, A_g, B_g$. Let $\zeta_1, \ldots, \zeta_g$ denote a basis of $H^0(X,\mathscr{K})$ dual to the chosen marking and let Ω denote the g by g matrix defined by

$$\Omega = \left(\int_{B_j} \zeta_k \right).$$

The Jacobian $J(X)$ can be realized as the complex torus obtained by $\mathbf{C}^g$ modulo the lattice $L(\Omega)$ generated by the period matrix (I_g, Ω). Let M_X denote the matrix associated to the hermitian form $\mathbf{H}_X$. Then

$$M_X = (\operatorname{Im}\Omega)^{-1}.$$

This is proved in Chapter 13, Proposition 3.1 of [La 83]. By definition,

$$\det M_X = \det \mathbf{H}_X.$$

The sheaf $\mathscr{S}$ corresponds to the degree $g-1$ divisor class determined by the vector of Riemann constants (see [FK 80], page 298, or [Fy 73], page 10). The *Riemann theta function* is defined for $z \in \mathbf{C}^g$ by the series

(4.4) $$\theta(z, \Omega) = \sum_{n \in \mathbf{Z}^g} \exp(\pi i^t n \Omega n + 2\pi i^t nz).$$

Let us introduce the notation $\|\theta\|^2$ to denote

$$\|\theta\|^2(z, \Omega) = (\det(\operatorname{Im}(\Omega)))^{\frac{1}{2}} \exp(-2\pi^t y \operatorname{Im}(\Omega) y) |\theta(z, \Omega)|^2$$

where $y = \operatorname{Im} z$. The function $\|\theta\|^2(z, \Omega)$ is well-defined for all $z \in J(X)$. Basically, the Riemann theta function (4.4) is a section of the bundle $\mathcal{O}(-\Theta)$ over $J(X)$, and $\|\theta\|^2$ is the square of the norm of that section with respect to a specific metric (the theta metric, see [Jo 92]). Beyond what is stated above, additional basic properties of the theta function will be assumed.

One can view the Jacobian variety $J(X)$ as the space of isomorphism classes of degree zero line sheaves. The Picard variety $\operatorname{Pic}_{g-1}(X)$ is the space of isomorphism classes of degree $g-1$ line sheaves ([GH 78], page 313). We have a canonical isomorphism

$$J(X) \to \operatorname{Pic}_{g-1}(X)$$

given by

$$\mathscr{L} \mapsto \mathscr{L} \otimes \mathscr{S}.$$

Let us use $[\mathscr{L}]$ to denote the point in $J(X)$ corresponding to the degree zero line sheaf $\mathscr{L}$. In particular, $[\mathscr{L}]$ can be used as an argument of the Riemann theta function (4.4). Classically, the argument $[\mathscr{L}]$ is written as follows. Let K_P denote the vector of Riemann constants relative to the base point P of the Abel-Jacobi map ψ_P. Let $\mathscr{L}$ denote a degree zero line sheaf isomorphic to

$$\mathscr{L} \cong \mathcal{O}(P_1 + \cdots + P_d - R_1 - \cdots - R_d).$$

Then

$$[\mathscr{L}] = \sum_{j=1}^{d} \left(\int_P^{P_j} \zeta - \int_P^{R_j} \zeta \right) \mod L(\Omega).$$

Let $\mathscr{M}$ denote the degree $g-1$ sheaf

$$\mathscr{M} = \mathcal{O}(P_1 + \cdots + P_{g-1}).$$

The point $[\mathscr{M} \otimes \mathscr{S}^{-1}]$ in $J(X)$ is classically given by

$$[\mathscr{M} \otimes \mathscr{S}^{-1}] = \sum_{j=1}^{g-1} \int_P^{P_j} \zeta + K_P \mod L(\Omega).$$

With all this, we have the following corollary of Theorem 4.3 which, in rough terms, states that the logarithm of the regularized product associated to a sheaf of degree $g-1$, when considered as a function on the Picard group, satisfies a natural normalization when viewing the Picard group as a complex torus.

Corollary 4.4. *Let X be an algebraic curve with Picard group $\mathrm{Pic}^0(X)$, viewed as a complex torus of complex dimension g. Let $\mathrm{vol}_{\mathrm{Pic}}$ denote a translation invariant volume element on $\mathrm{Pic}^0(X)$. Then*

$$\int_{\mathrm{Pic}^0(X)} \zeta(X,[\mathscr{L}])'(0)\,\mathrm{vol}_{\mathrm{Pic}}([\mathscr{L}]) = 0\,.$$

Proof. The formula follows directly from Theorem 4.3 by dividing by k^{2g} and letting k approach infinity. The sum on the left hand side of the equation is the appropriate Riemann sum that approaches the above integral. □

We shall call the formula in Corollary 4.4 the *generalized Kronecker limit formula*. We shall now show how Proposition 4.1 and Corollary 4.4 generalize the classical Jacobi derivative formula and Kronecker's second limit formula.

For now, assume X is of genus one, realized as the complex plane modulo the lattice generated by 1 and τ, where $\tau = a + ib$ and $b > 0$. The canonical metric on X is given by the $(1,1)$ form

$$\mu_{\mathrm{ca}}(z) = \frac{i}{2b}\, dz \wedge d\bar{z}\,,$$

with associated Laplacian

$$\Delta_{\mathrm{ca}} = -4b\left(\frac{\partial^2}{\partial z \partial \bar{z}}\right).$$

Let us define the operator dd^c to be

$$dd^c = \frac{i}{2\pi}\left(\frac{\partial^2}{\partial z \partial \bar{z}}\right) dz \wedge d\bar{z}\,.$$

For any point u in the Jacobian $J(X)$, let $\mathscr{L}_u$ denote the associated degree zero line sheaf. Given a degree zero line sheaf $\mathscr{L}$, let $[\mathscr{L}]$ denote the corresponding point in $J(X)$.

Lemma 4.5. *Consider the action of the Laplacian Δ_{ca} on C^∞ sections of $\mathscr{L}_u$. The eigenvalues of this action are*

$$\lambda_{n,m}(u) = \frac{(2\pi)^2}{b}|u - \tau n + m|^2\,,$$

for all n and m in $\mathbf{Z}$, with multiplicity one.

The reader is referred to page 166 of [RS 73] for a proof and further discussion.

For any positive $(1,1)$ form μ on X, also called a metric, and degree zero line sheaf $\mathscr{L}$, let $u = [L]$ in $J(X)$ and define

$$\zeta(s, u, \mu) = \zeta_{\mathscr{L},\mu}(s)\,.$$

By Lemma 4.5, if $\mu = \mu_{ca}$, the spectral zeta function $\zeta(s, u, \mu_{ca})$ is explicitly known as a function of s, u and τ. Specifically, if $u \neq 0$,

$$\zeta(s, u, \mu_{ca}) = \left(\frac{b}{(2\pi)^2}\right)^s \sum_{n,m} |u - \tau n + m|^{-2s}, \tag{4.5}$$

and

$$\zeta(s, 0, \mu_{ca}) = \left(\frac{b}{(2\pi)^2}\right)^s \sum_{(n,m) \neq (0,0)} |\tau n - m|^{-2s}. \tag{4.6}$$

It is important to notice that (4.5) and (4.6) are multiples of the Eisenstein series associated to the discrete subgroup $SL(2, \mathbf{Z})$ of the group $SL(2, \mathbf{R})$ (see [La 87], page 273).

Lemma 4.6. *For any s in $\mathbf{C}$,*

$$d_u d_u^c \zeta(s, u, \mu_{ca}) = 4\pi s^2 \zeta(s+1, u, \mu_{ca}) \mu_{ca}.$$

Proof. By Lemma 4.5, it is immediate that for any n and m,

$$\frac{\partial^2}{\partial u \partial \bar{u}} \lambda_{n,m}(u) = \frac{(2\pi)^2}{b}, \tag{4.7}$$

and

$$\frac{\partial}{\partial u} \lambda_{n,m}(u) \cdot \frac{\partial}{\partial \bar{u}} \lambda_{n,m}(u) = \frac{(2\pi)^2}{b} \lambda_{n,m}(u). \tag{4.8}$$

A simple computation yields the identity

$$\frac{\partial^2}{\partial u \partial \bar{u}} \left(\lambda_{n,m}^{-s}(u)\right) \tag{4.9}$$

$$= s\left((s+1)\frac{\partial \lambda_{n,m}(u)}{\partial u} \cdot \frac{\partial \lambda_{n,m}(u)}{\partial \bar{u}} \lambda_{n,m}^{-1}(u) - \frac{\partial^2 \lambda_{n,m}(u)}{\partial u \partial \bar{u}}\right) \lambda_{n,m}^{-s-1}(u).$$

By combining (4.7), (4.8) and (4.9), the proposition is proved when $\mathrm{Re}(s)$ is sufficiently large. By analytic continuation, the stated result follows. $\square$

Theorem 4.7. *For any metric μ on X,*

$$d_u d_u^c \zeta'(0, u, \mu) = \mu_{ca}(u),$$

and

$$\zeta'(0, u, \mu) = -\log|u|^2 + C^\infty \text{ function in } u$$

as $|u|$ approaches zero. Also,

$$\int_X \zeta'(0, u, \mu_{ca}) \mu_{ca}(u) = 0.$$

Proof. Since $\text{vol}_{ca}(X) = 1$, it follows from the analytic continuation of the spectral zeta function that $\zeta(s, u, \mu_{ca})$ has the following expansion near $s = 0$:

(4.10) $$\zeta(s, u, \mu_{ca}) = \frac{1}{4\pi}\left(\frac{1}{s-1}\right) + \text{holomorphic at } s = 0.$$

Combining Lemma 4.6 and (4.10), we have that

$$\frac{\partial^2}{\partial u \partial \bar{u}} \zeta(s, u, \mu_{ca}) = \frac{(2\pi)^2 s}{4\pi b} + s^2 \cdot \text{holomorphic at } s = 0,$$

which proves the theorem in the case $\mu = \mu_{ca}$.

For a general metric μ, there is a smooth function ϕ such that $\mu = e^\phi \mu_{ca}$. The Polyakov formula (see §1 of [JoLu92], for example, as well as references therein) asserts the existence of some functional $F(\phi)$ such that

$$\zeta'(0, u, \mu) = \zeta'(0, u, \mu_{ca}) + F(\phi).$$

Since the operator $d_u d_u^c$ annihilates the constant (in u) $F(\phi)$, the proof of the first assertion is complete. In the case $\mu = \mu_{ca}$, the second assertion follows from Lemma 4.5. The Polyakov formula applies to complete the proof for general μ. Finally, the last assertion is simply a restatement of Corollary 4.4. □

Using the results of [Qu86], one can prove the first assertion of Theorem 4.7 *without* explicit knowledge of the eigenvalues, as we have used.

Let us recall various special functions associated to the elliptic curve X with complex modulus τ. The (analytic) discriminant $\Delta(\tau)$ is expressible as the infinite product

$$\Delta(\tau) = q_\tau \prod_{n=1}^{\infty} (1 - q_\tau^n)^{24}$$

where

$$q_\tau = \exp(2\pi i \tau).$$

The function $\Delta(\tau)$ is, up to constant multiple, the unique cusp form of weight 12 for $SL(2, \mathbf{Z})$. Kronecker's eta function $\eta(\tau)$ is a particular 24th root of $\Delta(\tau)$. The norm of $\Delta(\tau)$ is defined as

$$\|\Delta\|(\tau) = b^6 |\Delta(\tau)|,$$

and

$$\|\eta\|(\tau) = \|\Delta\|^{\frac{1}{24}}(\tau).$$

The (analytic) discriminant $\Delta(\tau)$ is related to the theta function via the expression

$$\left(q_\tau^{1/8} \frac{\partial \theta}{\partial z}([\mathscr{S}], \tau)\right)^8 = (2\pi)^8 \Delta(\tau).$$

The theta divisor Θ on the elliptic curve X, viewed as its own Jacobian, is the 2-torsion point $\frac{1}{2}(\tau + 1)$; hence the sheaf $\mathscr{S}$ is isomorphic to $\mathscr{O}\left(\frac{1}{2}(\tau + 1) - (0)\right)$, and

$$[\mathscr{S}] = \frac{1}{2}(\tau + 1) - (0).$$

Lemma 4.8. *Let X denote an elliptic curve with complex modulus τ, and let $\mathscr{S}$ denote the square root of the canonical sheaf. Let θ denote the Riemann theta function associated to X, with theta divisor Θ. The following identities hold:*

$$d_u d_u^c \log \|\theta\|^2 (u - [\mathscr{S}], \tau) = -\mu_{ca}(u),$$

$$\log \|\theta\|^2 (u - [\mathscr{S}], \tau) = -\log |u|^2 + C^\infty \text{ in } u,$$

$$\int_{J(X)} \log \|\theta\|^2 (u, \tau) \mu_{ca}(u) = \log \|\eta\|^2 (\tau).$$

Proof. The reader is referred to [Mu 83], pages 68 and 123, for the appropriate formulas from which the above identities follow. □

Theorem 4.9.

$$\zeta(0, u, \mu_{ca}) = -\log \|\theta\|^2 (u - [\mathscr{S}], \tau) + \log \|\eta\|^2 (\tau).$$

Proof. Combining the above results, we get

$$d_u d_u^c [\zeta(0, u, \mu_{ca}) + \log \|\theta\|^2 (u - [\mathscr{S}], \tau)] = 0$$

for all $u \in J(X)$. Hence, this function is a constant that depends solely on τ. To finish, use the integrals given in Theorem 4.7 and Lemma 4.8. □

If we combine Theorem 4.3 and Theorem 4.9 with $n = 2$, we obtain the relation

(4.11) $$\|\theta\|(0, \tau) \|\theta\|\left(\frac{1}{2}, \tau\right) \|\theta\|\left(\frac{\tau}{2}, \tau\right) = 2 \|\eta\|^3 (\tau).$$

Up to a constant of modulus 1, this is the classical Jacobi derivative formula as stated on page 64 of [Mu 83]. This constant is easily determined by considering the limiting behavior of (4.9) when $\tau = ib$ and b tends to infinity.

Let V denote a projective variety and D a Cartier divisor on V. By a *complex Weil function associated to D* we mean a function

$$\lambda_D : V \backslash \text{Supp}(D) \to \mathbf{R}$$

satisfying the following condition. If D is locally represented by $\{(U,f)\}$ where U is a Zariski open set, and f is a rational function, then there exists a continuous function α on U such that

$$\lambda_D(P) = -\log|f(P)|^2 + \alpha(P)$$

for all $P \in U \setminus \mathrm{Supp}(D)$. The function α is then uniquely determined by λ_D and the pair (U, f). A *Green's function* for a divisor D with respect to a positive $(1,1)$ form μ is function

$$g_D : X \setminus \mathrm{supp}(D) \to \mathbf{R}$$

such that

$$g_D(P) = -\log|f(P)|^2 + \alpha(P),$$

in notation stated above, and

$$dd^c g_D = (\deg D)\mu.$$

With this language, we can summarize the above calculations as follows.

Theorem 4.10. *For any metric μ, $\zeta'(0, u, \mu)$ is a Weil function for the point $u = 0$ on $J(X)$. Also, $\zeta(0, u, \mu_{\mathrm{ca}})$ is the Green's function $g(0, u)$ for the divisor (0) with respect to the form μ_{ca}.*

To conclude this section, we will combine results obtained in [Jo 91] and [Jo 92] to present a Jacobi formula for algebraic curves. The following proposition from [Jo 91] relates the regularized products for two degree zero line sheaves on any non-singular algebraic curve X to classical theta functions. Let us set the notation

$$\log \det \Delta(X, \mathscr{L}) = -\zeta(X, \mathscr{L})'(0) \quad \text{and} \quad c_A(X) = -\zeta(X, \mathcal{O})'(0) - \log \mathrm{vol}(X).$$

Proposition 4.11. *Let X denote a non-singular algebraic curve of genus g, let $\mathscr{L}$ denote a non-trivial degree zero line sheaf on X. Let $P = P_g$ denote a fixed point on X and let $P_1, \ldots, P_{g-1}$ denote generic, distinct points on X. Let $\eta_1, \ldots, \eta_{g-1}$ denote a basis of $H^0(X, \mathscr{K} \otimes \mathscr{L})$ and $\omega_1, \ldots, \omega_g$ denote a basis of $H^0(X, \mathscr{K})$. Let*

$$\mathscr{N} = \mathscr{L} \otimes \mathcal{O}(-P_1 - \cdots - P_{g-1})$$

and $\mathscr{M} = \mathscr{N} \otimes \mathscr{L}$. Then for any metric μ on X,

$$\log \det \Delta(X, \mathscr{L}) = c_A(X) + \log F(X, \mathscr{L})$$

where

$$\log F(X, \mathscr{L}) = 2\log 2\pi + \log \|\theta\|^2([\mathscr{M}], \Omega) + \log\left[\frac{\det(\langle \eta_i, \eta_j \rangle)}{\det(\langle \omega_i, \omega_j \rangle)}\right]$$

$$- \log\left(\left|\frac{\det((\eta_i)(P_j))}{\det((\omega_i)(P_j))} \sum_{j=1}^{g} \frac{\partial \|\theta\|}{\partial z_j}([\mathscr{N}], \Omega)\zeta_j(P_g)\right|^2\right).$$

The key point of Proposition 4.11 is that one has an explicit relation between $\det \Delta(X, \mathscr{L})$ and $c_A(X)$ through the theta functions as given by $F(X, \mathscr{L})$.

The following result, proved in §1 of [Jo 92], relates the regularized product associated to any degree $g - 1$ line sheaf to that of the trivial line sheaf.

Proposition 4.12. *Let $\mathscr{L}$ denote a degree zero line sheaf. There is a constant $a(g)$ of the form $a(g) = a(0)(1 - g)$ such that*

$$\log \det \Delta(X, \mathscr{S} \otimes \mathscr{L}) + \frac{1}{2} c_A(X) = \frac{a(g)}{4} + \log \|\theta\|^2([\mathscr{L}], \Omega).$$

The proof of Proposition 4.12 comes from considering the Mumford isomorphism

$$M : \det H(\mathscr{S}) \to \left(\det H(\mathcal{O})\right)^{-\frac{1}{2}}$$

and computing Quillen metrics of generators of this isomorphism. The reader is referred to [Jo 92], §1, for a proof.

By combining the formulas in Proposition 4.11 and Proposition 4.12, we arrive at the equation

$$(4.12) \quad \log \det \Delta(X, \mathscr{L}) + 2 \log \det \Delta(X, \mathscr{S} \otimes \mathscr{L}) - \frac{a(g)}{2} = \log L(X, \mathscr{L})$$

where

$$\log L(X, \mathscr{L}) = \log F(X, \mathscr{L}) + 2 \log \|\theta\|^2([\mathscr{L}], \Omega).$$

The terms $\log \det \Delta(X, \mathscr{S} \otimes \mathscr{L})$, $a(g)$ and $\log \det \Delta(X, \mathscr{L})$ all satisfy Artin's formalism. Therefore, the right hand side of (4.12) also satisifies the Artin formalism.

Theorem 4.13. *Let Y denote a compact Riemann surface of genus with fixed point free action by the abelian group G and quotient surface X. Let $\mathscr{L}$ denote a degree zero line sheaf on Y. Then*

$$\prod_{\chi \in \hat{G}} L(X, \mathscr{L}(\chi)_{X,Y}) = L(Y, \mathscr{L}).$$

Proof. Immediate by applying Proposition 4.1, once with $\mathscr{E} = \mathscr{L}$ and once with $\mathscr{E} = \mathscr{S}$, and using (4.12). □

As above, if one considers the case of an elliptic curve X covering itself via multiplication by 2 with $\mathscr{L} = \mathcal{O}$, Theorem 4.13 reduces to the classical Jacobi derivative formula. Also, Theorem 4.13 contains the Green's function identity stated in §1 of [Sz 90] when using formulas from [Jo 90]. It would be interesting to combine Theorem 4.13 with the Green's function formulas from [Jo 90] to address the questions stated in Vojta's appendix to [La 88] (see page 172). In addition, it remains to relate these results to the theta function identities studied in chapter 4 of Fay's book [Fy 73]. This work will be left for future discussions.

§ 5. Riemann theta identities

In this section we will give short proofs of the Riemann theta identities and the functional equation of the Riemann theta function. As before, the underlying ideas only involve the existence and uniqueness of the heat kernel together with elementary representation theory of finite groups. For example, the classical Riemann theta identity for the Riemann theta function in one variable becomes a statement involving the heat kernel for the manifold $\mathbf{R}/2\pi\mathbf{Z}$ and the representations of the group $\mathbf{Z}/2\mathbf{Z}$. For the sake of brevity, we will give complete proofs only for the theta function of one variable and simply outline proofs for theta functions of several variables.

Throughout this section we will consider the theta function in the variable $z \in \mathbf{C}$ and modulus $\tau \in \mathfrak{h}$. Recall that the theta function with real characteristics α and β is defined as

$$(5.1) \quad \theta\begin{bmatrix}\alpha\\ \beta\end{bmatrix}(z,\tau) = \sum_{m \in \mathbf{Z}^g} \exp(\pi i(m+\alpha)^2\tau + 2\pi i(m+\alpha)(z+\beta)).$$

Lemma 5.1. *Consider the manifold* $Y = \mathbf{R}/2\pi\mathbf{Z}$ *with translation invariant metric under which the volume of* Y *is* 2π *and Laplacian* $\Delta = -d^2/dx^2$. *Let* $\mathcal{O}(\alpha)$ *denote the sheaf on* Y *represented by the space of functions on* $\mathbf{R}$ *for which*

$$f(x+2\pi) = \exp(2\pi i\alpha)f(x).$$

Then, relative to the action of Δ *on smooth sections on* $\mathcal{O}(\alpha)$,

$$K(Y, \mathcal{O}(\alpha))(y_1, y_2, t) = \frac{1}{2\pi}\theta\begin{bmatrix}\alpha\\ 0\end{bmatrix}((y_1 - y_2)/2\pi, it/\pi).$$

Proof. The orthonormal eigensections of $\mathcal{O}(\alpha)$ are given by the functions (defined on $\mathbf{R}$)

$$(5.2) \quad \phi_k(x) = \frac{1}{\sqrt{2\pi}} e^{i(k+\alpha)x} \quad \text{for } k \in \mathbf{Z}$$

with associated eigenvalue $(k+\alpha)^2$. The result follows by substituting this data into the spectral expression for the heat kernel (1.3) and the defining equation for the theta function (5.1). □

Proposition 5.2. *Let* $K(\mathbf{R}, \mathcal{O})(\tilde{x}_1, \tilde{x}_2, t)$ *be the heat kernel on* $\mathbf{R}$ *associated to the operator* $\Delta = -d^2/dx^2$. *Then*

$$K(Y, \mathcal{O}(\alpha))(x_1, x_2, t) = \sum_{k=-\infty}^{\infty} K(\mathbf{R}, \mathcal{O})(\tilde{x}_1, \tilde{x}_2 + 2\pi k, t) e^{2\pi i k\alpha},$$

where $\tilde{x}_1$ *and* $\tilde{x}_2$ *are points in* $\mathbf{R}$ *that project to* x_1 *and* x_2, *respectively.*

Proof. Proposition 5.2 follows directly from the existence and uniqueness of the heat kernel, and is a direct consequence of Corollary 2.7. □

Recall that the heat kernel on $\mathbf{R}$ associated to the Laplacian $\Delta = -d^2/dx^2$ which acts on the space of C^∞ functions is

$$(5.3) \qquad K(\mathbf{R}, \mathcal{O})(x, y, t) = \frac{1}{\sqrt{4\pi t}} \exp\left(-\frac{(x-y)^2}{4t}\right)$$

(see page 142 of [Ch 84]). Let $u = x_1 - x_2$ and substitute (5.3) into the equation given in Proposition 5.2 to obtain

$$(5.4) \qquad \frac{1}{2\pi} \sum_{k=-\infty}^{\infty} e^{-(k+\alpha)^2 t + i(k+\alpha)u} = \frac{1}{\sqrt{4\pi t}} \sum_{k=-\infty}^{\infty} e^{-\frac{1}{4t}(u - 2\pi k)^2 + 2\pi i k \alpha}.$$

Specializing to the case $\alpha = 0$, (5.4) becomes the statement

$$(5.5) \qquad \frac{1}{2\pi} \theta\begin{bmatrix}0\\0\end{bmatrix}(u/2\pi, it/\pi) = \frac{1}{\sqrt{4\pi t}} e^{-u^2/4t} \theta\begin{bmatrix}0\\0\end{bmatrix}(u/2it, i\pi/t).$$

As discussed on page 32 of [Mu 83], equation (5.5) yields the general functional equation for the Riemann theta function for a restricted class of moduli τ and arguments u, namely $u \in \mathbf{R}$ and $\tau \in i\mathbf{R}$ (see also page 269 of [La 87]). By analytic continuation, the equality in (5.5) extends for all $u \in \mathbf{C}$ and all moduli, provided one is careful when defining $\sqrt{\tau}$ for complex τ. This square root will introduce an eight-root of unity, thus explaining the appearance of this factor (see page 33 of [Mu 83]).

In summary, the equality of heat kernels as asserted in Proposition 5.2, which is a special case of Corollary 2.7, implies the functional equation of the one-variable Riemann theta function in (5.5).

Similarly, one can obtain a proof of the functional equation for a theta function in several variables (see page 189 of [Mu 83]). One approach is to let Y be the compact real torus $(S^1)^n$ and to consider a self-adjoint second-order elliptic operator with constant coefficients. As in Lemma 5.1, the associated heat kernel will be a type of theta function with modulus corresponding to the matrix of coefficients of the operator. Then one would need the appropriate version of (5.3), which also would depend on the coefficients of the operator. A second approach is to consider the usual Laplacian on a real torus given by $\mathbf{R}^n$ modulo a lattice which is represented by the columns of a positive definite symmetric real matrix. In either case, a proof would come from an existence and uniqueness result, as in Proposition 5.2, and analytic continuation, as discussed above.

We shall now prove the classical Riemann theta relations. Let us begin with following elementary result.

Theorem 5.3. *Let G be an abelian group, and let Y be a G-manifold with G-sheaf $\mathscr{E}$. Then*

$$\sum_{\pi \in \hat{G}} \sum_{g \in G} \chi_\pi(g) K(Y, \mathscr{E})(y_1, g y_2, t) = [G:1] K(Y, \mathscr{E})(y_1, y_2, t).$$

Furthermore, for any fixed $g \in G$,

$$K(Y, \mathscr{E})(y_1, y_2, t) = K(Y, \mathscr{E})(gy_1, gy_2, t).$$

Proof. This first equality follows by interchanging the order of summation, the fact

$$\sum_{\pi \in \hat{G}} \chi_\pi(g) = \begin{cases} 0, & g \neq \mathrm{id}, \\ [G:1], & g = \mathrm{id}, \end{cases}$$

and, as always, the existence and uniqueness of the heat kernel. The second equality follows from the G-invariance of the metrics on Y and $\mathscr{E}$. □

We call the equation in Theorem 5.3 the *generalized Riemann theta identity*. To justify this label, we need to show that Theorem 5.3 contains the classical result by the same name. For the sake of brevity, we shall do so in the simplest and most often cited example of the formula ([Ig72], page 137 and [Mu83], page 20). The classical Riemann theta identity is the following result.

Corollary 5.4. *Let M denote the rational orthogonal matrix*

$$M = \frac{1}{2} \begin{pmatrix} 1 & 1 & 1 & 1 \\ 1 & 1 & -1 & -1 \\ 1 & -1 & 1 & -1 \\ 1 & -1 & -1 & 1 \end{pmatrix}$$

and $z = (z_1, \ldots, z_4) \in \mathbf{C}^4$. Then with notation as above, we have

$$\sum_{\alpha, \beta \in \{0, 1/2\}} \prod_{j=1}^{4} \left(\theta \begin{bmatrix} \alpha \\ \beta \end{bmatrix} (z_j, \tau) \right) = 2 \prod_{j=1}^{4} \left(\theta \begin{bmatrix} 0 \\ 0 \end{bmatrix} (Mz_j, \tau) \right).$$

Proof. Let L be the lattice $\mathbf{Z}^4 \cap M\mathbf{Z}^4$ and Y be the manifold $\mathbf{R}^4/L$. Since Y is a real torus and M is orthogonal, one can use the lattice L to construct a self-adjoint second-order elliptic operator with constant coefficients such that, by Lemma 5.1, the heat kernel relative to this operator is

$$\frac{1}{(2\pi)^4} \prod_{k=1}^{4} \left(\theta \begin{bmatrix} 0 \\ 0 \end{bmatrix} (My_k/2\pi, it/\pi) \right),$$

where $y = (y_1, \ldots, y_4) \in \mathbf{R}^4$. Since $\mathbf{Z}^4 \cap M\mathbf{Z}^4$ has index 2 in $M\mathbf{Z}^4$, Y admits an action by the group $\mathbf{Z}/2\mathbf{Z}$. The left-hand-side of the equation in Theorem 5.3 is computed using the coset representatives of $\mathbf{Z}^4 \cap M\mathbf{Z}^4$ in $M\mathbf{Z}^4$ (see [Mu83], page 212). By analytically continuing in the variables y and t, as discussed above, the result is proved. □

From Lemma 5.1 one sees that, in rough terms, the theta identities from Theorem 5.3 are such that an α characteristic corresponds to an element of G and a β characteristic corresponds to an element of $\hat{G}$.

The general Riemann theta relation (see page 136 of [Ig72] or [Mu83], page 211) can be proved as follows. Let M be an $h \times h$ rational unitary matrix, and let Δ_P be the constant coefficient second order elliptic operator naturally associated to a $g \times g$ positive definite symmetric matrix P. The operator Δ_P acts on smooth functions on $\mathbf{R}^g$. Let

$$L = (\mathbf{Z}^g)^h \cap M(\mathbf{Z}^g)^h$$

and

$$Y = (\mathbf{R}^g)^h / L.$$

The general Riemann theta relation follows from Theorem 5.3 by considering the heat kernel on Y associated to the Laplacian $\Delta_P \oplus \cdots \oplus \Delta_P$. Further details are left for the reader.

Finally, let us point out the similarity between our approach to the Riemann theta identities via Theorem 5.3 and the results in [Ko76], specifically Lemma 1.1, where the Riemann theta relations (general addition laws) are considered and shown to be a consequence of this basic lemma and various calculations involving coset representatives.

Bibliography

[Ar23]   E. *Artin*, Über eine neue Art von *L*-Reihen, Abh. math. Semin. Univ. Hamburg **3** (1923), 89–108. (Collected Papers, Springer Verlag, Berlin–Heidelberg–New York 1965.)

[Ar30]   E. *Artin*, Zur Theorie der *L*-Reihen mit allgemeinen Gruppencharakteren, Abh. math. Semin. Univ. Hamburg **8** (1930), 292–306. (Collected Papers, Springer Verlag, Berlin–Heidelberg–New York 1965.)

[At67]   M. *Atiyah*, *K*-Theory, Benjamin, New York 1967.

[BGV92] N. *Berline*, E. *Getzler* and M. *Vergne*, Heat Kernels and Dirac Operators, Grundl. math. Wiss. **298**, Springer Verlag, New York–Berlin–Heidelberg 1992.

[CV90] P. *Cartier* and A. *Voros*, Une nouvelle interprétation de la formulae des traces de Selberg, Progr. Math. **87** (1990), 1–68.

[Ch84] I. *Chavel*, Eigenvalues in Riemannian Geometry, Academic Press, New York 1984.

[DP86] E. *d'Hoker* and D. *Phong*, On determinants of Laplacians on Riemann surfaces, Comm. Math. Phys. **105** (1986), 537–545.

[Fa92] G. *Faltings*, Lectures on the Arithmetic Riemann-Roch theorem, Ann. Math. Stud. **127**, Princeton University Press 1992.

[FK80] H. *Farkas* and I. *Kra*, Riemann Surfaces, Grad. Texts Math. **71**, Springer Verlag, New York–Berlin–Heidelberg 1980.

[Fy73] J. *Fay*, Theta functions on Riemann surfaces, Lect. Notes Math. **352**, Springer Verlag, New York–Berlin–Heidelberg 1973.

[Ga77] R. *Gangolli*, Zeta functions of Selberg's type for compact space forms of symmetric spaces of rank one, Ill. J. Math. **21** (1977), 1–42.

[GH78] P. *Griffiths* and J. *Harris*, Principles of Algebraic Geometry, Wiley, New York 1978.

[He76] D. A. *Hejhal*, The Selberg Trace Formula for PSL(2, $\mathbb{Z}$), volume 1, Lect. Notes Math. **548**, Springer Verlag, New York–Berlin–Heidelberg 1976.

[Ig72] J. *Igusa*, Theta Functions, Grundl. math. Wiss. **194**, Springer Verlag, New York–Berlin–Heidelberg 1972.

[Jo90] J. *Jorgenson*, Asymptotic behavior of Faltings's delta function, Duke Math. J. **61** (1990), 221–254.

[Jo91] J. *Jorgenson*, Analytic torsion for line bundles on Riemann surfaces, Duke Math. J. **62** (1991), 527–549.

[Jo92] J. *Jorgenson*, Degenerating hyperbolic Riemann surfaces and an evaluation Deligne's constant, Yale University Preprint 1992.

[JoL92] J. *Jorgenson* and S. *Lang*, Complex analytic properties of regularized products, Yale University Preprint 1992, Lect. Notes Math. **1564**, Springer Verlag, New York–Berlin–Heidelberg 1993.

[JoLu92] J. Jorgenson and R. Lundelius, Continuity of relative hyperbolic spectral theory through metric degeneration, Yale University Preprint 1992.
[KaS87] A. Katsuda and T. Sunada, Homology and closed geodesics in a compact Riemann surface, Am. J. Math. **109** (1987), 145–156.
[Ko76] S. Koizumi, Theta relations and projective normality of abelian varieties, Am. J. Math. **98** (1976), 865–889.
[La56] S. Lang, L-series of a covering, Proc. Natl. Acad. Sci. **42** (1956), 442–424.
[La82] S. Lang, Introduction to Algebraic and Abelian functions, Grad. Texts Math. **89**, Springer Verlag, New York–Berlin–Heidelberg 1982.
[La83] S. Lang, Fundamentals of Diophantine Geometry, Springer Verlag, New York–Berlin–Heidelberg 1983.
[La86] S. Lang, Algebraic Number Theory, Grad. Texts Math. **110**, Springer Verlag, New York–Berlin–Heidelberg 1986.
[La87] S. Lang, Elliptic Functions, second edition, Grad. Texts Math. **112**, Springer Verlag, New York–Berlin–Heidelberg 1987.
[La88] S. Lang, Introduction to Arakelov Theory, Springer Verlag, New York–Berlin–Heidelberg 1988.
[La93] S. Lang, Algebra, second edition, Addison-Wesley, Menlo Park, CA, 1984, third edition 1993.
[Mc72] H. McKean, Selberg's trace formula as applied to a compact Riemann surface, Comm. pure appl. math. **25** (1972), 225–246.
[Mu83] D. Mumford, Tata Lectures on Theta I, Progr. Math. **28**, Birkhäuser, Boston 1983.
[PS88] R. Phillips and P. Sarnak, Geodesics in homology classes, Duke Math. J. **55** (1988), 287–297.
[Qu86] D. Quillen, Determinants of Cauchy-Riemann operators over a Riemann surface, Func. Anal. Appl. **19** (1986), 31–34.
[RS73] D. Ray and I. Singer, Analytic torsion for complex manifolds, Ann. Math. **98** (1973), 154–177.
[Sa81] P. Sarnak, Prime Geodesic Theorems, Ph.D dissertation, Stanford University 1981.
[Sa87] P. Sarnak, Determinants of Laplacians, Comm. Math. Phys. **110** (1987), 113–120.
[Sel56] A. Selberg, Harmonic analysis and discontinuous groups in weakly symmetric Riemannian spaces with applications to Dirichlet series, J. Indian Math. Soc. B. **20** (1956), 47–87.
[Su85] T. Sunada, Riemannian coverings and isospectral manifolds, Ann. Math. **121** (1985), 169–186.
[Sz90] L. Szpiro, Sur les propriétés numériques du dualisant relatif d'une surface arithmétique, Progr. Math. **88** (1990), 229–246.
[Ve79] A. B. Venkov, The Artin-Takagi formula for Selberg's zeta function and the Roelcke conjecture, Soviet Math. Dokl. **20** (1979), 745–748.
[VeZ81] A. B. Venkov and P. G. Zograf, On analogues of the Artin factorization formulas in the spectral theory of automorphic functions connected with induced representations of Fuchsian groups, Soviet Math. Dokl. **21** (1981), 84–96.
[VeZ83] A. B. Venkov and P. G. Zograf, On analogues of the Artin factorization formulas in the spectral theory of automorphic functions connected with induced representations of Fuchsian groups, Math. USSR-Izv. **21** (1983), 435–443.
[Vo87] A. Voros, Spectral functions, special functions, and the Selberg zeta function, Comm. Math. Phys. **110** (1987), 439–465.
[Wi92] F. L. Williams, A factorization of the Selberg zeta function attracted to a rank 1 space form, Manuscr. Math. **77** (1992), 17–39.

Department of Mathematics, Yale University, Box 2155 Yale Station, New Haven, CT 06520

Eingegangen 11. Mai 1992, in revidierter Fassung 25. Mai 1993

EXPLICIT FORMULAS FOR REGULARIZED PRODUCTS AND SERIES

Jay Jorgenson and Serge Lang

EXPLICIT FORMULAS FOR REGULARIZED PRODUCTS AND SERIES

Jay Jorgenson and Serge Lang

Introduction	3
I Asymptotic estimates of regularized harmonic series	**11**
1. Regularized products and harmonic series	14
2. Asymptotics in vertical strips	20
3. Asymptotics in sectors	22
4. Asymptotics in a sequence to the left	24
5. Asymptotics in a parallel strip	34
6. Regularized product and series type	36
7. Some examples	39
II Cramér's Theorem as an Explicit Formula	**43**
1. Euler sums and functional equations	45
2. The general Cramér formula	47
3. Proof of the Cramér theorem	51
4. An inductive theorem	57
III Explicit Formulas under Fourier Assumptions	**61**
1. Growth conditions on Fourier transforms	62
2. The explicit formulas	66
3. The terms with the q's	73
4. The term involving Φ	78
5. The Weil functional and regularized product type	79
IV From Functional Equations to Theta Inversions	**85**
1. An application of the explicit formulas	87
2. Some examples of theta inversions	92

V From Theta Inversions to Functional Equations — 97

    1. The Weil functional of a Gaussian test function — 99
    2. Gauss transforms — 101
    3. Theta inversions yield zeta functions — 109
    4. A new zeta function for compact quotients of $\mathbf{M}_3$ — 113

VI A Generalization of Fujii's Theorem — 119

    1. Statement of the generalized Fujii theorem — 122
    2. Proof of Fujii's theorem — 125
    3. Examples — 128

Bibliography — 131

Introduction

Explicit formulas in number theory were originally motivated by the counting of primes, and Ingham's exposition of the classical computations is still a wonderful reference [In 32]. Typical of these formulas is the Riemann-von Mangoldt formula

$$\sum_{p^n \leq x} \log p = x - \sum_\rho \frac{x^\rho}{\rho} - \zeta'_\mathbf{Q}/\zeta_\mathbf{Q}(0) - \frac{1}{2}\log(1 - x^{-2}).$$

Here the sum on the left is taken over all prime powers, and the sum on the right is taken over the non-trivial zeros of the Riemann zeta function.

Later, Weil [We 52] pointed out that these formulas could be expressed much more generally as stating that the sum of a suitable test function taken over the prime powers is equal to the sum of the Mellin transform of the function taken over the zeros of the zeta function, plus an analytic term "at infinity", viewed as a functional evaluated on the test function.

It is the purpose of these notes to carry through the derivation of the analogous so-called "explicit formulas" for a general zeta function having an Euler sum and functional equation whose fudge factors are of regularized product type. As a result, our general theorem applies to many known examples, some of which are listed in §7 of [JoL 93c]. The general Parseval formula from [JoL 93b] provides an evaluation of the "term at infinity", which we call the Weil functional. Also, as an example of our results, let us note that even in the well-studied case of the Selberg zeta function of a compact Riemann surface, our computations show that one may deal with a larger class of test functions than previously known.

For some time, analogies between classical analytic number theory and spectral theory have been realized. Minakshisundaram-Pleijel defined a zeta function in connection with the Laplacian on an arbitrary Riemannian manifold [MiP 49], and subsequently Selberg defined his zeta function [Se 56]. In [JoL 93a,b] we developed a general theory of regularized products and series applicable equally to the classical analytic number theory and to some of these analogous spectral situations. In particular, we proved the basic properties of the Weil functional at infinity in the context of regularized products and series, with a view to using the functional for the explicit formulas in this general context.

A fundamental class of zeta functions. In [JoL 93c] we defined a fundamental class of functions to which we could apply these properties and carry out analogues of results in analytic number theory. Roughly speaking, the functions Z in our class are those which satisfy the three conditions:

- there is a functional equation;
- the logarithm of the function admits a generalized Dirichlet series converging in some half plane (we call this Dirichlet series an Euler sum for Z);
- the fudge factors in the functional equation are of regularized product type.

The precise definition of our class of functions is recalled in Chapter II, §1. The explicit formula can be formulated and proved for functions in this class. In Chapter II, §1, we discuss the extent to which this class is a much broader class than a certain class defined by Selberg [Se 91]. Furthermore, certain applications require an even broader class of functions to which all the present techniques can be applied. We shall describe the need for such a class in greater detail below.

Just as we did for the analogue of Cramér's theorem proved in [JoL 93c], we emphasize that the explicit formula involves an inductive step which describes a relation between some of the zeros and poles of the fudge factors and some of those of the principal zeta function Z. Such a step can be viewed as a step in the ladder of regularized products, because our generalized Cramér theorem insures that a function Z in our class is also of regularized product type provided the fudge factors are of regularized product type.

If Z is a function in our class, and, for $\text{Re}(s)$ sufficiently large, the expression

$$\log Z(s) = \sum_{\mathbf{q}} \frac{c(\mathbf{q})}{\mathbf{q}^s}$$

is the Euler sum for $\log Z(s)$, with a sequence $\{\mathbf{q}\}$ of real numbers > 1 tending to infinity, and complex coefficients $c(\mathbf{q})$, then such $\mathbf{q}$ play the role of prime powers. However, readers should keep in mind cases when $\mathbf{q}$ does not look at all like a prime power. For example, the general theory applies to the case when $Z(s)$ is a general

Dirichlet polynomial, up to an exponential fudge factor; a precise definition is given in Chapter II, §4. Such polynomials contain as special cases the local factors of more classical zeta functions and L-functions. In examples having to do with Riemannian geometry, $\log \mathbf{q}$ is the Riemannian distance between two points in the universal covering space.

The general version of Cramér's theorem in [JoL 93c] was carried out for the original Cramér's test function $\phi_z(s) = e^{sz}$. One can also view this version as a special case of an explicit formula with more general test functions. This is carried out in Chapter II. In [JoL 93c], §7 we gave a number of examples for our Cramér-type theorem. To these we are adding not only the general Dirichlet polynomials as mentioned above, but also Fujii-type L-functions, obtained from a zeta function by inserting what amounts to a generalized character as coefficient of the Dirichlet series defining the zeta function (see the papers by Fujii listed in the bibliography). In Chapter V we show both how to recover Fujii's theorems for the functions he considered, namely the Riemann zeta function and the Selberg zeta function for $PSL(2, \mathbf{Z})$, as well as an analogous theorem for the general zeta functions in our class, all as corollaries of our Cramér's theorem. Similarly, a result of Venkov, which relates the eigenvalues of the Laplacian relative to $PSL(2, \mathbf{Z})$ to the classical von Mangoldt function, will be generalized to any non-compact finite volume hyperbolic Riemann surface in [JoL 94]. The generalization involves another inductive type argument, using the fact that the fudge factor in the functional equation of the non-compact Selberg zeta function involves the determinant of the scattering matrix, which itself is in our class of functions since it has an Euler sum and a simple functional equation with constant fudge factors. In this case, the Euler sum exists whereas a classical Euler product does not. Thus, the general theory simultaneously contains previous results and gives new ones which were not proved previously by authors using such tools as the Selberg trace formula.

Analytic estimates for the proof. In addition to the Parseval formula of [JoL 93b], the proof of the general explicit formula relies on certain analytic estimates for regularized harmonic series, including the logarithmic derivatives of regularized products in strips. We gave such estimates already in [JoL 93a,b], but we need further such estimates which we present in Chapter I of the present work, using the technique of our generalized Gauss formula. Hard-core analytic estimates having thus been put out of the way, the rest of the work is then relatively formal. It is noteworthy that to each regularized product we associate naturally two non-negative integers determined directly from the definition. Then the fundamental estimates of Chapter I show that the order of growth of the logarithmic derivatives of such products in strips is determined by these two integers. In the application to the evaluation of certain integrals involving test functions, one can then see that the order of decay of these test functions, needed to insure that the integrals converge, is also determined by these two integers. Our systematic approach both improves known estimates for the Selberg zeta function (cf. Chapter I, §4), and provides estimates for functions in our class which had not been considered previously.

Theta inversions. We shall postpone to still another work the application of explicit formulas to the counting of those objects which play the role of prime powers. Here we shall emphasize an entirely different type of application, obtained by taking Gaussian type functions as the test functions instead of other test functions which lead to the counting. Applying the general explicit formula to such Gaussians gives rise to relations which are vast generalizations of the classical Jacobi inversion formula for the classical Jacobi theta function, where t on one side gets inverted to $1/t$ on the other side of the formula. The classical Jacobi inversion formula is the relation

$$\frac{1}{2\pi}\sum_{n=-\infty}^{\infty} e^{-n^2 t} = \frac{1}{\sqrt{4\pi t}} \sum_{n=-\infty}^{\infty} e^{-(2\pi n)^2/4t},$$

which holds for all $t > 0$. Here, $\log \mathbf{q} = 2\pi n$ where n is a positive integer. The zeta function $Z(s)$ giving rise to the above theta series is essentially the special Dirichlet polynomial

$$\sin(\pi i s) = -\frac{1}{2i} e^{\pi s}(1 - e^{-2\pi s}).$$

Thus, the most classical theta series appears in a new context, associated to a "zeta function" which looks quite different from those visualized classically.

The general context of Chapter IV and Chapter V allows a formulation of a theta inversion when the theta series is of type

$$\sum_k a_k e^{-\lambda_k t}$$

with various coefficients a_k. Theta inversion applies in certain cases when the sequence $\{\lambda_k\}$ is the sequence of eigenvalues of an operator. For example, as we will show in Chapter V, §4, such an inversion formula comes directly from considering the heat kernel on the compact quotient of an odd dimensional hyperbolic space which has metric with constant negative sectional curvature.

For certain manifolds, the theta inversion already gives rise to an extended class of zeta functions, which instead of an Euler sum may have a Bessel sum. For manifolds of even dimension, the class of functions having an Euler sum or Bessel sum is still not adequate, and it is necessary to define an even further extended class, which we shall describe briefly below. At this moment, it is not yet completely clear just how far an extension we shall need, but so far, whatever the extension of the fundamental class we have met, the techniques of [JoL 93a,b,c] and of Chapter I apply.

In [JoL 94], we show how the general explicit formula also applies to the scattering determinant of Eisenstein series. Here, the Euler sum exists, and scattering determinants are in the fundamental class.

An additive theory rather than multiplicative theory, and an extended class of functions. The conditions defining our fundamental class of functions are phrased in a manner still relatively close to the classical manner, involving the functions multiplicatively. However, it turns out that many essential properties of these functions involve only their logarithmic derivative, and thus give rise to an additive theory. For a number of applications, it is irrelevant that the residues are integers, and in some applications we are forced to deal with the more general notions of a regularized harmonic series (suitably normalized Mittag-Leffler expansions, with poles of order one) whose definition is recalled in Chapter I, §1. In general, the residues of such a series are not integers, so one cannot integrate back to realize this series as a logarithmic derivative of a meromorphic function. Even for the Artin L-functions, although they can be defined by an Euler product, it was natural for Artin to define them via their logarithmic derivative, and at the time, Artin could only prove that the residues were rational numbers. It took many years before the residues were finally proved to be integers. The systematic approach of [JoL 93a,b,c] in fact has been carried out so that it applies to this additive situation. The example of Chapter V, §4, shows why such an additive theory is essential.

Thus we are led to define not only the fundamental class of functions whose logarithmic derivative admits a Dirichlet series expression as mentioned above, but an extended class of functions where this condition is replaced by another one which will allow applications to more situations, starting with applications to various spectral theories as in [JoL 94]. Nevertheless, we still defined the fundamental class of functions having Euler sums, and we phrase some results multiplicatively, partly because at the present time, we feel that a complete change of notation with existing works would only make the present work less accessible, and partly because the class of functions admitting Euler sums is still a very important one including the classical functions of algebraic number theory and representation theory. However, we ask readers to keep in mind the additive rather than multiplicative formalism. Many sections, e.g. Chapter I and §1 and §2 of Chapter V, are written so that they apply directly to the additive situation.

Functions in the multiplicative fundamental class are obtained as Mellin transforms of theta functions having an inversion formula. Functions in the extended additive class are obtained as regularized harmonic series which are Gaussian transforms of such theta functions. For example, the (not regularized) harmonic series obtained from the heat kernel theta function in the special case of compact quotients of the three dimensional, complete, simply connected, hyperbolic manifold is essentially

$$\sum_k \frac{\phi_k(x)\phi_k(y)}{s(s-2)+\lambda_k}.$$

Observe how the presence of $s(s-2)$ in the series formally insures a trivial functional equation, that is invariance under $s \mapsto 2-s$.

Conversely, given a function in our extended additive class, one may go in reverse and see that the original theta inversion is only a special case of the general explicit formula valid for much more general test functions. The existence of an explicit formula with a more general test function will then allow us to obtain various counting results in subsequent publications.

Finally, let us note that many examples of explicit formulas using various test functions involving many examples of zeta functions have been treated in the literature, providing a vast number of papers on the subject. Most of the papers dealing with such explicit formulas are not directly relevant for what we do here, which is to lay out a general inductive "ladder principle" for explicit formulas in line with our treatment of Cramér's theorem. For instance, Deninger in [Den 93] emphasizes the compatibility of an explicit formula for the Riemann zeta function with a conjectural formalism of a Lefschetz trace formula. Such a formalism might occur in the presence of an operator whose eigenvalues are zeros of the zeta function. Our inductive hypotheses cover a wider class of functions than in [Den 93], and our treatment emphasizes another direction in the study of regularized products and series. Factors of regularized product type behave as if there were an operator, but no operator may be available.

We also mention Gallagher's attempt to unify a treatment of Selberg's trace formula with treatments of ordinary analytic number theory [Ga 84]. However, the conditions under which Gallagher proves his results are very restrictive compared to ours, and, in particular, are too restrictive to take into account the inductive ladder principle which we are following.

Acknowledgement: During the preparation of this work, the first author received support from NSF grant DMS-93-07023. Both authors benefited from visits to the Max-Planck-Institut in Bonn.

CHAPTER I

Asymptotic estimates of regularized harmonic series.

The proof of the general explicit formulas for functions whose fudge factors are of regularized product type will require a number of asymptotic estimates of general regularized harmonic series. The purpose of this chapter is to establish and tabulate these estimates in convenient form. The main definitions and results of this chapter are stated in §1 and §6.

These asymptotic formulas are needed just as one needs the asymptotic behavior of the gamma function and the zeta function in classical analytic number theory (see, for example, Chapter XVII of [La 70]). However, classical arguments which estimate this behavior cannot be applied in general, and must be replaced by more powerful tools, such as our extension of Cramér's theorem, proved in [JoL 93c], as well as our systematic analysis of the regularized harmonic series, given in [JoL 93a] and [JoL 93b].

Following the notation of [JoL 93a] and [JoL 93b], we let $R(z)$ be the regularized harmonic series associated to the theta function $\theta(t) = \sum a_k e^{-\lambda_k t}$; in other words

$$R(z) = \mathrm{CT}_{s=1} \mathbf{LM}\theta(s, z)$$

where **LM** is the Laplace-Mellin transform, and $\mathrm{CT}_{s=1}$ is the constant term of the power series in s at $s = 1$. As is shown in §4 of [JoL 93a], the function $R(z)$ has a meromorphic continuation to all $z \in \mathbf{C}$ whose singularities are simple poles located at $z = -\lambda_k$ with corresponding residue a_k. In what we call **the spectral case**, meaning $a_k \in \mathbf{Z}$ for all k, we have a regularized product $D(z)$ which is a meromorphic function defined for all $z \in \mathbf{C}$ and which satisfies the relation

$$R(z) = D'/D(z).$$

However, in this chapter, we will work in the more general situation by considering a regularized harmonic series which is not necessarily the logarithmic derivative of a regularized product. Our basic tool for estimating $R(z)$ is our general Gauss formula, which we shall recall at the end of §1.

From the general Gauss formula, we shall determine the asymptotic behavior of $R(z)$ as $z \to \infty$ in each of the following cases:

1) in a vertical strip obtained by restricting $\text{Re}(z)$ to a compact interval;
2) in a sector $|\text{Im}(z)| \ll \text{Re}(z)$;
3) in a sequence of vertical line segments $z = -T_m + iy$ with $T_m \to \infty$ and y in a compact interval so that $R(z)$ grows as slowly as possible;
4) in a strip parallel and disjoint from a strip which contains all $-\lambda_k$, assuming that $\lambda_k \to \infty$ in a horizontal strip.

In all cases, the asymptotic behavior of $R(z+w)$ will be determined from the general Gauss formula through judicious choices of z and w.

In §6 we apply these results to functions which are obtained from regularized products by a suitable change of variables $z \mapsto \alpha z + \beta$ with $\alpha, \beta \in \mathbf{C}$. Such a change of variables is needed, for example, because zeta functions usually have their zeros in vertical strips and not in horizontal strips. Power products of regularized products subject to such changes of variables will be said to be of regularized product type, formally defined in §6.

We conclude this chapter by comparing our results to various examples that exist elsewhere in the literature.

Recall that a regularized harmonic series has a natural "reduced order M" which is closely related to the exponent of the leading term in the asymptotic expansion of the associated theta function near $t = 0$. We recall the precise definition at the end of §1. There is another characterization of the reduced order in the spectral case, since we can then write the regularized product as a Weierstrass product

$$D(z) = e^{P(z)} E(z)$$

where P has degree $m+1$ and E is a canonical Weierstrass product of order $\leq m$. The smallest integer m for which such an expression is possible is the reduced order of D. In the classical case of analytic number theory, the gamma and zeta functions of number fields have reduced order 0.

From §1 to §5, we let R be a regularized harmonic series whose definition will be recalled in §1, as well as other basic definitions used throughout the chapter.

§1. Regularized products and harmonic series.

Let us briefly recall necessary background material from the theory of regularized products and series, as established in [JoL 93a] and [JoL 93b], to which we refer for details and further results.

We let $L = \{\lambda_k\}$ and $A = \{a_k\}$ be sequences of complex numbers which may be subject to the following conditions.

DIR 1. For every positive real number c, there is only a finite number of k such that $\operatorname{Re}(\lambda_k) \leq c$.

We use the convention that $\lambda_0 = 0$ and $\lambda_k \neq 0$ for $k \geq 1$. Under condition **DIR 1** we delete from the complex plane $\mathbf{C}$ the horizontal half lines going from $-\infty$ to $-\lambda_k$ for each k, together, when necessary, the horizontal half line going from $-\infty$ to 0. We define the open set:

$\mathbf{U}_L$ = the complement of the above half lines in $\mathbf{C}$.

If all λ_k are real and positive, then we note that $\mathbf{U}_L$ is simply $\mathbf{C}$ minus the negative real axis $\mathbf{R}_{\leq 0}$.

DIR 2.
  (a) The Dirichlet series
  $$\sum_k \frac{a_k}{\lambda_k^\sigma}$$
  converges absolutely for some real σ, say σ_0.

  (b) The Dirichlet series
  $$\sum_k \frac{1}{\lambda_k^\sigma}$$
  converges absolutely for some real σ, say σ_1.

DIR 3. There is a fixed $\epsilon > 0$ such that for all k sufficiently large, we have
$$-\frac{\pi}{2} + \epsilon \leq \arg(\lambda_k) \leq \frac{\pi}{2} - \epsilon.$$

We will consider a **theta series** or **theta function**, which is defined by

$$\theta_{A,L}(t) = \theta(t) = a_0 + \sum_{k=1}^{\infty} a_k e^{-\lambda_k t},$$

and, for each integer $N \geq 1$, we define the **asymptotic exponential polynomials** by

$$Q_N(t) = a_0 + \sum_{k=1}^{N-1} a_k e^{-\lambda_k t}.$$

We are also given a sequence of complex numbers $\{p\} = \{p_j\}$ with

$$\mathrm{Re}(p_0) \leq \mathrm{Re}(p_1) \leq \cdots \leq \mathrm{Re}(p_j) \leq \cdots$$

increasing to infinity, and, to every p in this sequence, we associate a polynomial B_p of degree n_p and set

$$b_p(t) = B_p(\log t).$$

We then define the **asymptotic polynomials at** 0 by

$$P_q(t) = \sum_{\mathrm{Re}(p) < \mathrm{Re}(q)} b_p(t) t^p.$$

We define

$$m(q) = \max \deg B_p \quad \text{for} \quad \mathrm{Re}(p) = \mathrm{Re}(q)$$
$$n(q) = \max \deg B_p \quad \text{for} \quad \mathrm{Re}(p) < \mathrm{Re}(q),$$
$$n(q') = \max \deg B_p \quad \text{for} \quad \mathrm{Re}(p) \leq \mathrm{Re}(q).$$

We shall use the term **special case** to describe the instance when $n(q) = 0$ for all q. The **principal part** of $\theta(t)$ is defined to be

$$P_0 \theta(t) = \sum_{\mathrm{Re}(p) < 0} b_p(t) t^p.$$

Let $\mathbf{C}\langle T \rangle$ be the algebra of polynomials in T^p with arbitrary complex powers $p \in \mathbf{C}$. Then, with this notation, $P_q(t) \in \mathbf{C}[\log t]\langle t \rangle$.

The function θ on $(0, \infty) = \mathbf{R}_{>0}$ is subject to **asymptotic conditions:**

AS 1. Given a positive number C and $t_0 > 0$, there exists N and $K > 0$ such that

$$|\theta(t) - Q_N(t)| \leq Ke^{-Ct} \text{ for } t \geq t_0.$$

AS 2. For every q, we have

$$\theta(t) - P_q(t) = O(t^{\text{Re}(q)} |\log t|^{m(q)}) \text{ for } t \to 0,$$

which shall written as

$$\theta(t) \sim \sum_p b_p(t) t^p.$$

AS 3. Given $\delta > 0$, there exists an $\alpha > 0$ and a constant $C > 0$ such that for all N and $0 < t \leq \delta$ we have

$$|\theta(t) - Q_N(t)| \leq C/t^\alpha.$$

We shall assume throughout that the theta series converges absolutely for $t > 0$. From **DIR 1** it follows that the convergence of the theta series is uniform for $t \geq \delta > 0$ for every δ.

The **Laplace-Mellin transform** of a measurable function f on $(0, \infty)$ is defined by

$$\mathbf{LM} f(s, z) = \int_0^\infty f(t) e^{-zt} t^s \frac{dt}{t}.$$

Theorem 1.1. *Let θ satisfy* **AS 1**, **AS 2** *and* **AS 3**. *Then* $\mathbf{LM}\theta$ *has a meromorphic continuation for $s \in \mathbf{C}$ and $z \in \mathbf{U}_L$. For each z, the function $s \mapsto \mathbf{LM}\theta(s, z)$ has poles only at the points $-(p+n)$ with $b_p \neq 0$ in the asymptotic expansion of θ at 0. A pole at $-(p+n)$ has order at most $n(p') + 1$. In the special* case when the asymptotic expansion at 0 has no log terms, the poles are simple.

We shall use a systematic notation for the coefficients of the Laurent expansion of $\mathbf{LM}\theta(s, z)$ near $s = s_0$. Namely we let $R_j(s_0; z)$ be the coefficient of $(s - s_0)^j$, so that

$$\mathbf{LM}\theta(s, z) = \sum R_j(s_0; z)(s - s_0)^j.$$

The constant term $R_0(s_0; z)$ is so important that we give it a special notation, namely,

$$\mathbf{CT}_{s=s_0} \mathbf{LM}\theta(s, z) = R_0(s_0; z).$$

In particular, we define the **regularized harmonic series** $R(z)$ to be the meromorphic function defined by

$$R(z) = \mathbf{CT}_{s=1} \mathbf{LM}\theta(s, z) = R_0(1; z).$$

Theorem 1.2. Let θ satisfy **AS 1**, **AS 2** and **AS 3**.
  a) For every $z \in \mathbf{U}_L$ and s near 0, the function $\mathbf{LM}\theta(s, z)$ has a pole at $s = 0$ of order at most $n(0') + 1$, and the function $\mathbf{LM}\theta(s, z)$ has the Laurent expansion

$$\mathbf{LM}\theta(s, z) = \frac{R_{-n(0')-1}(0; z)}{s^{n(0')+1}} + \cdots + R_0(0; z) + R_1(0; z)s + \ldots$$

  where, for each $j < 0$, $R_j(0; z) \in \mathbf{C}[z]$ is a polynomial of degree $\leq -\mathrm{Re}(p_0)$.
  b) One has the differential equation

$$\partial_z \mathbf{LM}\theta(s, z) = -\mathbf{LM}\theta(s + 1, z),$$

and hence the relation

$$-\partial_z R_0(0; z) = R(z).$$

The regularized harmonic series is a particular meromorphic function that has simple poles at $z = -\lambda_k$ with residue a_k.

Next we recall conditions when $R(z)$ is the logarithmic derivative of a meromorphic function. We define the **spectral case** to be when all $a_k \in \mathbf{Z}$.

Theorem 1.3. *In the spectral case, there exists a unique meromorphic function $D(z)$, called the **regularized product**, such that*
$$-\log D(z) = \mathrm{CT}_{s=0}\mathbf{LM}\theta(s,z) = R_0(0;z).$$
We have the relation
$$D'/D(z) = R(z).$$
This regularized product is meromorphic of finite order having zeros at the points $z = -\lambda_k$ with multiplicity a_k.

To make the situation more explicit, and to compare with results that exist elsewhere in the literature, let us record the following formula. In the spectral case, define
$$\zeta_\theta(s,z) = \frac{1}{\Gamma(s)}\mathbf{LM}\theta(s,z)$$
If we assume that
$$\theta(t) = \sum_{k=1}^{\infty} a_k e^{-\lambda_k t}$$
satisfies the asymptotic conditions **AS 1**, **AS 2** and **AS 3**, then we have the equality
$$\zeta_\theta(s,z) = \sum_{k=1}^{\infty} \frac{a_k}{(z+\lambda_k)^s}$$
for $\mathrm{Re}(z)$ and $\mathrm{Re}(s)$ sufficiently large. By Theorem 1.1 and Theorem 1.2, $\zeta_\theta(s,z)$ is holomorphic at $s=0$ for $z \in \mathbf{U}_L$ and
$$\zeta'_\theta(0,z) = \mathrm{CT}_{s=0}\mathbf{LM}\theta(s;z) + \gamma R_{-1}(0;z) = R_0(0;z) + \gamma R_{-1}(0;z).$$

Thus, in the spectral and special case, $-R_0(0;z)$ amounts to a normalization of the analytic torsion $-\zeta'(0,z)$. We have the Lerch formula
$$D'/D(z) = -\zeta'(0,z) + \quad \text{a constant.}$$
However, even in the most general case, for many applications one can work just as well with the regularized harmonic series $R(z)$ or

$R_0(0; z)$ even through $\zeta(s, z)$ is not holomorphic at $s = 0$, and D does not exist. For further comments, see §4 of Chapter V.

For any regularized harmonic series R, or regularized product D in the spectral case, we define the **reduced order** to be the pair of integers (M, m) where, in the notation of the asymptotic condition **AS 2**:

M is the largest integer $< -\text{Re}(p_0)$;

$m = m(p_0) + 1$ if there is a complex index p with

$$\text{Re}(p_0) = \text{Re}(p) \in \mathbf{Z}_{<0} \quad \text{and} \quad b_p \neq 0,$$

otherwise simply set $m = m(p_0)$.

Finally we recall the Gauss formula of [JoL 93a], §4. For any complex index q with $\text{Re}(q) > 0$, and variables z and w with $\text{Re}(w) > 0$ and

$$\text{Re}(w) > -\text{Re}(z) - \text{Re}(\lambda_k) \quad \text{for all } k$$

we then have the equality

$$R(z + w) = I_w(z; q) + S_w(z; q)$$

where

$$I_w(z; q) = \int_0^\infty [\theta_z(t) - P_q \theta_z(t)] e^{-wt} dt,$$

with

$$\theta_z(t) = e^{-zt} \theta(t)$$

and

$$S_w(z; q) = \sum_{\text{Re}(p)+k<\text{Re}(q)} \frac{(-z)^k}{k!} \text{CT}_{s=0} B_p(\partial_s) \left[\frac{\Gamma(s+p+k+1)}{w^{s+p+k+1}} \right].$$

§2. Asymptotics in vertical strips.

In this section we will determine the asymptotic behavior of a regularized harmonic series in a vertical strip. We shall see that the asymptotics in a vertical strip are given by the term $S_w(z;q)$.

We let
$$\Lambda_n = \max_{k \geq n}\{-\text{Re}(\lambda_k)\}.$$

Note that Λ_n is bounded above for all n and $\Lambda_n \to -\infty$ as $n \to \infty$.

Theorem 2.1. *Let $x_1, x_2 \in \mathbf{R}$ with $x_1 < x_2$. Select n sufficiently large so that $\Lambda_n < x_1 + 1$. Then for all q with $\text{Re}(q) > 0$ and uniformly for $x \in [x_1, x_2]$, we have the asymptotic relation*
$$R(x+1+iy) = S_{1+iy}(x;q) + \sum_{k=0}^{n-1} \frac{a_k}{x+1+iy+\lambda_k} + o\left(|y|^{-[\text{Re}(q)]}\right)$$

for $|y| \to \infty$.

Proof. With n chosen as above, let $L_n = \{\lambda_n, \ldots\}$. We decompose $R(z)$ as
$$R(z) = \sum_{k=0}^{n-1} \frac{a_k}{z+\lambda_k} + R_n(z),$$

where $R_n(z)$ is the regularized harmonic series formed from the sequence Λ_n with coefficients $\{a_k\}$, $k \geq n$. The desired estimate is obvious for the finite sum, so we may assume that we are working with the sequence L_n, and we will suppress this subscript. We apply the general Gauss formula by setting $z = x$ and $w = 1+iy$ with $y \in \mathbf{R}$. It suffices to prove the estimate
$$I_{1+iy}(x;q) = o(|y|^{-\text{Re}(q)}) \qquad \text{for } |y| \to \infty.$$

The integer n has been chosen sufficiently large so that, as a function of t,
$$\theta_x(t)e^{-t} = O\left(e^{-(x_1+1-\Lambda_n)t}\right) \qquad \text{for } x \in [x_1, x_2] \text{ and } t > 1.$$

In particular, this implies, by **AS 1** and **DIR 2**, that we have
$$\theta_x(t)e^{-t} \in L^1[1, \infty) \cap C^\infty[1, \infty).$$

Since $P_q\theta_x(t)$ has polynomial growth in t,
$$P_q\theta_x(t)e^{-t} \in L^1[1,\infty) \cap C^\infty[1,\infty).$$
Directly from **AS 2**, we have
$$[\theta_x(t) - P_q\theta_x(t)]\, e^{-t} \in L^1[0,1] \cap C^{[\mathrm{Re}(q)]}[0,1],$$
and
$$[\theta_x(t) - P_q\theta_x(t)]\, e^{-t} = O(t^{\mathrm{Re}(q)}|\log t|^{m(q)}) \quad \text{for } t \to 0.$$
We now apply the Riemann-Lebesgue lemma to obtain the bound
$$I_{1+iy}(x;q) = \int_0^\infty [\theta_x(t) - P_q\theta_x(t)]\, e^{-t} e^{-iyt}\, dt = o(|y|^{-[\mathrm{Re}(q)]})$$
which holds uniformly for all $x \in [x_1, x_2]$ as $|y| \to \infty$. With this, the proof of the theorem is complete. $\square$

For many purposes, it suffices simply to know the lead asymptotic of $R(z)$, which is obtained from the bound
$$\sum_{k=0}^{n-1} \frac{a_k}{x+1+iy+\lambda_k} = o(1) \quad \text{for } |y| \to \infty,$$
and the following result.

Corollary 2.2. *Let* $m = m(p_0) + 1$ *if there is an index p with* $\mathrm{Re}(p_0) = \mathrm{Re}(p) \in \mathbf{Z}_{<0}$ *and* $b_p \neq 0$, *otherwise set* $m = m(p_0)$. *Then for any q with* $\mathrm{Re}(q) > 0$, *we have*
$$S_{1+iy}(x;q) = O\left(|y|^{-\mathrm{Re}(p_0)-1}(\log|y|)^m\right) \quad \text{for } |y| \to \infty.$$
Since $-\mathrm{Re}(p_0)-1 \leq M < -\mathrm{Re}(p_0)$, *this estimate can be written as*
$$S_{1+iy}(x;q) = O\left(|y|^M (\log|y|)^m\right) \quad \text{for } |y| \to \infty.$$

Proof. Immediate from the definition of $S_w(z;q)$ as applied to $S_{1+iy}(x;q)$. $\square$

§3. Asymptotics in sectors.

In this section we will determine the asymptotic behavior of the regularized harmonic series $R(w)$ as $w \to \infty$ in a sector of the form

$$\text{Sec}_\epsilon = \{w \in \mathbf{C} | -\frac{\pi}{2} + \epsilon < \arg(w) < \frac{\pi}{2} - \epsilon\}$$

for some $\epsilon > 0$. As in the previous section, the asymptotics are determined by the term $S_w(0; q)$ in the general Gauss formula.

Theorem 3.1. *Let $x = \text{Re}(w)$. For all q with $\text{Re}(q) > 0$, we have*

$$R(w) = S_w(0; q) + O(x^{-\text{Re}(q)-1}(\log x)^{m(q)})$$

as $w \to \infty$ in Sec_ϵ where, as above,

$$S_w(0; q) = \sum_{\text{Re}(p) < \text{Re}(q)} \text{CT}_{s=0} B_p(\partial_s) \left[\frac{\Gamma(s+p+1)}{w^{s+p+1}} \right].$$

Proof. We apply the general Gauss formula with $z = 0$. The proof follows from estimating the integrals

$$I_w(0; q) = \int_0^1 [\theta(t) - P_q\theta(t)] e^{-wt} dt + \int_1^\infty [\theta(t) - P_q\theta(t)] e^{-wt} dt.$$

By **AS 2**, we have

$$\left| \int_0^1 [\theta(t) - P_q\theta(t)] e^{-wt} dt \right| \ll \int_0^1 t^{\text{Re}(q)} |\log t|^{m(q)} e^{-xt} dt$$

$$\leq \left[(\partial_s)^{m(q)} \frac{\Gamma(s+1)}{x^{s+1}} \right]_{s=\text{Re}(q)} + \int_1^\infty t^{\text{Re}(q)} (\log t)^{m(q)} e^{-xt} dt$$

$$= O\left(x^{-\text{Re}(q)-1} (\log x)^{m(q)} \right)$$

as $w \to \infty$ in Sec_ϵ. For $t \geq 1$, $\theta(t) - P_q\theta(t) = O(e^{ct})$ for some $c > 0$ and so, for $w \in \text{Sec}_\epsilon$ with x sufficiently large, we have

$$\left| \int_1^\infty [\theta(t) - P_q\theta(t)] e^{-wt} dt \right| \leq K \int_1^\infty e^{-(x-c)t} dt = O\left(e^{-x}/x\right),$$

thus yielding the stated estimate. □

As in the case of asymptotics in vertical strips, one often needs to know only the lead asymptotics of $R(z)$. Using Corollary 2.2, we can state this result.

Corollary 3.2. *Let $m = m(p_0) + 1$ if there is an index p with $\text{Re}(p_0) = \text{Re}(p) \in \mathbb{Z}_{<0}$ and $b_p \neq 0$, otherwise set $m = m(p_0)$. Then, with notation as above,*

$$R(w) = O\left(x^{-\text{Re}(p_0)-1}(\log x)^m\right) \quad \text{for } w \to \infty \text{ in } \text{Sec}_\epsilon.$$

In particular, since $-\text{Re}(p_0) - 1 \leq M < -\text{Re}(p_0)$, this estimate can be written as

$$R(w) = O\left(x^M (\log x)^m\right) \quad \text{for } w \to \infty \text{ in } \text{Sec}_\epsilon.$$

It is interesting to note that the asymptotic relation given in Corollary 3.2 is identical to the result given in Corollary 2.2 even though the directions in which $w \to \infty$ are quite different.

We can integrate the result in Theorem 3.1 in order to obtain an alternative proof of the general Stirling's formula for a regularized product, first established in §5 of [JoL 93a]. For completeness, we shall simply state this result.

Theorem 3.3. *Let D be a regularized product, so that we have $R = D'/D$. Let*

$$\mathbf{B}_q(w) = \sum_{\text{Re}(p) < \text{Re}(q)} \text{CT}_{s=0} B_p(\partial_s) \left[\frac{\Gamma(s+p)}{w^{s+p}}\right] \in \mathbf{C}\langle w\rangle[\log w],$$

and set $x = \text{Re}(w)$. Then for all q with $\text{Re}(q) > 0$, we have

$$\log D(w) = \mathbf{B}_q(w) + O(x^{-\text{Re}(q)}(\log x)^{m(q)})$$

as $w \to \infty$ in Sec_ϵ.

§4. Asymptotics in a sequence to the left.

In this section we will determine the asymptotic behavior of the regularized harmonic series $R(z)$ for a particular sequence of vertical line segments going to the left so that $R(z)$ grows as slowly as possible. Roughly, these line segments must pass between certain consecutive pairs of poles of $R(z)$ that are "sufficiently far apart".

We shall state and prove results for general regularized harmonic series and for regularized harmonic series which are the logarithmic derivatives of regularized products. The main result of this section is the following theorem.

Theorem 4.1. *There is a sequence of real numbers $T_n \to \infty$ such that for all y in any compact interval of $\mathbf{R}$ the following asymptotic relations hold.*

a) *In the notation of* **DIR 2**, *we have*

$$R(-T_n + iy) = o(T_n^{\sigma_1 + \sigma_0 - 1} \log T_n) \quad \text{as } T_n \to \infty.$$

b) *Let $m = m(p_0) + 1$ if there is a complex index p with $\mathrm{Re}(p_0) = \mathrm{Re}(p) \in \mathbf{Z}_{<0}$ and $b_p \neq 0$, otherwise simply set $m = m(p_0)$. Then for any q with $\mathrm{Re}(q) > 0$, we have*

$$R(-T_n + iy) = O(T_n^{-\mathrm{Re}(p_0)-1}(\log T_n)^m) \quad \text{as } T_n \to \infty.$$

In particular, since $-\mathrm{Re}(p_0) - 1 \leq M < -\mathrm{Re}(p_0)$, we have

$$R(-T_n + iy) = O(T_n^M (\log T_n)^m) \quad \text{as } T_n \to \infty.$$

The proof of Theorem 4.1 comes from considering the general Gauss formula for $R(u+w)$ with

$$w = -T - 1 + iy \quad \text{and} \quad u \in [0, 1],$$

for various choices of T and u in $\mathbf{R}$ which are to be made later. For T sufficiently large, let us write

$$(1) \quad R(u+w) = \sum_{|\lambda_k| \leq 2T} \frac{a_k}{u + w + \lambda_k} + S_w(u; 0) + I_w^{(2T)}(u; 0),$$

where

$$\theta_u^{(2T)}(t) = \sum_{|\lambda_k|>2T} a_k e^{-\lambda_k t} \cdot e^{-ut}$$

$$= \theta_u(t) - \sum_{|\lambda_k|\leq 2T} a_k e^{-\lambda_k t} \cdot e^{-ut}$$

and

$$I_w^{(2T)}(u;0) = \int_0^\infty \left[\theta_u^{(2T)}(t) - P_0\theta_u^{(2T)}(t)\right] e^{-wt} dt.$$

In (1) we have used the relation

$$P_0\theta_u^{(2T)}(t) = P_0\theta_u(t),$$

which, in particular, gives the identity

$$S_w^{(2T)}(u;0) = S_w(u;0).$$

Theorem 4.1 will follow by estimating each term in (1) separately. In a manner similar to that of Corollary 2.2, we can estimate the term $S_w(u;0)$. Specifically, with the above choice of variables u and $w = -T - 1 + iy$, we immediately have

(2) $\quad S_w(u;0) = O\left(T^{-\text{Re}(p_0)-1}(\log T)^m\right) \quad$ for $|T| \to \infty.$

This bound fits the result stated in Theorem 4.1(a) after noting the relations

$$\sigma_0 > 0 \quad \text{and} \quad \sigma_1 > -\text{Re}(p_0).$$

Throughout this section, which is devoted to the proof of Theorem 4.1, we will consider both general regularized harmonic series and those regularized harmonic series which are logarithmic derivatives of regularized products. In this latter case, we will, for this section only, assume that all coefficients a_k are equal to ± 1 and count the zeros λ_k with multiplicities.

In order to estimate the remaining two terms in (1), we need the following bounds.

Lemma 4.2. *With the notation as above, the following asymptotic relations hold:*

a) *In the notation of* **DIR 2**, *we have*
$$k = o(|\lambda_k|^{\sigma_1}) \quad \text{and} \quad |a_k| = o(|\lambda_k|^{\sigma_0}) = o\left(k^{\sigma_0/\sigma_1}\right).$$

b) *In the spectral case, there is a positive constant C' such that*
$$k \sim C' |\lambda_k|^{-\text{Re}(p_0)} (\log |\lambda_k|)^{m(p_0)},$$

or, equivalently, there is a positive constant C such that
$$|\lambda_k| \sim C \left(k/(\log k)^{m(p_0)}\right)^{-1/\text{Re}(p_0)}.$$

Proof. Define the **absolute zeta function** associated to the sequence L by
$$\zeta_{\text{abs}}(s) = \sum_{k=1}^{\infty} |\lambda_k|^{-s}.$$

By the inverse Mellin transform (see §7 of [JoL 93a]), we have, for all $\sigma > \sigma_1$
$$t^{\sigma_1} \cdot \sum_{k=1}^{\infty} e^{-|\lambda_k|t} = \frac{1}{2\pi i} \int_{\sigma-i\infty}^{\sigma+i\infty} \zeta_{\text{abs}}(s)\Gamma(s) t^{-(s-\sigma_1)} ds.$$

Since $\text{Re}(-s + \sigma_1) < 0$, we can apply the dominated convergence theorem to conclude

(3) $$\lim_{t \to 0} \left[t^{\sigma_1} \cdot \sum_{k=1}^{\infty} e^{-|\lambda_k|t} \right] = 0.$$

Similarly, one shows that

(4) $$\lim_{t \to 0} \left[t^{\sigma_0} \cdot \sum_{k=1}^{\infty} |a_k| e^{-|\lambda_k|t} \right] = 0.$$

In the spectral case, we have, by **AS 2**, the existence of a positive constant c such that

(5) $$\lim_{t \to 0} \left[t^{-\text{Re}(p_0)} (\log t)^{-m(p_0)} \cdot \sum_{k=1}^{\infty} e^{-|\lambda_k|t} \right] = c.$$

At this point, we apply the following general result.

The Karamata Theorem. *Let $\alpha \in \mathbf{R}_{>0}$ and $\beta \in \mathbf{Z}_{\geq 0}$. Let μ be a positive measure on $\mathbf{R}^+$ such that the integral*

$$\int_0^\infty e^{-t\lambda} \mu(\lambda)$$

converges for $t > 0$, and such that

$$\lim_{t \to 0} \left[t^\alpha (-\log t)^\beta \int_0^\infty e^{-t\lambda} \mu(\lambda) \right] = C$$

for some positive constant C. If $f(x)$ is a continuous function on the interval $[0, 1]$, then

$$\lim_{t \to 0} \left[t^\alpha (-\log t)^\beta \int_0^\infty f(e^{-t\lambda}) e^{-t\lambda} \mu(\lambda) \right]$$
$$= \frac{C}{\Gamma(\alpha)} \int_0^\infty f(e^{-t}) t^\alpha e^{-t} \frac{dt}{t}.$$

The proof of the Karamata theorem in the above form with the factor $(-\log t)^\beta$ follows the proof given on page 94 of [BGV 92].

We apply the Karamata theorem as on page 95 of [BGV 92], namely by considering a decreasing sequence of continuous test functions converging to the function x^{-1} on the interval $[1/e, 1]$ and zero on the interval $[0, 1/e)$. In the setup of (a), we use the spectral measure that places unit mass at the points $|\lambda_k| \in \mathbf{R}^+$, taking into account multiplicities. From (3), we obtain the bound

(6) $$N(\lambda) = \#\{k : |\lambda_k| \leq \lambda\} = o(\lambda^{\sigma_1}).$$

Since
$$N(\lambda_k) \geq k,$$
and we conclude

(7) $$k = o(|\lambda_k|^{\sigma_1}),$$

as was claimed in the first assertion of part (a).

To continue, let us again apply the Karamata theorem, this time with the measure that places a mass of $|a_k|$ at the points $|\lambda_k| \in \mathbf{R}^+$. The Karamata theorem then yields the estimate

$$\sum_{|\lambda_n| \leq \lambda} |a_n| = o(\lambda^{\sigma_0}),$$

which, in particular, implies the bound

$$|a_k| = o(|\lambda_k|^{\sigma_0}).$$

Combining this estimate with (7), we have

$$|a_k| = o(k^{\sigma_0/\sigma_1}),$$

which completes the proof of (a).

In the spectral case, we apply the Karamata theorem to (5) with the family of test functions described above and

$$\alpha = -\operatorname{Re}(p_0) \quad \text{and} \quad \beta = -m(p_0),$$

from which we obtain the estimate

$$(8) \qquad N(\lambda) = \#\{k : |\lambda_k| \leq \lambda\} \sim C'\lambda^\alpha (\log \lambda)^{-\beta}$$

for some positive constant C'. This estimate gives the relation

$$k = N(\lambda_k) \sim C'|\lambda_k|^\alpha (\log |\lambda_k|)^{-\beta},$$

as asserted in the statement of the lemma. Then

$$k^{1/\alpha} \sim C'^{1/\alpha} |\lambda_k| (\log |\lambda_k|)^{-\beta/\alpha}$$

and

$$\log k \sim \alpha \cdot \log |\lambda_k|.$$

Hence, for some positive constant C, we obtain the asymptotic relation

$$|\lambda_k| \sim C'^{-1/\alpha} k^{1/\alpha} (\log |\lambda_k|)^{\beta/\alpha} \sim C k^{1/\alpha} (\log k)^{\beta/\alpha},$$

which completes the proof of the lemma. $\square$

Lemma 4.2 allows us to estimate the term $I_w^{(2T)}(u;0)$ from (1) in the following proposition.

Proposition 4.3. Let $w = -T - 1 + iy$ and $u \in [0, 1]$. For all y in any compact interval of $\mathbf{R}$ the following asymptotic relations hold.

a) In the notation of **DIR 2**, we have

$$I_w^{(2T)}(u; 0) = o\left(T^{\sigma_1 + \sigma_0 - 1} \log T\right) \quad \text{for } T \to \infty,$$

b) Let $m = m(p_0) + 1$ if there is a p with $\mathrm{Re}(p_0) = \mathrm{Re}(p) \in \mathbf{Z}_{<0}$ and $b_p \neq 0$, otherwise set $m = m(p_0)$. Then, in the spectral case, we have

$$I_w^{(2T)}(u; 0) = O\left(T^{-\mathrm{Re}(p_0) - 1} (\log T)^m\right) \quad \text{for } T \to \infty.$$

Proof. For n sufficiently large, one can write

$$\partial_u^n I_w^{(2T)}(u; 0) = (-1)^n \Gamma(n+1) \sum_{|\lambda_k| > 2T} \frac{a_k}{(u + w + \lambda_k)^{n+1}}.$$

Since y lies in a compact interval, we have the estimate

$$|u + w| \sim T,$$

so we have the bound

$$\partial_u^n I_w^{(2T)}(u; 0) = O\left(\sum_{|\lambda_k| > 2T} \frac{|a_k|}{(|\lambda_k| - T)^{n+1}}\right).$$

Using Lemma 4.2(a) and an integral comparison, there exist positive constants c_1 and c_2 such that

$$\partial_u^n I_w^{(2T)}(u; 0) = o\left(\int_{c_1 T^{\sigma_1}}^{\infty} \frac{x^{\sigma_0/\sigma_1}}{(c_2 x^{1/\sigma_1} - T)^{n+1}} dx\right)$$

$$= o\left(\int_{c_1 T^{\sigma_1}}^{\infty} x^{\sigma_0/\sigma_1 - (n+1)/\sigma_1} dx\right)$$

$$= o\left(T^{\sigma_0 + \sigma_1 - (n+1)}\right).$$

To finish one integrates n-times. An extra power of $\log T$ occurs precisely in the case when $\sigma_0 + \sigma_1 \in \mathbf{Z}_{<0}$.

The proof of (b) is similar to that of (a). As above, let

$$\alpha = -\text{Re}(p_0) \quad \text{and} \quad \beta = -m(p_0).$$

Lemma 4.2(b) and an integral comparison yield the estimate

$$\partial_u^n I_w^{(2T)}(u;0) = O\left(\int_{c_1 T^\alpha (\log T)^{-\beta}}^{\infty} \frac{dx}{\left(c_2 \left(x(\log x)^\beta\right)^{1/\alpha} - T\right)^{n+1}}\right)$$

$$= O\left(\int_{c_1 T^\alpha (\log T)^{-\beta}}^{\infty} x^{-(n+1)/\alpha} (\log x)^{-\beta(n+1)/\alpha} dx\right).$$

If we integrate by parts, we obtain

$$\partial_u^n I_w^{(2T)}(u;0) = O\left(\left. x^{-(n+1)/\alpha+1}(\log x)^{-\beta(n+1)/\alpha} dx \right|_{c_1 T^\alpha (\log T)^{-\beta}}^{\infty}\right)$$

$$= O\left(\left(T^\alpha (\log T)^{-\beta}\right)^{-(n+1)/\alpha+1} (\log T)^{-\beta(n+1)/\alpha}\right)$$

$$= O\left(T^{\alpha-(n+1)}(\log T)^{-\beta}\right)$$

$$= O\left(T^{-\text{Re}(p_0)-(n+1)}(\log T)^{m(p_0)}\right).$$

To finish one integrates n-times. An extra power of $\log T$ occurs precisely in the case when $\text{Re}(p_0) \in \mathbf{Z}_{<0}$. $\square$

The results stated in (2) and Proposition 4.3 are for $T \to \infty$. To estimate the remaining term in (1), namely the finite sum

$$\sum_{|\lambda_k| \leq 2T} \frac{a_k}{u + w + \lambda_k},$$

it is necessary to choose carefully a sequence T_n. Roughly speaking, the sequence must be as far from any $-\lambda_k$ as possible. The following lemma makes this statement precise.

Lemma 4.4. *Assume the notation as above.*
a) *Let $r_n = |\lambda_n|^{\sigma_1}$. There is a positive constant c and a sequence $\{n_k\}$ of positive integers with $n_k \to \infty$ such that*

$$r_{n_k+1} - r_{n_k} \geq c \quad \text{for all } n_k.$$

Equivalently, there is a positive constant c and a sequence $\{n_k\}$ of positive integers with $n_k \to \infty$ such that

$$|\lambda_{n_k+1}| - |\lambda_{n_k}| \geq c \cdot n_k^{1/\sigma_1 - 1} \quad \text{for all } n_k.$$

b) *Let $r_n = |\lambda_n|^{\alpha}(\log |\lambda_n|)^{-\beta}$. There is a positive constant c and a sequence $\{n_k\}$ of positive integers with $n_k \to \infty$ such that*

$$r_{n_k+1} - r_{n_k} \geq c \quad \text{for all } n_k.$$

Equivalently, there is a positive constant c and a sequence $\{n_k\}$ of positive integers with $n_k \to \infty$ such that

$$|\lambda_{n_k+1}| - |\lambda_{n_k}| \geq c \cdot n_k^{1/\alpha - 1}(\log n_k)^{-\beta/\alpha} \quad \text{for all } n_k.$$

Proof. For the first statement in (a), one has, by (6), the bound $k = o(r_k)$, which would be contradicted if no such constant c or subsequence $\{n_k\}$ would exist. The second assertion in (a) and both assertions in (b) are established similarly. □

By assumption **DIR 3**, there is a constant C such that for all k, we have the inequalities

$$\operatorname{Re}(\lambda_k) \leq |\lambda_k| \leq C\operatorname{Re}(\lambda_k).$$

One then has results analogous to Lemma 4.2 and Lemma 4.4 for the sequence $\{\operatorname{Re}(\lambda_k)\}$. Specifically, we shall need the following version of Lemma 4.4.

Lemma 4.5. *Assume the notation as above.*
a) *Let $r_n = \operatorname{Re}(\lambda_n)^{\sigma_1}$. There is a positive constant c and a sequence $\{n_k\}$ of positive integers with $n_k \to \infty$ such that for all n_k,*

$$r_{n_k+1} - r_{n_k} \geq c.$$

Equivalently, there is a positive constant c and a sequence $\{n_k\}$ of positive integers with $n_k \to \infty$ such that for all n_k,

$$\mathrm{Re}(\lambda_{n_k+1}) - \mathrm{Re}(\lambda_{n_k}) \geq c \cdot n_k^{1/\sigma_1 - 1}.$$

b) *Let $r_n = \mathrm{Re}(\lambda_n)^\alpha (\log \mathrm{Re}(\lambda_n))^{-\beta}$. There is a positive constant c and a sequence $\{n_k\}$ of positive integers with $n_k \to \infty$ such that for all n_k,*

$$r_{n_k+1} - r_{n_k} \geq c.$$

Equivalently, there is a positive constant c and a sequence $\{n_k\}$ of positive integers with $n_k \to \infty$ such that for all n_k,

$$\mathrm{Re}(\lambda_{n_k+1}) - \mathrm{Re}(\lambda_{n_k}) \geq c \cdot n_k^{1/\alpha - 1} (\log n_k)^{-\beta/\alpha}.$$

The proof of Lemma 4.5 is identical to that of Lemma 4.4, hence will be omitted.

The following proposition estimates the finite sum in (1) for a particular sequence T_n of real numbers with $T_n \to \infty$ so that the finite sum grows as slowly as possible. With the proposition, the proof of Theorem 4.1 is complete.

Proposition 4.6. *There is a sequence of real numbers T_n with $T_n \to \infty$ such that for all y in any compact interval of $\mathbf{R}$, the following asymptotic relations hold.*

a) *In the notation of* **DIR 2**, *we have*

$$\sum_{|\lambda_k| \leq 2T_n} \frac{a_k}{-T_n + iy + \lambda_k} = o\left(T_n^{\sigma_1 + \sigma_0 - 1} \log T_n\right),$$

b) *Let $m = m(p_0) + 1$. Then, in the spectral case, we have*

$$\sum_{|\lambda_k| \leq 2T_n} \frac{a_k}{-T_n + iy + \lambda_k} = O\left(T_n^{-\mathrm{Re}(p_0) - 1} (\log T_n)^m\right).$$

Proof. For any z with $\text{Re}(z)$ sufficiently large and $|z| \neq |\lambda_k|$, we can estimate the finite sum in (a) by

$$\left| \sum_{|\lambda_k| \leq 2T_n} \frac{a_k}{z + \lambda_k} \right| = O\left(\sum_{|\lambda_k| \leq 2T_n} \frac{|a_k|}{|\text{Re}(\lambda_k) + \text{Re}(z)|} \right)$$

(9)
$$= o\left(\sum_{|\lambda_k| \leq 2T_n} \frac{|\lambda_k|^{\sigma_0}}{||\text{Re}(\lambda_k)| - |\text{Re}(z)||} \right).$$

Let $\{\lambda_{n_k}\}$ be the sequence determined by Lemma 4.5(a), and set

$$T_k = \frac{1}{2} \left(\text{Re}(\lambda_{n_k}) + \text{Re}(\lambda_{n_k+1}) \right).$$

For any fixed positive number h and any number T, there is a uniform bound on the number of elements of the sequence $\{r_k\}$ in the interval $[T, T+h]$. Also, there is a positive constant c such that $|r_k - T_n| \geq c$ for all k and n. With these facts and Lemma 4.2, (9) can be bounded by

$$o\left(\sum_{k=1}^{T_n^{\sigma_1} - cT_n^{\sigma_1 - 1}} \frac{k^{\sigma_0/\sigma_1}}{T_n - k^{1/\sigma_1}} \right),$$

which we can estimate by the integral

$$o\left(\int_0^{T_n^{\sigma_1} - cT_n^{\sigma_1 - 1}} \frac{x^{\sigma_0/\sigma_1}}{T_n - x^{1/\sigma_1}} dx \right) = o\left(\int_0^{T_n - c} \frac{u^{\sigma_0}}{T_n - u} d[u^{\sigma_1}] \right)$$

$$= o\left(T_n^{\sigma_0 + \sigma_1 - 1} \log T_n \right).$$

Similarly, for part (b), we can bound the finite sum in the spectral case by the integral

$$O\left(\int_0^{T_n - c} \frac{1}{T_n - x} d\left[x^\alpha (\log x)^{-\beta} \right] \right) = O\left(T_n^{\alpha - 1} (\log T_n)^{-\beta + 1} \right)$$

$$= O\left(T_n^{-\text{Re}(p_0) - 1} (\log T_n)^{m(p_0) + 1} \right).$$

With this, the proof of Proposition 4.6, and Theorem 4.1, is complete. $\square$

§5. Asymptotics in a parallel strip.

In this section we will consider a regularized harmonic series $R(z)$ corresponding to a sequence $\{\lambda_k\}$ that converges to infinity in a strip, meaning there exists a constant C such that for all k, we have the inequality

$$|\text{Im}(\lambda_k)| < C.$$

Recall that the sequence $\{\lambda_k\}$ is such that $\text{Re}(\lambda_k) \to \infty$. An individual term $1/(z+\lambda_k)$ may become arbitrarily large if z approaches λ_k. Under the assumption $|\text{Im}(\lambda_k)| < C$, we will show that by bounding z to a parallel horizontal strip given by

$$0 < c < |\text{Im}(z) - \text{Im}(\lambda_k)| < c' < \infty,$$

we shall be able to determine the asymptotic behavior of the regularized harmonic series $R(z)$.

Theorem 5.1. *Assume there is a constant C such that for all k we have $|\text{Im}(\lambda_k)| \leq C$. Let $z = -T + iy$ and assume there are constants $c, c' > 0$ such that for all k, we have*

$$0 < c < |\text{Im}(z) - \text{Im}(\lambda_k)| < c' < \infty$$

a) *In the notation of* **DIR 2**, *we have*

$$R(-T + iy) = o(T^{\sigma_1 + \sigma_0 - 1} \log T)$$

uniformly as $T \to \infty$.

b) Let $m = m(p_0) + 1$ if there is a complex index p with $\text{Re}(p_0) = \text{Re}(p) \in \mathbf{Z}_{<0}$ and $b_p \neq 0$, otherwise simply set $m = m(p_0)$. Then for any q with $\text{Re}(q) > 0$, we have

$$R(-T + iy) = O(T^{-\text{Re}(p_0) - 1} (\log T)^m)$$

uniformly as $T \to \infty$. Since $-\text{Re}(p_0) - 1 \leq M < -\text{Re}(p_0)$, we have

$$R(-T + iy) = O(T^M (\log T)^m)$$

uniformly as $T \to \infty$.

Proof. The proof follows that of Theorem 5.1. The only aspect of the proof which needs to be reconsidered is the asymptotic behavior of the finite sum

$$\sum_{|\lambda_k|\leq 2T} \frac{a_k}{u+w+\lambda_k}.$$

However, we immediately have the estimate

$$\left|\sum_{|\lambda_k|\leq 2T} \frac{a_k}{z+\lambda_k}\right| = O\left(\sum_{|\lambda_k|\leq 2T} \frac{|a_k|}{c+|\mathrm{Re}(\lambda_k)+\mathrm{Re}(z)|}\right)$$

$$= o\left(\sum_{|\lambda_k|\leq 2T} \frac{|\lambda_k|^{\sigma_0}}{c+||\mathrm{Re}(\lambda_k)|-|\mathrm{Re}(z)||}\right),$$

which can be bounded by the sum

$$o\left(\sum_{k=1}^{c_1 T^{\sigma_1}} \frac{k^{\sigma_0/\sigma_1}}{c+|c_2 k^{1/\sigma_1}-T|}\right).$$

As in the proof of Proposition 4.6, this sum can be estimated by the integral

$$o\left(\int_0^{c_1 T^{\sigma_1}} \frac{x^{\sigma_0/\sigma_1}}{|T-x^{1/\sigma_1}|+c}dx\right) = o\left(\int_0^{c_1' T} \frac{u^{\sigma_0}}{|T-u|+c}d\left[u^{\sigma_1}\right]\right)$$

$$= o\left(T^{\sigma_0+\sigma_1-1}\log T\right).$$

The proof of the remaining parts of the theorem follows the same pattern. □

§6. Regularized product and series type.

In §1 we defined what is meant for a function to be a regularized product. In this section, we define functions which are of regularized product type. After this, we will summarize the asymptotic formulae of the previous sections for this new class of functions.

Definition 6.1. A meromorphic function of finite order $G(z)$ is said to be of **regularized product type** if

$$G(z) = Q(z)e^{P(z)} \prod_{j=1}^{n} D_j(\alpha_j z + \beta_j)^{k_j}$$

where:
   i) $Q(z)$ is a rational function;
   ii) $P(z)$ is a polynomial;
   iii) $D_j(z)$ is a regularized product, with $\alpha_j, \beta_j \in \mathbf{C}$ and $k_j \in \mathbf{Z}$;
   iv) the α_j, β_j are restricted so that the zeros and poles of $D_j(\alpha_j z + \beta_j)$ lie in the the union of regions of the form:
   - a sector opening to the right, meaning

$$\{z \in \mathbf{C} : -\frac{\pi}{2} + \epsilon < \arg(z) < \frac{\pi}{2} - \epsilon\} \text{ for some } \epsilon > 0,$$

   - a sector opening to the left, meaning

$$\{z \in \mathbf{C} : \frac{\pi}{2} + \epsilon < \arg(z) < \frac{3\pi}{2} - \epsilon\} \text{ for some } \epsilon > 0,$$

   - a vertical strip, meaning

$$\{z \in \mathbf{C} : a < \mathrm{Re}(z) < b\} \text{ for some } a, b \in \mathbf{R}.$$

If G is of regularized product type, then any vertical strip which contains at most a finite number of zeros of G is called an **admissible strip**. Observe that if the zeros and poles of the function G are in sectors, as described in the first two conditions of Definition 6.1(iv) above, then every vertical strip is admissible. The main results of [JoL 93c] will be assumed (see in particular Theorem 1.5), and will not be repeated in these notes. Many examples of functions of regularized product type are given in §7 of [JoL 93c].

Corresponding to the multiplicative group of regularized product types, we have the additive group of their logarithmic derivatives, namely

$$G'/G(s) = \alpha_0 D'_0/D_0(\alpha_0 z + \beta_0) + \sum_{j=1}^{n} k_j \alpha_j D'_j/D_j(\alpha_j z + \beta_j).$$

This leads us to define the additive notion for its own sake.

A function R will be said to be of **regularized harmonic series type** if it is a finite linear combination

$$R(z) = \sum c_j R_j(\alpha_j z + \beta_j) + P'(z) + \sum \frac{c'_k}{z - \beta'_k},$$

where $c_j, c'_k, \alpha_j, \beta_j, \beta'_k \in \mathbf{C}$, each R_j is a regularized harmonic series as defined in §1, P' is a polynomial, and α_j and β_j are subject to condition (iv) with respect to the poles of $R_j(\alpha_j z + \beta_j)$. Sections §2 to §5 provide systematic estimates for functions of regularized harmonic series type in vertical and horizontal strips, not just for those which happen to be logarithmic derivatives of regularized product types. Indeed, a transformation $z \mapsto \alpha z + \beta$ amounts to a dilation, rotation and translation, so the estimates of §2 and §5 apply to each term in the above linear combination for a function of regularized harmonic series type. The resulting estimates will be summarized in Theorem 6.2 below.

Let $D_0(z) = Q(z)e^{P(z)}$ and define the **reduced order** of D_0 to be $(M_0, 0)$ where $M_0 = \max\{0, \deg P - 1\}$. Suppose G is of regularized product type of the form

$$G(z) = D_0(\alpha_0 z + \beta_0) \prod_{j=1}^{n} D_j(\alpha_j z + \beta_j)^{k_j}.$$

Let (M_j, m_j) be the reduced order of D_j, as defined at the end of Chapter I, §1. then we define the **reduced order** of G to be the pair of integers (M, m) where $M = \max\{M_j\}$ and m is the maximum over all m_j for which $M = M_j$.

We make a similar definition for the reduced order (M, m) of a regularized harmonic series type. In this case, the reduced order of P' is $(\deg P', 0)$ if $P' \neq 0$.

We shall use for a vertical strip the notation

$$\text{Str}(x_1, x_2) = \{z \in \mathbf{C} : x_1 \leq \text{Re}(z) \leq x_2\}.$$

The asymptotic formulas of the previous sections can now be easily summarized for any function R of regularized harmonic series type of reduced order (M, m).

Theorem 6.2. *Let R be of regularized harmonic series type of reduced order (M, m).*

a) *Let $S = \text{Str}(x_1, x_2)$ be an admissible strip. Then uniformly for $x \in [x_1, x_2]$, we have the asymptotic relation*

$$R(x + iy) = O(y^M (\log y)^{m+1}) \quad \text{as } |y| \to \infty.$$

b) *Let $S = \text{Str}(x_1, x_2)$ be a vertical strip which contains an infinite number of poles of the function R. Then there is a sequence of real numbers $T_n \to \infty$ such that for all $x \in [x_1, x_2]$, we have the following uniform asymptotic relation*

$$R(x \pm iT_n) = O(T_n^M (\log T_n)^{m+1}) \quad \text{as } T_n \to \infty.$$

Proof. Assertion (a) follows directly from the definition of regularized harmonic series type, admissible strip, and Corollary 2.2, Corollary 3.2, and Theorem 5.1. Assertion (b) follows from the same results together with Theorem 4.1. Note that in order to obtain the symmetry as asserted, one needs to consider the union of the sequences in the regularized harmonic series decomposition of G, which a possible change of sign. □

Correction added 2000. Theorem 6.2 requires M, m to be sufficiently large, or an adjusted definition of reduced order. A complete exposition of this and an extension of the theory will be given in a forthcoming book.

§7. Some examples.

To conclude this chapter, we shall briefly discuss how the results of the previous sections contain various classical asymptotic formula for the gamma function and the Riemann zeta function, and improve an existing bound for the Selberg zeta function associated to a compact hyperbolic Riemann surface.

Example 1: The gamma function. The classical gamma function $\Gamma(s)$ is, up to a factor of the form e^{as+b}, the regularized product associated to the sequence $\mathbf{Z}_{\geq 0}$. In the notation used above, we have $-\mathrm{Re}(p_0) = 1$ and $m(p_0) = 0$, hence the gamma function is of regularized product type of reduced order $(0,0)$. Theorem 2.3 yields the classical Stirling's formula for $\log \Gamma(s)$. As for asymptotics in vertical strips, Corollary 2.2 yields the equally classical asymptotic formula

$$\Gamma'/\Gamma(s) = \log|s| + O(1)$$

as $s \to \infty$ in any vertical strip of finite width. This result is an important ingredient in the proof of the classical explicit formula for zeta functions of number fields (see, for example, Chapter XVII of [La 70]).

Example 2: Dirichlet polynomials. We define a **Dirichlet polynomial** to be a holomorphic function of the form

$$P(s) = \sum_{n=1}^{N} a_n b_n^s$$

where $\{a_n\}$ is a finite sequence of complex numbers, $\{b_n\}$ is a finite sequence of positive real numbers. In Chapter II we will apply the general Cramér theorem from [JoL 93c] to show that P is of regularized product type of reduced order $(0,1)$.

Example 3: The Riemann zeta function. Let $\zeta_{\mathbf{Q}}(s)$ be the Riemann zeta function and consider the sequences

$$\Lambda_+ = \{\rho/i \in \mathbf{C} \mid \zeta_{\mathbf{Q}}(\rho) = 0, \ 0 \leq \mathrm{Re}(\rho) \leq 1, \ \mathrm{Im}(\rho) > 0\}$$

and

$$\Lambda_- = \{\rho/(-i) \in \mathbf{C} \mid \zeta_{\mathbf{Q}}(\rho) = 0, \ 0 \leq \mathrm{Re}(\rho) \leq 1, \ \mathrm{Im}(\rho) < 0\}.$$

By Corollary 1.3 of [JoL 93c], the theta function θ_{Λ_+} associated to the sequence Λ_+ satisfies the asymptotic conditions **AS 1**, **AS 2**, and **AS 3** with $-\mathrm{Re}(p_0) = 1$ and $m(p_0) = 1$. Similarly, the theta function θ_{Λ_-} associated to the sequence Λ_- satisfies the asymptotic conditions **AS 1**, **AS 2**, and **AS 3**, again with $-\mathrm{Re}(p_0) = 1$ and $m(p_0) = 1$. Theorem 1.6 of [JoL 93a] then implies that the functions

$$D_+(z) = \exp\left(-CT_{s=0}\mathbf{LM}\theta_{\Lambda_+}(s,z)\right)$$

and

$$D_-(z) = \exp\left(-CT_{s=0}\mathbf{LM}\theta_{\Lambda_-}(s,z)\right)$$

are holomorphic functions of finite order with

$$D_+(z) = 0 \quad \text{if and only if } -z \in \Lambda_+,$$

and

$$D_-(z) = 0 \quad \text{if and only if } -z \in \Lambda_-.$$

The argument of §6 of [JoL 93c] produces the relation

$$\zeta_{\mathbf{Q}}(s)\Gamma(s/2) = D_+(-s/i)D_-(s/i)(s(s-1))^{-1}e^{as+b}$$

for some constants a and b. Therefore, the Riemann zeta function is of regularized product type of reduced order $(0,1)$.

Example 3 and Theorem 5.1 combine to assert the existence of a sequence of numbers $\{T_n\}$ with $T_n \to \infty$ such that

$$|\zeta'_{\mathbf{Q}}/\zeta_{\mathbf{Q}}(x \pm iT_n)| = O\left((\log T_n)^2\right)$$

for $x \in [x_1, x_2]$ and $T_n \to \infty$. This result is proved classically in a different manner, see, for instance [In 32], pages 71-73.

Example 4: The Selberg zeta function $Z(s)$ associated to a compact Riemann surface. The results of [JoL 93c] show that the Selberg zeta function $Z(s)$ associated to any compact hyperbolic Riemann surface is an entire function of regularized product type and reduced order $(1,1)$, since, in this case, $-\mathrm{Re}(p_0) = 2$ and $m(p_0) = 1$. Theorem 5.1 asserts the existence of a sequence of numbers $\{T_n\}$ with $T_n \to \infty$ such that

$$|Z'/Z(x \pm iT_n)| = O\left(T_n(\log T_n)^2\right)$$

for $x \in [x_1, x_2]$ and $T_n \to \infty$. This improves a result stated on page 80 of [He 76], giving an upper bound of the form $O(T_n^2)$. As we will see in subsequent chapters, this improvement is significant since it allows us to apply the explicit formula for the Selberg zeta function to any test function that satisfies the three basic conditions to order one, which essentially amounts to requiring that the test function and its first derivative satisfies certain smoothness conditions. With the weaker bound from page 80 of [He 76], one would be forced to require smoothness conditions on a test function and its first two derivatives.

Example 5: The Selberg zeta function $Z(s)$ associated to a non-compact Riemann surface. One can apply the results of this section to the Selberg zeta function associated to any non-compact hyperbolic Riemann surface of finite volume X since, by Theorem 7.1 of [JoL 93c], this Selberg zeta function is of regularized product type of reduced order $(1,1)$. Associated to X there is a meromorphic function called the scattering determinant, which is the determinant of the constant terms in the Fourier expansions in the cusps of the Eisenstein series (see the discussion starting on page 498 of [He 83] or page 49 of [Sel 56]). Theorem 7.1 of [JoL 93c] shows that the scattering determinant is of regularized product type of reduced order $(0, 1)$.

In [JoL 94] we will give a more complete discussion of the application of our results to scattering determinants and to Eisenstein series on non-compact hyperbolic Riemann surfaces of finite volume.

A larger but non-exhaustive list of further examples is given at the end of [JoL 93c], including zeta functions arising in representation theory and the theory of modular forms, and zeta functions associated with certain higher dimensional manifolds. We shall return to these specific applications in a subsequent publication devoted exclusively to them.

CHAPTER II

Cramér's theorem as an explicit formula.

In [Cr 19] Cramér showed that if $\{\rho_k\}$ ranges over the non-trivial zeros of the Riemann zeta function with $\text{Im}(\rho_k) > 0$, then the series

$$V(z) = \sum e^{\rho_k z}$$

converges for $\text{Im}(z) > 0$ and has a singularity at the origin of the type $\log z/(1 - e^{-z})$, by which we mean that the function

$$F(z) = 2\pi i V(z) - \frac{\log z}{1 - e^{-z}},$$

which is defined for $\text{Im}(z) > 0$, has a meromorphic continuation to all $\mathbf{C}$, with simple poles at the points $\pm \pi i n$, where n ranges over the integers, and at the points $\pm \log p^m$, where p^m ranges over the prime powers. In [JoL 93c] we proved an analogous theorem for any meromorphic function with an Euler sum and functional equation whose fudge factors which are of regularized product type. In this chapter we will prove a Cramér-type theorem by considering the same contour integral as analyzed in [JoL 93c] for a more general class of test functions. Specifically, we consider the contour integral

$$\frac{1}{2\pi i} \int_{\mathcal{R}_a} \phi(s) Z'/Z(s) ds$$

over a semi-infinite vertical rectangle $\mathcal{R}_a$ which is assumed to contain the top half of the "critical strip" of Z. The test function ϕ is assumed to be holomorphic on the closure of $\mathcal{R}_a$ and to have reasonably weak growth conditions, essentially what is needed to make the proof go through. In the Cramér theorem from [JoL 93c], the test function ϕ depended on a complex parameter z, namely

$$\phi(s) = \phi_z(s) = e^{sz}.$$

For $\text{Im}(z) > 0$, the function ϕ_z has exponential decay when s lies in a finite strip and $\text{Im}(s) \to \infty$. As a result, the analysis needed in [JoL 93c] required a very weak growth result, which we proved for general meromorphic functions of prescribed order (see, in particular, Lemma 2.1 of [JoL 93c]).

In §1 we recall the fundamental class of functions which we have defined, and discuss its relation to the Selberg class of functions defined in [Sel 91]. The definitions of §1 are used throughout, but the subsequent sections may be logically omitted for the rest of this work. Taking into account the asymptotic estimates of Chapter I, our method of proof from [JoL 93c] applies to the more general class of test functions considered here. In §2 we establish notation and state the main result of this chapter. The proof is given in §3, and various applications are discussed in §4. As remarked on page 397 of [JoL 93c], our proof does differ from the original proof due to Cramér, which is one of the reasons why we can easily generalize the theorems to the class of functions which have Euler sums and functional equations with fudge factors which are of regularized product type.

§1. Euler sums and functional equations.

We shall say that the functions Z and $\tilde{Z}$ have an **Euler sum and functional equation** if the following properties are satisfied:

1. **Meromorphy.** The functions Z and $\tilde{Z}$ are meromorphic functions of finite order.

2. **Euler Sum.** There are sequences $\{\mathbf{q}\}$ and $\{\tilde{\mathbf{q}}\}$ of real numbers > 1 that depend on Z and $\tilde{Z}$, respectively, and that converge to infinity, such that for every $\mathbf{q}$ and $\tilde{\mathbf{q}}$, there exist complex numbers $c(\mathbf{q})$ and $c(\tilde{\mathbf{q}})$ and $\sigma_0' \geq 0$ such that for all $\text{Re}(s) > \sigma_0'$,

$$\log Z(s) = \sum_{\mathbf{q}} \frac{c(\mathbf{q})}{\mathbf{q}^s} \quad \text{and} \quad \log \tilde{Z}(s) = \sum_{\tilde{\mathbf{q}}} \frac{c(\tilde{\mathbf{q}})}{\tilde{\mathbf{q}}^s}.$$

The series are assumed to converge uniformly and absolutely in any half-plane of the form $\text{Re}(s) \geq \sigma_0' + \epsilon > \sigma_0'$.

3. **Functional Equation.** There are functions G and $\tilde{G}$, meromorphic and of finite order, and there exists σ_0 with $0 \leq \sigma_0 \leq \sigma_0'$ such that

$$Z(s)G(s) = \tilde{Z}(\sigma_0 - s)\tilde{G}(\sigma_0 - s).$$

We let
$$\Phi(s) = G(s)/\tilde{G}(\sigma_0 - s),$$
so the functional equation can be written in the form
$$Z(s)\Phi(s) = \tilde{Z}(\sigma_0 - s).$$

We call G, $\tilde{G}$ or Φ the **fudge factors** of the functional equation.

Remark 1. We inadvertently took $\sigma_0 = \sigma_0'$ in [JoL 93c], but it is important to allow $\sigma_0 \neq \sigma_0'$ for some applications, e.g. the scattering determinants which we considered in §7 of [JoL 93c]. No change is needed in the proof of the Cramér theorem except for choosing $a > 0$ such that $\sigma_0 + a > \sigma_0'$ as in §2 below. We dealt with the scattering determinant in [JoL 93c], in the context of Cramér's theorem.

The Euler sum for Z implies that Z is uniformly bounded for $\text{Re}(s) \geq \sigma_0' + \epsilon$ for every $\epsilon > 0$. Notice that Z has no zeros or poles for $\text{Re}(s) > \sigma_0'$, and all zeros and poles of Z in the region $\text{Re}(s) < -a$ agree in location and order with poles and zeros of Φ.

Remark 2. A Dirichlet series expression is assumed only for Z and $\tilde{Z}$, so the fudge factors do not occur symetrically for the zeta function in the above conditions, although they might appear to do so in the functional equation. For example, in the most classical case of the Riemann zeta function, the fudge factor is essentially the gamma function, which does not have a Dirichlet series expansion but is of regularized product type.

We define a triple $(Z, \tilde{Z}, \Phi)$ to be in **the fundamental class** if Z, $\tilde{Z}$ have an Euler sum and functional equation, and the fudge factors are of regularized product type. Selberg has defined a "Selberg class" of functions in analytic number theory (see [Se 91], [CoG 93]). Our class is much wider than Selberg's class in several major respects:

1) Selberg's fudge factors are of gamma type, i.e. $\Gamma(\alpha s + \beta)$.
2) Selberg assumes a Ramanujan-Petersson estimate on the coefficients of the Euler sum, but we do not.
3) Selberg's Euler sum involves ordinary integers and ordinary prime powers. We allow arbitrary positive numbers $\{\mathbf{q}\}$.

Our conditions allow for a much wider domain of applicability in spectral theory as illustrated by our varied examples, and many more to be treated in subsequent papers. For example, our conditions allow the fudge factors to include Γ_2, Γ_d for general d, or for ζ and L functions themselves, or any function of regularized product type.

Remark 3. Even so, the Euler sum condition is still not sufficiently general for our purposes and will be ultimately be generalized to a Bessel sum condition. For further comments on this point of view, see §3 and §4 of Chapter V. In the same vein, the functional equation will also be replaced by an additive relation where the additive fudge factor will be assumed to be a regularized harmonic series, or regularized harmonic series type, in analogy with regularized product type.

§2. The general Cramér formula.

Let $a > 0$ be such that $\sigma_0 + a > \sigma_0'$. We define the following regions in the complex plane:

$\mathcal{R}_a^+$ = semi-infinite open rectangle bounded by the lines

$$\text{Re}(s) = -a, \quad \text{Re}(s) = \sigma_0 + a, \quad \text{Im}(s) = 0.$$

$\mathcal{R}_a^+(T)$ = the portion of $\mathcal{R}_a^+$ below the line $\text{Im}(s) = T$.

We allow Z and Φ to have zeros or poles on the finite real segment $[-a, \sigma_0 + a]$, but we assume that Φ and Z have no zeros or poles on the vertical edges with $\text{Re}(s) = -a$ and $\text{Re}(s) = \sigma_0 + a$ and $\text{Im}(s) > 0$.

Let ϕ be any function which is holomorphic on the closure of the semi-infinite rectangle $\bar{\mathcal{R}}_a^+$, and let H be a meromorphic function on this closure. We are interested in studying the (formal) sums

$$(1) \qquad \text{div}_{H,a}^+(\phi) = \sum_{z \in \mathcal{R}_a^+} v_H(z) \phi(z)$$

where

$$v_H(z) = \text{ord}_H(z)$$

is the order of the zero or pole of H at z, so the sum (1) is actually over the divisor of H which lies in $\mathcal{R}_a^+$. Such sums do not converge a priori, so we need to define them as limits in a suitable sense, and for suitable functions ϕ. On a space of functions decaying sufficiently fast (depending on H), the divisor

$$\text{div}_{H,a}^+ = \sum_{z \in \mathcal{R}_a^+} v_H(z)(z)$$

gives rise to the functional defined by the sum (1). The functional itself may be denoted by $[\text{div}_{H,z}^+]$ to distinguish the functional from the divisor. To determine such a space of functions, we proceed as follows.

Let $H_1, \ldots, H_r$ be a finite number of functions which are meromorphic on the closure of $\mathcal{R}_a^+$, and let ϕ be a function which is holomorphic on this closure. We say that a sequence $\{T_m\}$ of positive real numbers tending to infinity is ϕ-**admissible** for $\{H_1, \ldots, H_r\}$ if for any k, H_k has no zero or pole on the segment

$$S_m = [-a + iT_m, \sigma_0 + a + iT_m]$$

and

$$\phi(s)H'_k/H_k(s) \to 0 \quad \text{for } s \in S_m \text{ with } m \to \infty.$$

When $\{H_1, \ldots, H_r\}$ is the set of functions $\{Z, \tilde{Z}, \Phi\}$, we say simply that $\{T_m\}$ is **admissible**. With respect to a ϕ-admissible sequence $\{T_m\}$, we define the **divisor functional for H**, which we denote by $\text{div}^+_H(\phi)$, to be the limit

$$(2) \qquad \text{div}^+_{H,a}(\phi) = \lim_{m \to \infty} \sum_{z \in \mathcal{R}^+_a(T_m)} v_H(z)\phi(z),$$

if such a limit exists. In particular, for $H = Z$, we let

$$\{\rho\} = \text{set of zeros and poles of } Z \text{ in } \mathcal{R}^+_a$$

so the sum (2) can be written as

$$\text{div}^+_{Z,a}(\phi) = \lim_{m \to \infty} \sum_{\rho \in \mathcal{R}^+_a(T_m)} v_Z(\rho)\phi(\rho).$$

We define other functionals as follows. Here we do not try to give subtle conditions on what amounts to a half-Fourier transform, so we simply assume that the derivative $\phi'(s)$ is in L^1 of each vertical half line $[-a, -a + i\infty]$ and $[\sigma_0 + a, \sigma_0 + a + i\infty]$. We define the **positive Cramér functional for q**, which we denote by $\text{Cr}^+_{\mathbf{q},a}(\phi)$, to be the integral

$$\text{Cr}^+_{\mathbf{q},a}(\phi) = \int_{\sigma_0+a}^{\sigma_0+a+i\infty} \phi'(s)\mathbf{q}^{-s}ds.$$

Similarly, the **negative Cramér functional for q** is defined by the integral

$$\text{Cr}^-_{\mathbf{q},a}(\phi) = \int_{\sigma_0+a-i\infty}^{\sigma_0+a} \phi'(\sigma_0 - s)\mathbf{q}^{-s}ds.$$

Also, with respect to an H-admissible sequence $\{T_m\}$, where H is a meromorphic function which is holomorphic on $\text{Re}(s) = -a$, we define the functional

$$U^+_{H,-a}(\phi) = \lim_{m \to \infty} \int_{-a}^{-a+iT_m} \phi(s) H'/H(s) ds.$$

When all the above functionals are defined, we then consider

The Cramér formula.

$$2\pi i \text{div}^+_{Z,a}(\phi) = \sum_{\tilde{\mathbf{q}}} c(\tilde{\mathbf{q}}) \text{Cr}^-_{\tilde{\mathbf{q}},a}(\phi) - \sum_{\mathbf{q}} c(\mathbf{q}) \text{Cr}^+_{\mathbf{q},a}(\phi)$$

$$+ U^+_{\Phi,-a}(\phi) + \int_{-a}^{\sigma_0+a} \phi'(s) \log Z(s) ds$$

$$+ \phi(-a) \left(\log \tilde{Z}(\sigma_0 + a) - \log Z(-a) \right).$$

Such a formula is derived formally by considering the contour integral

$$\frac{1}{2\pi i} \int_{\partial \mathcal{R}^+_a} \phi(s) Z'/Z(s) ds,$$

which can be evaluated in one way by using the residue theorem, and in another way by using the Euler sum and functional equation for Z.

We are interested in conditions on ϕ for which the Cramér formula holds. For this purpose, we consider the following growth assumptions on ϕ.

GR 1. $\phi(s)\Phi'/\Phi(s)$ is in L^1 on the vertical ray $[-a, -a+i\infty]$.

GR 2. The derivative $\phi'(s)$ is in L^1 of each vertical half line

$$[-a, -a+i\infty] \quad \text{and} \quad [\sigma_0+a, \sigma_0+a+i\infty].$$

The first condition compares the decay of $\phi(s)$ with the exponential growth of $\Phi(s)$. As in [JoL 93c], one often considers the situation when Φ is of regularized product type. In this case, the above growth conditions can be verified as follows.

Proposition 2.1.
   a) *If Φ is of regularized product type of reduced order (M, m) and the ray $[-a, -a+i\infty]$ lies in an admissible strip (defined in Chapter I, §6), then*

$$\Phi'/\Phi(s) = O(|s|^M (\log|s|)^m)$$

   for $|s| \to \infty$ and s on the ray $[-a, -a+i\infty]$.
   b) *If for some $\delta > 0$, ϕ has the decay*

$$\phi(s) = O(1/|s|^{M+1} (\log|s|)^{m+1+\delta}) \quad \text{for } |s| \to \infty \text{ on the ray,}$$

then **GR 1** *is satisfied.*

The hypotheses in the above criterion have been shown to be satisfied in several cases which are of direct interest in our theory of regularized products, for instance Theorem 5.2 of Chapter I. The next theorem asserts that the divisor functional $[\text{div}_{Z,a}^+]$ is defined on the vector space of functions satisfying **GR 1** and **GR 2**, and satisfies the Cramér formula.

Theorem 2.2. *Assume ϕ and Φ satisfy the two growth conditions* **GR 1** *and* **GR 2**. *Then all the functionals $\text{div}_{Z,a}^+(\phi)$, $U_{\Phi,-a}^+(\phi)$, $\text{Cr}_{q,a}^+(\phi)$, and $\text{Cr}_{q,a}^-(\phi)$ are defined, and the following formula holds:*

$$2\pi i \text{div}_{Z,a}^+(\phi) = \sum_{\tilde{q}} c(\tilde{q}) \text{Cr}_{\tilde{q},a}^-(\phi) - \sum_{q} c(q) \text{Cr}_{q,a}^+(\phi)$$

$$+ U_{\Phi,-a}^+(\phi) + \int_{-a}^{\sigma_0 + a} \phi'(s) \log Z(s) ds$$

$$+ \phi(-a) \left(\log \tilde{Z}(\sigma_0 + a) - \log Z(-a) \right).$$

As stated above, the proof of Theorem 2.2 will be given in the following section, and various applications of the theorem will be discussed in §4.

§3. Proof of the Cramér theorem.

The pattern of proof of Theorem 2.2 follows §2 of [JoL 93c], which, as we shall remark below, contains one significant technical improvement over the proof of the original theorem given by Cramér [Cr 19] for the Riemann zeta function.

Choose an $\epsilon > 0$ sufficiently small so that Z has no zeros or poles in the open rectangle with vertices

$$-a, \quad -a + i\epsilon, \quad \sigma_0 + a + i\epsilon, \quad \sigma_0 + a$$

or on the line segment $[-a+i\epsilon, \sigma_0+a+i\epsilon]$. Note that the function Z may have zeros or poles on the horizontal line segment $[-a, \sigma_0 + a]$. For T sufficiently large, we shall study the contour integral

(1)
$$2\pi i V_Z(z, \epsilon; T) = \int_{-a+iT}^{-a+i\epsilon} \phi(s) Z'/Z(s) ds + \int_{-a+i\epsilon}^{\sigma_0+a+i\epsilon} \phi(s) Z'/Z(s) ds$$
$$+ \int_{\sigma_0+a+i\epsilon}^{\sigma_0+a+iT} \phi(s) Z'/Z(s) ds \quad \int_{\sigma_0+a+iT}^{-a+iT} \phi(s) Z'/Z(s) ds.$$

We may assume that Z has no zeros or poles on the line segment connecting the points $-a + iT$ and $\sigma_0 + a + iT$, because we will pick $T = T_m$ for m sufficiently large. Let:

$\mathcal{R}_T(\epsilon) = $ the finite rectangle with vertices

$$-a + iT, \quad -a + i\epsilon, \quad \sigma_0 + a + i\epsilon, \quad \sigma_0 + a + iT.$$

By the residue theorem, we have

$$2\pi i V_Z(z, \epsilon; T) = \sum_{\rho \in \mathcal{R}_T(\epsilon)} v(\rho) \phi(\rho).$$

Theorem 2.2 will be established by studying each of the four integrals in (1). For simplicity, let us call these integrals the left, bottom, right, and top integrals, respectively. We begin with the top integral which will be shown to be arbitrarily small upon letting $T = T_m$ approach infinity.

Lemma 3.1. *Let $\{T_m\}$ be an admissible sequence relative to Z. Then we have*

$$\lim_{m \to \infty} \left[\int_{\sigma_0+a+iT_m}^{-a+iT_m} \phi(s) Z'/Z(s) ds \right] = 0.$$

The proof of Lemma 3.1 follows directly from the definition of an admissible sequence.

To continue, we have, from the growth assumption **GR 1**, the limit

$$\lim_{m \to \infty} \int_{-a+iT_m}^{-a+i\epsilon} \phi(s) Z'/Z(s) ds = \int_{-a+i\infty}^{-a+i\epsilon} \phi(s) Z'/Z(s) ds$$

and

$$\lim_{m \to \infty} \int_{\sigma_0+a+i\epsilon}^{\sigma_0+a+iT_m} \phi(s) Z'/Z(s) ds = \int_{\sigma_0+a+i\epsilon}^{\sigma_0+a+i\infty} \phi(s) Z'/Z(s) ds.$$

By combining these equations with Lemma 3.1, we have the following preliminary result.

Proposition 3.2. *With notation as above, we have*

$$2\pi i V_Z(z, \epsilon) = \lim_{m \to \infty} 2\pi i V_Z(z, \epsilon; T_m)$$

$$= \int_{-a+i\infty}^{-a+i\epsilon} \phi(s) Z'/Z(s) ds + \int_{-a+i\epsilon}^{\sigma_0+a+i\epsilon} \phi(s) Z'/Z(s) ds$$

$$+ \int_{\sigma_0+a+i\epsilon}^{\sigma_0+a+i\infty} \phi(s) Z'/Z(s) ds.$$

As before, let us call the integrals in Proposition 3.2 the left, bottom and right integrals, respectively. By the above stated assumption on ϵ, we have

$$V_Z(z, \epsilon) = V_Z(z).$$

To continue our proof of Theorem 2.2, we will compute the three integrals in Proposition 3.2 using the axioms of Euler sum and functional equation. After these computations, we will let ϵ approach 0, which will complete the proof.

Let us use the functional equation to re-write the left integral as the sum of three integrals involving $\tilde{Z}$ and Φ. Specifically, we have

$$\int_{-a+i\infty}^{-a+i\epsilon} \phi(s) Z'/Z(s) ds = \int_{-a+i\infty}^{-a+i\epsilon} \phi(s) \left[-\Phi'/\Phi(s) - \tilde{Z}'/\tilde{Z}(\sigma_0 - s)\right] ds$$

(2)
$$= \int_{-a+i\epsilon}^{-a+i\infty} \phi(s) \Phi'/\Phi(s) ds$$

(3)
$$+ \int_{\sigma_0+a-i\infty}^{\sigma_0+a-i\epsilon} \phi(\sigma_0 - s) \tilde{Z}'/\tilde{Z}(s) ds.$$

After we let $\epsilon \to 0$, the integral in (2) appears in the statement of Theorem 2.2 as the functional $U_\Phi^+(\phi)$. Note that letting $\epsilon \to 0$ is justified since Φ was assumed to be holomorphic and non-zero on the vertical lines of integration. As for (3), we can re-write this integral using the Euler sum of $\tilde{Z}$, yielding

$$\int_{\sigma_0+a-i\infty}^{\sigma_0+a-i\epsilon} \phi(\sigma_0 - s) \tilde{Z}'/\tilde{Z}(s) ds$$

$$= \phi(\sigma_0 - s) \log \tilde{Z}(s) \Big|_{\sigma_0+a-i\infty}^{\sigma_0+a-i\epsilon} + \int_{\sigma_0+a-i\infty}^{\sigma_0+a-i\epsilon} \phi'(\sigma_0 - s) \log \tilde{Z}(s) ds$$

$$= \phi(-a + i\epsilon) \log \tilde{Z}(\sigma_0 + a - i\epsilon)$$

$$+ \sum_{\tilde{q}} c(\tilde{q}) \int_{\sigma_0+a-i\infty}^{\sigma_0+a-i\epsilon} \phi'(\sigma_0 - s) \tilde{q}^{-s} ds.$$

By the Euler sum condition and the fact that $a > 0$, we can let ϵ

approach zero to get the equality

$$\int_{\sigma_0+a-i\infty}^{\sigma_0+a} \phi(\sigma_0-s)\tilde{Z}'/\tilde{Z}(s)ds$$

(4)
$$= \phi(-a)\log \tilde{Z}(\sigma_0+a) + \sum_{\tilde{q}} c(\tilde{q})\mathrm{Cr}_{\tilde{q},a}^{-}(\phi).$$

Both terms in (4) appear in the statement of Theorem 2.2.

In the same manner as above, the right integral can be re-written using the Euler sum of Z, yielding

$$\int_{\sigma_0+a+i\epsilon}^{\sigma_0+a+i\infty} \phi(s)Z'/Z(s)ds$$

$$= \phi(s)\log Z(s)\Big|_{\sigma_0+a+i\epsilon}^{\sigma_0+a+i\infty} - \int_{\sigma_0+a+i\epsilon}^{\sigma_0+a+i\infty} \phi'(s)\log Z(s)ds$$

$$= -\phi(\sigma_0+a+i\epsilon)\log Z(\sigma_0+a+i\epsilon)$$

$$- \sum_{q} c(q) \int_{\sigma_0+a+i\epsilon}^{\sigma_0+a+i\infty} \phi'(s)q^{-s}ds$$

Again, we can let ϵ approach zero to obtain the equality

$$\int_{\sigma_0+a}^{\sigma_0+a+i\infty} \phi(s)Z'/Z(s)ds$$

(5)
$$= -\phi(\sigma_0+a)\log Z(\sigma_0+a) - \sum_{q} c(q)\mathrm{Cr}_{q,a}^{+}(\phi).$$

The second term in (5) appears in Theorem 2.2. The first term in (5) does not appear in Theorem 2.2 because this term cancels with a term that appears in the evaluation of the bottom integral, as we shall now see.

In the evaluation of the bottom integral, we see the importance of choosing $\epsilon > 0$ before integrating by parts. By the choice of ϵ,

Z has no zeros or poles on the line segment $[-a + i\epsilon, \sigma_0 + a + i\epsilon]$, so we have

$$\int_{-a+i\epsilon}^{\sigma_0+a+i\epsilon} \phi(s) Z'/Z(s) ds$$

(6)
$$= \phi(s) \log Z(s) \Big|_{-a+i\epsilon}^{\sigma_0+a+i\epsilon} - \int_{-a+i\epsilon}^{\sigma_0+a+i\epsilon} \phi'(s) \log Z(s) ds.$$

Now let $\epsilon \to 0$ to get the equality

$$\int_{-a}^{\sigma_0+a} \phi(s) Z'/Z(s) ds$$

$$= \phi(\sigma_0 + a) \log Z(\sigma_0 + a) - \phi(-a) \log Z(-a)$$

(7)
$$- \int_{-a}^{\sigma_0+a} \phi'(s) \log Z(s) ds.$$

To complete the proof of Theorem 2.2, simply combine equations (2) through (7). Note the cancellation of one term in (5) with a term in (7).

Remark 1. The value of $\log Z(-a)$ is obtained by the analytic continuation of the Euler sum of Z along the horizontal line segment $[\sigma_0 + a + i\epsilon, -a + i\epsilon]$, followed by the continuation along the vertical line segment $[-a, -a + i\epsilon]$, which is equivalent to the analytic continuation along the top of the horizontal line segment $[-a, \sigma_0 + a]$ for small ϵ. To be precise, one should write the integral in (7) as

$$\int_{-a}^{\sigma_0+a} \phi'(s) \log Z(s) ds$$

(8)
$$= \int_{-a}^{\sigma_0+a} \phi'(s) \log |Z(s)| ds - \int_{-a}^{\sigma_0+a} \phi'(s) \arg(Z(s)) ds.$$

Remark 2. In the case that $Z(s)$ is real on the real axis, then $\arg(Z(s))$ is a step function on $[-a, \sigma_0 + a]$ and takes on values in $\mathbf{Z} \cdot \pi i$, except at the zeros and poles of Z, where the argument is undetermined. In this case the integral with $\arg Z(s)$ in (8) can be evaluated directly and trivially, as an elementary integral.

§4. An inductive theorem.

When comparing our work with that of Cramér in [Cr 19], the reader should note that we have overcome a point of substantial technical difficulty that Cramér encountered when proving Theorem 2.2 for the Riemann zeta function $\zeta_\mathbf{Q}(s)$. By choosing a suitably, we have avoided having to consider the convergence of the Euler sum of Z on the line $\mathrm{Re}(s) = \sigma'_0$. Cramér used the fact that $\zeta_\mathbf{Q}(s)$ does not vanish on the vertical line $\mathrm{Re}(s) = 1$ as well as specific knowledge about the distribution of prime numbers, namely

$$\sum_{p \leq x} \frac{1}{p} = O(\log \log x) \quad \text{as } x \to \infty$$

and the Landau theorem which states that the limit

$$\lim_{x \to \infty} \sum_{p \leq x} \frac{1}{p^{1+it}}$$

converges uniformly for t in compact subsets of $\mathbf{R} \setminus \{0\}$. By following Cramér's original proof exactly, we would have greatly increased the complexity of the axioms of meromorphy, Euler sum, and functional equation.

The simplest example is that of the original Cramér theorem, for the Riemann zeta function $\zeta_\mathbf{Q}$. The gamma function obviously satisfies the growth conditions vis a vis the test function

$$\phi(s) = e^{sz} \quad \mathrm{Im}(z) > 0.$$

In fact, for the zeta functions coming from modular forms, with gamma factors as fudge factors, the same remark applies.

When applying Theorem 2.2 to the test function $\phi_z(s) = e^{sz}$, we can then set $z = it$, and we determined the asymptotic behavior of $\sum e^{i\rho t}$ as t approaches zero in complete detail in [JoL 93c] under the assumption that Φ is of regularized product type. The result of these calculations is the following theorem.

Theorem 4.1. *Let* $(Z, \tilde{Z}, \Phi)$ *be in the fundamental class, and assume that* Φ *has reduced order* (M, m). *Then* Z *and* $\tilde{Z}$ *are* of regularized product type and of reduced order $(M, m+1)$ if $-\mathrm{Re}(p_0) \in \mathbf{Z}$, otherwise the reduced order is (M, m).

Examples. In §7 of [JoL 93c] we applied Theorem 4.1 to give many examples of zeta functions which are thus sknown to be of regularized product type. We add to this list any **Dirichlet polynomial**, which we define to be any holomorphic function of the form

$$P(s) = \sum_{n=1}^{N} a_n b_n^s$$

where $\{a_n\}$ is a finite sequence of complex numbers, $\{b_n\}$ is a finite sequence of positive real numbers, which we may assume, without loss of generality, to satisfy the inequalities

$$0 < b_1 < b_2 < \cdots < b_N.$$

Let

$$Q(s) = P(-s) = \sum_{n=1}^{N} a_n b_n^{-s},$$

and write

$$P(s) = a_N b_N^s \left[1 + \sum_{n=1}^{N-1} \frac{a_n}{a_N} \left(\frac{b_n}{b_N} \right)^s \right] = a_N b_N^s \cdot Z(s)$$

and

$$Q(s) = a_1 b_1^{-s} \left[1 + \sum_{n=2}^{N} \frac{a_n}{a_1} \left(\frac{b_n}{b_1} \right)^{-s} \right] = a_1 b_1^{-s} \cdot \tilde{Z}(s).$$

It is immediate that there exists some $\sigma_0' > 0$ such that for all s with $\mathrm{Re}(s) > \sigma_0'$ we have

$$\left| \sum_{n=1}^{N-1} \frac{a_n}{a_N} \left(\frac{b_n}{b_N} \right)^s \right| < 1 \quad \text{and} \quad \left| \sum_{n=2}^{N} \frac{a_n}{a_1} \left(\frac{b_n}{b_1} \right)^{-s} \right| < 1.$$

Therefore, Z and $\tilde{Z}$ have Euler sums, which means there exist sequences $\{\mathbf{q}\}, \{\tilde{\mathbf{q}}\}$ and $\{c(\mathbf{q})\}, \{c(\tilde{\mathbf{q}})\}$ such that for $\mathrm{Re}(s) > \sigma_0'$ we have

$$\log Z(s) = \sum_{\mathbf{q}} \frac{c(\mathbf{q})}{\mathbf{q}^s} \quad \text{and} \quad \log \tilde{Z}(s) = \sum_{\tilde{\mathbf{q}}} \frac{c(\tilde{\mathbf{q}})}{\tilde{\mathbf{q}}^s}.$$

If we set
$$G(s) = a_N b_N^s \quad \text{and} \quad \tilde{G}(s) = a_1 b_1^{-s},$$
then the trivial relation $P(s) = Q(-s)$ can be written as
$$G(s)Z(s) = \tilde{G}(-s)\tilde{Z}(-s),$$
so we also have a functional equation with $\sigma_0 = 0$. Notice that the functional equation implies that all the zeros of P lie in some vertical strip. Further, we can apply Theorem 4.1 to conclude that the Dirichlet polynomial P is of regularized product type, with reduced order $(0,1)$. As a result, the estimates from Chapter I, specifically Theorem 6.2, hold for any Dirichlet polynomial.

Finally, observe that the local factors of the more classical zeta functions are Dirichlet polynomials. Indeed, such factors are of the form $\text{Pol}_p(p^{-s})^{\pm 1}$ where Pol_p is a polynomial with constant term 1, and p is a prime number. For the Riemann zeta function, this local factor is simply $\text{Pol}_p(T) = 1 - T$, so we have $\text{Pol}_p(p^{-s}) = 1 - p^{-s}$. In the representation theory of $GL(n)$, the polynomial Pol_p has degree n. For representations in $GL(2)$ associated to an elliptic curve, say, we have $\text{Pol}_p(T) = 1 - a_p T + pT^2$, so in terms of p^{-s}, the local factor is
$$1 - a_p p^{-s} + pp^{-2s}.$$
Thus the local factors of classical zeta functions are themselves of regularized product type.

CHAPTER III
Explicit formulas under Fourier Assumptions

The classical "explicit formulas" of analytic number theory show that the sum of a certain function taken over the prime powers is equal to the sum of the Mellin transform taken over the zeros of the zeta function. Historically, only very special functions were used until Weil pointed out that the formulas could be proved for a much wider class of test functions (see [We 52]). We shall give here a version of these explicit formulas applicable to a wide class of test functions in connection with general zeta functions which have an Euler sum and functional equation whose fudge factor is of regularized product type. As a result, our general theorem contains the known explicit formulas for zeta functions of number fields and Selberg type zeta functions as well as new examples of explicit formulas such as that corresponding to the scattering determinant and Eisenstein series associated to any non-compact finite volume hyperbolic Riemann surface.

Various facts from analysis which we shall use in this chapter have been proved in our papers [JoL 93a] and [JoL 93b], as well as Chapter I. As a result, most of the steps taken here are relatively formal. We carry out the steps by integrating over a rectangle in the classical manner, but one aspect of this classical procedure emerges more clearly than in the case of classical zeta functions, namely the inductive procedure arising from a functional equation of the type

$$Z(s)\Phi(s) = \widetilde{Z}(\sigma_0 - s),$$

with zeta functions Z and $\widetilde{Z}$ and fudge factor Φ which is of regularized product type. For instance, for the Selberg zeta function of compact Riemann surfaces, these factors involve the Barnes double gamma function, and for the non-compact case, these fudge factors may involve the Riemann zeta function itself at the very least. Ultimately, arbitrarily complicated regularized products will occur as fudge factors in such a functional equation.

§1. Growth conditions on Fourier transforms.

We shall consider growth conditions on Fourier transforms and logarithmic derivatives of regularized products, and we begin by estimating Fourier transforms.

Following Barner [Ba 81], we require the test functions g to satisfy the following two basic Fourier conditions.

FOU 1. $g \in BV(\mathbf{R}) \cap L^1(\mathbf{R})$.

FOU 2. g is **normalized**, meaning

$$g(x) = \frac{1}{2}(g(x+) + g(x-)) \quad \text{for all } x \in \mathbf{R}.$$

These will be the only relevant conditions in this section, but in the next section to apply a Parseval formula, we shall consider a third condition at the origin, namely:

FOU 3. There exists $\epsilon > 0$ such that

$$g(x) = g(0) + O(|x|^\epsilon) \quad \text{for } x \to 0.$$

If we let N be any integer ≥ 0, then we say that g satisfies the **basic Fourier conditions to order** N if g is N times differentiable and its first N derivatives satisfy the above three basic conditions.

Lemma 1.1. *Assume g satisfies* **FOU 1** *and* **FOU 2** *to order M. Then*

$$g^\wedge(t) = O(1/|t|^{M+1}) \quad \text{for } |t| \to \infty.$$

Proof. We integrate by parts M times to give

$$g^\wedge(t) = \frac{1}{\sqrt{2\pi}} \left(\frac{1}{it}\right)^M \int_{-\infty}^{\infty} g^{(M)}(x) e^{-itx} dx.$$

To finish, note that for any $h \in BV(\mathbf{R}) \cap L^1(\mathbf{R})$, we have the Stieltjes integration by parts formula

$$h^\wedge(t) = \frac{1}{\sqrt{2\pi}} \frac{1}{it} \int_{-\infty}^{\infty} e^{-itx} dh(x),$$

from which we obtain the estimate

$$|h^\wedge(t)| \le \frac{1}{\sqrt{2\pi}|t|} V_{\mathbf{R}}(h),$$

whence the proposition follows. □

Let f be a measurable function on $\mathbf{R}^+$ so that under certain convergence conditions we have the **Mellin transform**

$$\mathbf{M}f(s) = \int_0^\infty f(u) u^s \frac{du}{u}.$$

Let $\sigma_0 \in \mathbf{R}$. We define $\mathbf{M}_{\sigma_0/2} f$ to be the translate of $\mathbf{M}f$ by $\sigma_0/2$, meaning

$$\mathbf{M}_{\sigma_0/2} f(s) = \mathbf{M}f(s - \sigma_0/2).$$

We put

$$F(x) = f(e^{-x}) \quad \text{so} \quad f(u) = F(-\log u).$$

Then letting $s = \sigma + it$, we find

$$\mathbf{M}_{\sigma_0/2} f(s) = \int_{-\infty}^\infty F(x) e^{x\sigma_0/2} e^{-sx} dx$$

$$= \int_{-\infty}^\infty F(x) e^{-(\sigma - \sigma_0/2)x} e^{-itx} dx,$$

which is $\sqrt{2\pi}$ times the Fourier transform of

$$F_\sigma(x) = F(x) e^{-(\sigma - \sigma_0/2)x}.$$

That is,

$$\mathbf{M}_{\sigma_0/2} f(\sigma + it) = \sqrt{2\pi} F_\sigma^\wedge(t).$$

In particular, if we let $\sigma = \sigma_0/2$, then we obtain:

On the line $\mathrm{Re}(s) = \sigma_0/2$, *the Mellin transform is a constant multiple of the Fourier transform, namely*

$$\mathbf{M}_{\sigma_0/2} f(\sigma_0/2 + it) = \sqrt{2\pi} F^\wedge(t).$$

Lemma 1.2. *Let F be a function on $\mathbf{R}$ and assume there is an $\epsilon > 0$ such that*
$$F(x)e^{(\sigma_0/2 - \sigma_1 + \epsilon)|x|}.$$
Assume F satisfies **FOU 1** *and* **FOU 2** *to order M and define the function f on $\mathbf{R}^+$ by*
$$f(u) = F(-\log u).$$
Let $\sigma_1, \sigma_2 \in \mathbf{R}$ be fixed real numbers with $\sigma_1 < \sigma_2$ and consider the strip consisting of all $s \in \mathbf{C}$ with $\mathrm{Re}(s) \in [\sigma_1, \sigma_2]$. Then for any s in the strip $\mathrm{Re}(s) \in [\sigma_1, \sigma_2]$, we have
$$\mathbf{M}_{\sigma_0/2} f(\sigma + it) = \sqrt{2\pi} F_\sigma^\wedge(t).$$
and
$$\mathbf{M}_{\sigma_0/2} f(s) = O(1/|s|^{M+1})$$
for $|s| \to \infty$.

The proof of Lemma 1.2 follows that of Lemma 1.1, hence will be omitted.

As in previous chapters, we consider a meromorphic function H, and we want to compare the rate of growth of the logarithmic derivative H'/H and $\mathbf{M}_{\sigma_0/2} f$ on vertical lines. Specifically, we consider the following growth condition.

GR. *There exists a sequence $\{T_m\}$ tending to ∞ such that*
$$\mathbf{M}_{\sigma_0/2} f(s) H'/H(s) \to 0 \quad \text{for } m \to \infty$$
for any s on the horizontal line segment $S_{\pm m}$ defined by $\sigma \pm iT_m$ with $\sigma_1 \leq \sigma \leq \sigma_2$.

Remark. Of course, we are interested in considering the growth condition **GR** when H is one of the functions Z, $\widetilde{Z}$, or Φ. In the present variation, we need only one growth condition of type **GR** to insure the existence of certain limiting integrals. The point is that under the additional Fourier theoretic conditions, the convergence of the integrals on the vertical lines will be reduced to Fourier inversion on the middle, or critical, line $\mathrm{Re}(s) = \sigma_0/2$, after shifting the line of integration, and picking up residues corresponding to zeros and poles of the factor Φ.

In this chapter, we shall apply the estimates summarized in §6 of Chapter I to obtain a criterion under which the growth condition **GR** is satisfied, as in the next theorem.

Theorem 1.3. *Let H be of regularized product type of reduced order (M,m), and assume that F satisfies the growth conditions of Lemma 1.2 to order M. Then H satisfies the growth condition* **GR** *whenever F satisfies the three basic conditions to order M*

The proof of Theorem 1.3 follows directly from Theorem 2.5 of Chapter I and Lemma 1.2 above.

Recall that the basic Fourier conditions **FOU 1** and **FOU 2** insure the possibility of applying one of the most classical inversion theorems of Fourier analysis, stemming from Dirichlet, and attributed more directly to Pringsheim, Prasad and Hobson by Titchmarsh [Ti 48], page 25 (see also Theorem 2.5 of Chapter X in [La 93b]). For completeness, let us recall this result.

Basic Fourier Inversion Formula. *Let $\beta \in \mathrm{BV}(\mathbf{R}) \cap L^1(\mathbf{R})$, and suppose β is normalized. Then*

$$\lim_{T \to \infty} \frac{1}{\sqrt{2\pi}} \int_{-T}^{T} \beta^{\wedge}(t) e^{itx} \, dt = \beta(x) \quad \text{for all } x \in \mathbf{R}.$$

§2. The explicit formulas.

Let $(Z, \tilde{Z}, \Phi)$ be in the fundamental class. Let $a > 0$ be such that $\sigma_0 + a > \sigma_0'$ and such that Z, $\tilde{Z}$ and Φ do not have a zero or pole on the lines $\text{Re}(s) = -a$ and $\text{Re}(s) = \sigma_0 + a$. We let f and F be measurable functions related by

$$F(x) = f(e^{-x}) \quad \text{so} \quad f(u) = F(-\log u).$$

For the moment, assume there is some $a' > a$ for which we have the bound

(1) $$|F(x)| \ll e^{-(\sigma_0/2 + a')|x|}.$$

We shall actually assume something stronger later, but for now we just want to deal with a region of absolute convergence for a Mellin integral. Assuming the bound stated in (1), the function $\mathbf{M}_{\sigma_0/2} f(s)$ is holomorphic in the closed strip $-a \leq \sigma \leq \sigma_0 + a$.

We let:

$\mathcal{R}_a$ be the infinite rectangle bounded by the vertical lines

$$\text{Re}(s) = -a \quad \text{and} \quad \text{Re}(s) = \sigma_0 + a.$$

$\mathcal{R}_a(T)$ be the finite rectangle bounded by the above vertical lines and the lines

$$\text{Im}(s) = -T \quad \text{and} \quad \text{Im}(s) = T.$$

We assume at first that Φ has no zeros or poles on the line $\text{Re}(s) = \sigma_0/2$. A variation without this restriction will also be treated. The line with $\sigma_0/2$ is especially useful for the more classical applications to counting **q**'s, or primes in the number theoretic case.

We let:

$\{\rho\}$ = the set of zeros and poles of Z in the *full* strip $-a \leq \sigma \leq \sigma_0 + a$;
$\{\alpha\}$ = the set of zeros and poles of Φ in the *half* strip $-a \leq \sigma \leq \sigma_0/2$;

Assume that T is chosen so that the functions Z, $\widetilde{Z}$, Φ have no zeros or poles on the horizontal lines that border $\mathcal{R}_a(T)$. Then, we may form the finite sum

$$V_{Z,a}(f,T) = \sum_{\rho \in \mathcal{R}_a(T)} v(\rho) \mathbf{M}_{\sigma_0/2} f(\rho)$$
$$= \mathrm{div}_{Z,a}(\mathbf{M}_{\sigma_0/2} f, T).$$

Similarly, we define

$$V_{\Phi,a,\sigma_0/2}(f,T) = \sum_{\alpha} v(\alpha) \mathbf{M}_{\sigma_0/2} f(\alpha).$$

Let $\mathcal{L}_a(T)$ be the boundary of the rectangle $\mathcal{R}_a(T)$ and consider the integral

$$\int_{\mathcal{L}_a(T)} \mathbf{M}_{\sigma_0/2} f(s) Z'/Z(s) ds.$$

By the residue theorem we have the equality

$$\int_{\mathcal{L}_a(T)} \mathbf{M}_{\sigma_0/2} f(s) Z'/Z(s) ds$$
$$= 2\pi i \sum_{\rho \in \mathcal{R}_a(T)} v(\rho) \mathbf{M}_{\sigma_0/2} f(\rho) = 2\pi i V_{Z,a}(f,T).$$

At this point, we want to take a limit as $T \to \infty$. For this, one must have a choice of $T \to \infty$ such that the integrand on the top and bottom segments of the rectangle tends to 0, as hypothesized in the growth condition **GR**. Roughly speaking, the more zeros and poles the function Z has in the strip $\mathcal{R}_a$, the larger $Z'/Z(s)$ could be on such horizontal segments, and so the smoother the function f must be so that its Mellin transform approaches 0 sufficiently fast, meaning faster than $Z'/Z(s)$ approaches infinity.

Assuming the growth condition **GR**, we are interested in the infinite sum

$$V_{Z,a}(f) = \sum_{\rho \in \mathcal{R}_a} v(\rho) \mathbf{M}_{\sigma_0/2} f(\rho),$$

271

which is understood in the limiting sense

$$\text{(2)} \quad \sum_{\rho \in \mathcal{R}_a} v(\rho) \mathbf{M}_{\sigma_0/2} f(\rho) = \lim_{m \to \infty} \sum_{\rho \in \mathcal{R}_a(T_m)} v(\rho) \mathbf{M}_{\sigma_0/2} f(\rho).$$

Since the similar sum over the family $\{\alpha\}$ is taken on the left half interval, we use the notation

$$V_{\Phi,a,\sigma_0/2}(f) = \sum_{\text{Re}(\alpha) \leq \sigma_0/2} v(\alpha) \mathbf{M}_{\sigma_0/2} f(\alpha)$$

$$= \lim_{m \to \infty} \sum_{\alpha \in \mathcal{R}_a(T_m), \text{Re}(\alpha) \leq \sigma_0/2} v(\alpha) \mathbf{M}_{\sigma_0/2} f(\alpha).$$

The Fourier conditions **FOU 1** and **FOU 2** are imposed in order to deal with questions of Fourier inversion in connection with sums over $\{\mathbf{q}\}$. As we shall see, the condition **FOU 3** is concerned with our evaluation of the **Weil functional** W_Φ, which is defined to be the limit

$$W_\Phi(F) = \lim_{m \to \infty} \frac{1}{\sqrt{2\pi}} \int_{-T_m}^{T_m} F^\wedge(t) \Phi'/\Phi(\sigma_0/2 + it) dt,$$

where Φ is assumed to be holomorphic on the line $\text{Re}(s) = \sigma_0/2$. In addition to these assumptions, we will require:

FOU 4. There exists a constant $a' > 0$ such that the function

$$x \mapsto F(x) e^{(\sigma_0/2 + a')|x|}$$

is in BV(**R**).

Under suitable conditions on the test function F, which will be expressed in terms of the above four Fourier conditions, we shall prove:

The Explicit Formula.

$$V_{Z,a}(f) + V_{\Phi,a,\sigma_0/2}(f) =$$

$$\sum_{\mathbf{q}} \frac{-c(\mathbf{q}) \log \mathbf{q}}{\mathbf{q}^{\sigma_0/2}} f(\mathbf{q}) + \sum_{\tilde{\mathbf{q}}} \frac{-c(\tilde{\mathbf{q}}) \log \tilde{\mathbf{q}}}{\tilde{\mathbf{q}}^{\sigma_0/2}} f(1/\tilde{\mathbf{q}}) + W_\Phi(F).$$

More specifically, the main result of this chapter is the following theorem.

Theorem 2.1. Let $(Z, \tilde{Z}, \Phi)$ be in the fundamental class, and assume that Φ has reduced order (M, m) with no zeros or poles on the line $\mathrm{Re}(s) = \sigma_0/2$. Then for any function F which satisfies the four Fourier conditions to order M, the functionals $W_\Phi(F)$, $V_{Z,a}(f)$ and $V_{\Phi,a,\sigma_0/2}(f)$ are defined and the explicit formula holds.

Observe that, as in our formulation of Cramér's theorem, the above theorem is an inductive one, expressing the sum $V_{Z,a}(f)$ in terms of a similar sum concerning $V_{\Phi,a,\sigma_0/2}$, the Weil functional, and terms involving the families $\{\mathbf{q}\}$ and $\{\tilde{\mathbf{q}}\}$.

Note that through a "change of notation", Theorem 2.1 can be used to express an explicit formula involving the zeros and poles of $\tilde{Z}$.

Remark 1. Suppose there are meromorphic functions G and $\tilde{G}$ such that
$$\Phi(s) = G(s)/\tilde{G}(\sigma_0 - s).$$
Then we may write the Weil functional as
$$W_\Phi(F) = W_G^+(F) + W_{\tilde{G}}^-(F)$$
where
$$W_G^+(F) = \lim_{m\to\infty} \frac{1}{\sqrt{2\pi}} \int_{-T_m}^{T_m} F^\wedge(t) G'/G(\sigma_0/2 + it) dt$$
and
$$W_{\tilde{G}}^-(F) = \lim_{m\to\infty} \frac{1}{\sqrt{2\pi}} \int_{-T_m}^{T_m} F^\wedge(-t) \tilde{G}'/\tilde{G}(\sigma_0/2 + it) dt.$$

Further, in the case where $G = \tilde{G}$, we can write the Weil functional as
$$W_\Phi(F) = \lim_{m\to\infty} \frac{1}{\sqrt{2\pi}} \int_{-T_m}^{T_m} [F^\wedge(t) + F^\wedge(-t)] G'/G(\sigma_0/2 + it) dt.$$

Remark 2. Observe that the sum $V_{Z,a} + V_{\Phi,a,\sigma_0/2}$ is independent of a even though neither term is independent of a.

For any u such that Φ has no zero or pole on the line $\text{Re}(s) = u$, we define

$$W^{\#}_{\Phi,u}(f) = \lim_{m \to \infty} \frac{1}{2\pi i} \int_{u-iT_m}^{u+iT_m} M_{\sigma_0/2}(f)\Phi'/\Phi(s)ds.$$

The proof of the explicit formula will go through the following intermediate stage.

Theorem 2.2. *Assume that Φ is of regularized product type of reduced order (M, m), and F satisfies the Fourier conditions to order M. Then*

$$V_{Z,a}(f) = \sum_{\mathbf{q}} \frac{-c(\mathbf{q})\log \mathbf{q}}{\mathbf{q}^{\sigma_0/2}} f(\mathbf{q}) + \sum_{\tilde{\mathbf{q}}} \frac{-c(\tilde{\mathbf{q}})\log \tilde{\mathbf{q}}}{\tilde{\mathbf{q}}^{\sigma_0/2}} f(1/\tilde{\mathbf{q}})$$
$$+ W^{\#}_{\Phi,-a}(f).$$

Theorem 2.2 will be proved in §3. After Theorem 2.2 has been proved, what will remain is to analyze the last term containing $\Phi'/\Phi(s)$. Different applications require different analyses of this term. For classical analytic number theory, one moves the line of integration from $-a$ to $\sigma_0/2$, and then one applies the general Parseval formula. This will be carried out in §4, thus yielding Theorem 2.1. In §5 we give a further determination of the Weil functional, based on the general Parseval formula from [JoL 93b].

However, for other applications and notably those occuring later in this book in Chapter V, §3 and §4, we shall move the line of integration far to the right. In these applications, one can thus completely bypass the Parseval formula, and the final result is therefore much simpler to prove. We now carry this out.

Any regularized product type can be expressed as a product of two such types, one of which has all of its zeros and poles in a left half plane and one of which has all of its zeros and poles in a right half plane, say

$$\Phi = \Phi_{\text{left}} \Phi_{\text{right}}.$$

We let $A > \sigma_0$ be a number such that all the zeros and poles of Φ_{left} are in the half plane $\text{Re}(s) \leq A - \delta$, for some $\delta > 0$. Similarly, select a is such that all the zeros and poles of Φ_{right} are in the half plane $\text{Re}(s) \geq -a + \delta$, for some $\delta > 0$.

In Theorem 2.2 we move the line of integration for Φ_{left} to $\text{Re}(s) = A$, thus picking up the residues of all the poles between these two lines. At the same time, we define

$$V_{\Phi_{\text{left}},a,A}(f) = \sum_{\zeta} v(\zeta) \mathbf{M}_{\sigma_0/2} f(\zeta)$$

where

$\{\zeta\}$ = the set of all zeros and poles of Φ_{left} such that $\text{Re}(\zeta) > -a$.

Then, from Theorem 2.2, we arrive at the following formula.

Theorem 2.3. *Let $(Z, \tilde{Z}, \Phi)$ be in the fundamental class, and assume that Φ is decomposed as above. Then for any function F which satisfies the four Fourier conditions to order M, all the functionals in the next formula are defined and the following formula holds:*

$$V_{Z,a}(f) + V_{\Phi_{\text{left}},a,A}(f) =$$
$$\sum_{\mathbf{q}} \frac{-c(\mathbf{q}) \log \mathbf{q}}{\mathbf{q}^{\sigma_0/2}} f(\mathbf{q}) + \sum_{\tilde{\mathbf{q}}} \frac{-c(\tilde{\mathbf{q}}) \log \tilde{\mathbf{q}}}{\tilde{\mathbf{q}}^{\sigma_0/2}} f(1/\tilde{\mathbf{q}})$$
$$+ W^{\#}_{\Phi_{\text{left}},A}(f) + W^{\#}_{\Phi_{\text{right}},-a}(f).$$

Steps which justify moving the line of integration are given in §4 in the case considered in Theorem 2.1. The same argument applies to Theorem 2.2, thus yielding Theorem 2.3.

Remark 3. In Chapter II, for the proof of Cramèr's theorem, we did not use a Mellin transform but worked directly on a half strip with a test function ϕ. One may do the same in a full strip to get the explict formula directly for such a function. In that case, the Weil functionals are expressed in terms of ϕ instead of the Mellin transform $\mathbf{M}_{\sigma_0/2} f$. For the applications to Chapter IV and Chapter V, this way of proceeding eliminates completely all Fourier conditions, and we could deal directly with the special test functions

$$\phi_t(s) = e^{(s-\sigma_0/2)^2 t}$$

72

or
$$\phi'_t(s) = (2s - \sigma_0)e^{(s-\sigma_0/2)^2 t},$$

just as we dealt with the function e^{sz} for the Cramér theorem in [JoL 93c].

§3. The terms with the q's.

In this and the next section we will begin our proof of the explicit formula based on the axioms **FOU 1** through **FOU 4**. In this section, we shall compute the sums over **q** that appear in Theorem 2.1. The remainder of the proof, namely the determination of the terms involving the fudge factor Φ and its set of zeros and poles $\{\alpha\}$ will be given in the next section.

Consider the rectangle $\mathcal{R}_a(T_m)$, as defined above, and integrate over the boundary of this rectangle. By the residue formula, we have

$$(1) \quad \sum_{\rho \in \mathcal{R}_a(T_m)} v(\rho) \mathbf{M}_{\sigma_0/2} f(\rho) = \frac{1}{2\pi i} \int_{\mathcal{L}_a(T_m)} \mathbf{M}_{\sigma_0/2} f(s) Z'/Z(s) ds.$$

Throughout we assume that Φ is of regularized product type of reduced order (M, m). By Theorem 4.1 of Chapter II, Z and $\widetilde{Z}$ are of regularized product type of reduced order $(M, m+1)$. We now can apply Lemma 1.2 to $H = Z$ to find, for $m \to \infty$,

$$\frac{1}{2\pi i} \int_{-a-iT_m}^{\sigma_0+a-iT_m} \mathbf{M}_{\sigma_0/2} f(s) Z'/Z(s) ds$$

$$(2) \quad + \frac{1}{2\pi i} \int_{\sigma_0+a+iT_m}^{-a+iT_m} \mathbf{M}_{\sigma_0/2} f(s) Z'/Z(s) ds = o(1).$$

Upon combining (1) and (2), we obtain

$$\sum_{\rho \in \mathcal{R}_a(T_m)} v(\rho) \mathbf{M}_{\sigma_0/2} f(\rho) + o(1)$$

$$= \frac{1}{2\pi i} \int_{-a+iT_m}^{-a-iT_m} \mathbf{M}_{\sigma_0/2} f(s) Z'/Z(s) ds$$

$$+ \frac{1}{2\pi i} \int_{\sigma_0+a-iT_m}^{\sigma_0+a+iT_m} \mathbf{M}_{\sigma_0/2} f(s) Z'/Z(s) ds.$$

Using the functional equation, we can express this equality as

$$= \frac{1}{2\pi i} \int_{-a+iT_m}^{-a-iT_m} \mathbf{M}_{\sigma_0/2} f(s) \left[-\widetilde{Z}'/\widetilde{Z}(\sigma_0 - s)\right] ds$$

(3)
$$+ \frac{1}{2\pi i} \int_{-a+iT_m}^{-a-iT_m} \mathbf{M}_{\sigma_0/2} f(s) [-\Phi'(s)/\Phi(s)] ds$$

$$+ \frac{1}{2\pi i} \int_{\sigma_0+a-iT_m}^{\sigma_0+a+iT_m} \mathbf{M}_{\sigma_0/2} f(s) Z'/Z(s) ds.$$

As stated above, in this section we will deal with the Z and $\widetilde{Z}$ integrals, and in the next section we will evaluate the Φ integral.

For the Z integral, we obtain

$$\frac{1}{2\pi i} \int_{\sigma_0+a-iT_m}^{\sigma_0+a+iT_m} \mathbf{M}_{\sigma_0/2} f(s) Z'/Z(s) ds$$

(4)
$$= \frac{-1}{2\pi} \int_{-T_m}^{T_m} dt \int_{-\infty}^{\infty} F(x) e^{-(\sigma_0/2+a+it)x} \sum_{\mathbf{q}} \frac{c(\mathbf{q}) \log \mathbf{q}}{\mathbf{q}^{\sigma_0+a+it}} dx.$$

This step follows by differentiating the Euler sum for $\log Z$ to obtain the series for Z'/Z. We shall give formal arguments to change this last expression (4), and then we give estimates to justify the calculations. We make a change of variables

$$x = y - \log \mathbf{q} \quad \text{and} \quad dx = dy$$

in the integral of each term with $\mathbf{q}$. Then the whole expression (4) becomes

$$= \frac{-1}{2\pi} \int_{-T_m}^{T_m} dt \sum_{\mathbf{q}} \int_{-\infty}^{\infty} \frac{c(\mathbf{q}) \log \mathbf{q}}{\mathbf{q}^{\sigma_0/2}} F(y - \log \mathbf{q}) e^{-(\sigma_0/2+a)y} e^{-ity} dy$$

$$= \frac{-1}{2\pi} \int_{-T_m}^{T_m} dt \sum_{\mathbf{q}} \int_{-\infty}^{\infty} H_{\mathbf{q}}(y) e^{-ity} dy$$

where
$$H_{\mathbf{q}}(y) = \frac{c(\mathbf{q})\log \mathbf{q}}{\mathbf{q}^{\sigma_0/2}} F(y - \log \mathbf{q}) e^{-(\sigma_0/2+a)y}.$$

We let
$$H(y) = \sum_{\mathbf{q}} H_{\mathbf{q}}(y).$$

With the definition of H, we may express our desired limit integral in (4) as being

$$= \frac{-1}{2\pi} \lim_{m \to \infty} \int_{-T_m}^{T_m} dt \int_{-\infty}^{\infty} H(y) e^{-ity} dy = -H^{\wedge\wedge}(0) = -H(0).$$

by the Basic Fourier Inversion Formula. We then see that

$$-H(0) = \sum_{\mathbf{q}} \frac{-c(\mathbf{q})\log \mathbf{q}}{\mathbf{q}^{\sigma_0/2}} F(-\log \mathbf{q}) = \sum_{\mathbf{q}} \frac{-c(\mathbf{q})\log \mathbf{q}}{\mathbf{q}^{\sigma_0/2}} f(\mathbf{q}),$$

which is a term in the statement of Theorem 2.1.

We shall now justify the above steps. First, by **FOU 4**, there is a constant C such that

$$|F(x)| \leq C e^{-(\sigma_0/2 + a')|x|},$$

from which we get the estimates

$$|F(y - \log \mathbf{q})| e^{-(\sigma_0/2+a)y} \leq \begin{cases} C\mathbf{q}^{-(\sigma_0/2+a')} e^{(a'-a)y} & y \leq \log \mathbf{q} \\ C\mathbf{q}^{\sigma_0/2+a'} e^{-(\sigma_0+a+a')y} & y \geq \log \mathbf{q}, \end{cases}$$

and, in particular,

$$|F(y - \log \mathbf{q})| e^{-(\sigma_0/2+a)y} \leq C\mathbf{q}^{-(\sigma_0/2+a)}$$

for all y. From this it follows that

(5) $$|H_{\mathbf{q}}(y)| \leq 2C \frac{|c(\mathbf{q})|\log \mathbf{q}}{\mathbf{q}^{\sigma_0+a}},$$

and after actually performing the integration, we obtain

(6) $$\int_{-\infty}^{\infty} |H_{\mathbf{q}}(y)|\, dy \leq 2C \frac{|c(\mathbf{q})|\log \mathbf{q}}{\mathbf{q}^{\sigma_0+a}} \left[\frac{1}{a'-a} + \frac{1}{\sigma_0+a+a'}\right].$$

Estimate (5) shows that the series

$$H(y) = \sum_{\mathbf{q}} H_{\mathbf{q}}(y)$$

is absolutely and uniformly convergent, and estimate (6) shows that this series defines a function $y \mapsto H(y)$ in $L^1(\mathbf{R})$.

Since each term $H_{\mathbf{q}}$ is of bounded variation and normalized, the uniform convergence of the series (6) shows that H is normalized. Furthermore, the total variation $V_{\mathbf{R}}$ satisfies the triangle inequality for an infinite sum, as one verifies directly from the Riemann-Steiltjes sums defining this variation. Similarly for a product, we have
$$V_{\mathbf{R}}(gh) \leq \|g\| V_{\mathbf{R}}(h) + \|h\| V_{\mathbf{R}}(g).$$
Then
$$V_{\mathbf{R}}(H) \leq \sum_{\mathbf{q}} \frac{|c(\mathbf{q})|\log \mathbf{q}}{\mathbf{q}^{\sigma_0+a}} V_{\mathbf{R}}\left(F(x)e^{-(\sigma_0/2+a)x}\right).$$

Hence, by **FOU 4**, H has bounded variation, and we have justified all the formal operations and the application of Fourier inversion in the evaluation of the Z integral as $-H(0)$.

One may carry out similar arguments for the $\widetilde{Z}$ integral

$$\frac{-1}{2\pi i} \int_{-a+iT_m}^{-a-iT_m} \mathbf{M}_{\sigma_0/2} f(s) \widetilde{Z}'/\widetilde{Z}(\sigma_0 - s)\, ds,$$

which we write as

$$= \frac{-1}{2\pi} \int_{-T_m}^{T_m} dt \int_{-\infty}^{\infty} F(x) e^{-(\sigma_0/2+a+it)x} \sum_{\widetilde{\mathbf{q}}} \frac{c(\widetilde{\mathbf{q}})\log \widetilde{\mathbf{q}}}{\widetilde{\mathbf{q}}^{\sigma_0+a+it}} dx.$$

In this case, one uses the inequalities

$$|F(y + \log \tilde{\mathbf{q}})| e^{(\sigma_0/2+a)y} \leq \begin{cases} C\tilde{\mathbf{q}}^{-(\sigma_0/2+a')} e^{(a-a')y} & y \geq -\log \tilde{\mathbf{q}} \\ C\tilde{\mathbf{q}}^{\sigma_0/2+a'} e^{(\sigma_0+a+a')y} & y \leq -\log \tilde{\mathbf{q}}, \end{cases}$$

and

$$|F(y + \log \tilde{\mathbf{q}})| e^{(\sigma_0/2+a)y} \leq C\tilde{\mathbf{q}}^{-(\sigma_0/2+a)},$$

which holds for all y. Inequalities similiar to (5) and (6) then follow when we define

$$H_{\tilde{\mathbf{q}}}(y) = \frac{c(\tilde{\mathbf{q}}) \log \tilde{\mathbf{q}}}{\tilde{\mathbf{q}}^{\sigma_0/2}} F(y + \log \tilde{\mathbf{q}}) e^{(\sigma_0/2+a)y}$$

and

$$H(y) = \sum_{\tilde{\mathbf{q}}} H_{\tilde{\mathbf{q}}}(y),$$

with the bounded variation

$$V_{\mathbf{R}}(H) \leq \sum_{\tilde{\mathbf{q}}} \frac{|c(\tilde{\mathbf{q}})| \log \tilde{\mathbf{q}}}{\tilde{\mathbf{q}}^{\sigma_0+a}} V_{\mathbf{R}}\left(F(x) e^{(\sigma_0/2+a)x}\right).$$

In this way, we obtain the term involving the $\widetilde{Z}$ integral in Theorem 2.1. This concludes the proof of the Theorem 2.1 as far as it involves the Z and $\widetilde{Z}$ integrals.

At this point, we have proved Theorem 2.2.

§4. The term involving Φ.

In this section we will compute the terms in the explicit formula containing the fudge factor Φ and its set of roots $\{\alpha\}$. To do so, we begin by considering the integral containing Φ'/Φ, namely

$$\frac{1}{2\pi i} \int_{-a-iT_m}^{-a+iT_m} \mathbf{M}_{\sigma_0/2} f(s) \Phi'/\Phi(s) ds.$$

We want to move the line of integration to the critical line $\sigma = \sigma_0/2$. Upon doing this, we pick up the residues of $\mathbf{M}_{\sigma_0/2} f(s)\Phi'/\Phi(s)$ at the points α, and, hence, by using **FOU 1**, we find

$$\frac{1}{2\pi i} \int_{-a-iT_m}^{-a+iT_m} \mathbf{M}_{\sigma_0/2} f(s) \Phi'/\Phi(s) ds$$

$$= \frac{1}{2\pi i} \int_{\sigma_0/2-iT_m}^{\sigma_0/2+iT_m} \mathbf{M}_{\sigma_0/2} f(s) \Phi'/\Phi(s) ds$$

$$- \sum_{\alpha \in \mathcal{R}_a(T_m)} v(\alpha) \mathbf{M}_{\sigma_0/2} f(\alpha) + o(1)$$

$$= \frac{1}{\sqrt{2\pi}} \int_{-T_m}^{T_m} F^{\wedge}(t) \Phi'/\Phi(\sigma_0/2 + it) dt$$

(1)
$$- \sum_{\alpha \in \mathcal{R}_a(T_m)} v(\alpha) \mathbf{M}_{\sigma_0/2} f(\alpha) + o(1).$$

Using **FOU 1** and **FOU 2**, we obtain the equality

$$\frac{1}{2\pi i} \lim_{m \to \infty} \int_{-a-iT_m}^{-a+iT_m} \mathbf{M}_{\sigma_0/2} f(s) \Phi'/\Phi(s) ds = W_\Phi(F) - V_{\Phi,a,\sigma_0/2}(f).$$

With this, as well as the calculations from the previous section, we have concluded the proof of Theorem 2.1, up to the evaluation of the Weil functional, which will now be dealt with.

§5. The Weil functional and regularized product type.

In this section we consider the evaluation of the Weil functional W_Φ when Φ is of regularized product type of reduced order (M, m) and F satisfies the Fourier conditions to order M. Essentially, we will apply the general Parseval formula of [JoL 93b].

Quite generally, for suitable functions g and H, and $u \in \mathbf{R}$, we shall consider the Weil functional, which we define to be

$$W_{H,u}(g) = \lim_{T \to \infty} \frac{1}{\sqrt{2\pi}} \int_{-T}^{T} g^\wedge(t) H'/H(u + it) dt.$$

Our analysis involves cases when H is one of the following three types of functions.

Case 1. $H = Q$ for some rational function Q.

Case 2. $H = e^P$ for some polynomial P.

Case 3. $H = D$ for some regularized product D.

In the remainder of this section, we will evaluate the Weil functional in each of the above three cases.

Case 1: $H = Q$ for some rational function Q. It suffices to consider the function $H(z) = z + \alpha$ for some complex number α.

Theorem 5.1. *Assume g satisfies* **FOU 1** *and* **FOU 2**. *Then*

$$\frac{1}{\sqrt{2\pi}} \int_{-\infty}^{\infty} g^\wedge(t) \frac{1}{t + \alpha} dt = \begin{cases} -i \int_0^\infty g(x) e^{i\alpha x} dx, & \operatorname{Im}(\alpha) > 0, \\ i \int_0^\infty g(-x) e^{-i\alpha x} dx, & \operatorname{Im}(\alpha) < 0. \end{cases}$$

Proof. Lemma 1.1 shows that $g^\wedge(t)/(t + \alpha)$ is in $L^1(\mathbf{R})$, so if

Im(α) > 0, we have

$$\frac{1}{\sqrt{2\pi}} \int_{-\infty}^{\infty} g^\wedge(t) \frac{1}{t+\alpha} dt = \frac{1}{\sqrt{2\pi}} \int_{-\infty}^{\infty} g^\wedge(t) \int_{0}^{\infty} -i e^{i(t+\alpha)x} dx\, dt$$

$$= -i \int_{0}^{\infty} g(x) e^{i\alpha x} dx,$$

by the Fubini Theorem and the Basic Fourier Inversion Formula. The case Im(α) < 0 is treated similarly. □

Case 2: $H = e^P$ for some polynomial P.

In this case, the evaluation of the Weil functional reduces to the Basic Fourier Inversion Formula as the following theorem demonstrates.

Theorem 5.2. *Assume g satisfies* **FOU 1** *and* **FOU 2** *to order M. Let P' be a polynomial of degree $\leq M$, and let $P'(-i\partial)$ be the corresponding constant coefficient partial differential operator. Then*

$$\lim_{T \to \infty} \frac{1}{\sqrt{2\pi}} \int_{-T}^{T} g^\wedge(t) P'(t) dt = P'(-i\partial) g(0).$$

Proof. The Basic Fourier Inversion Formula states

$$\lim_{T \to \infty} \frac{1}{\sqrt{2\pi}} \int_{-T}^{T} g^\wedge(t) P'(t) e^{itx} dt = P'(-i\partial) g(x).$$

The assertion follows after we set $x = 0$. □

Case 3: $H = D$ for some regularized product D.

This case is handled by the results from [JoL 92b], which generalizes Barner's formulation [Ba 81] of Weil's formula [We 52] in the special case H'/H is the logarithmic derivative of the classical gamma function. For completeness, let us recall the main theorem from [JoL 92b].

Let L and A be as in §1, and let $\theta(x)$ be the corresponding theta function
$$\theta(x) = \sum a_k e^{-\lambda_k t}$$
that satisfies the asymptotic axioms **AS 1**, **AS 2** and **AS 3**. For any n let
$$L_n = \{\lambda_{n+1}, \ldots, \},$$
so then we have

(1) $$D'_L/D_L(z) = \sum_{k=0}^{n} \frac{a_k}{z + \lambda_k} + D'_{L_n}/D_{L_n}(z).$$

Since Theorem 5.1 applies to the sum in (1), we may assume, without loss of generality, that L is such that
$$\max_{\lambda_k \in L}\{-\text{Re}(\lambda_k)\} < u.$$

The principal part of the theta function $\theta(x)$ is
$$P_0\theta(x) = \sum_{\text{Re}(p)<0} b_p(x) x^p,$$
which, by **AS 2**, is such that one has the asymptotic behavior
$$\theta(x) - P_0\theta(x) = O(|\log x|^{m(0)}) \quad \text{as } x \to 0.$$

For any $z \in \mathbf{C}$, let
$$\theta_z(x) = \theta(x)e^{-zx}.$$

By expanding e^{-zx} in a power series, we see that the principal part of $\theta_z(x)$ is
$$P_0\theta_z(x) = P_0\left[e^{-zx}\theta(x)\right] = \sum_{\text{Re}(p)+k<0} \frac{b_p(x)x^{p+k}}{k!}(-z)^k.$$

285

Recall from §1 (Theorem 4.1 of [JoL 92a]) that, for any constant $\alpha > 0$, the logarithmic derivative of the regularized product can be written as

$$D'/D(u+it) = I_\alpha(u-\alpha+it) + S_\alpha(u-\alpha+it)$$

where

$$I_w(z) = \int_0^\infty [\theta_z(x) - P_0\theta_z(x)] e^{-wx} dx$$

and

$$S_w(z) = \sum_{\text{Re}(p)+k<0} \frac{(-z)^k}{k!} CT_{s=0} B_p(\partial_s) \left[\frac{\Gamma(s+p+k+1)}{w^{s+p+k+1}} \right].$$

Note that, as a function of z, $S_\alpha(z)$ is a polynomial of degree $\leq M$, in which case Theorem 5.2 applies to show that $S_\alpha(z)$ satisfies **GR** for any test function g that satisfies the basic conditions to order M. Therefore, in order to evaluate

$$\lim_{T\to\infty} \frac{1}{\sqrt{2\pi}} \int_{-T}^T g^\wedge(t) D'/D(u+it) dt,$$

it suffices to evaluate

$$\lim_{T\to\infty} \frac{1}{\sqrt{2\pi}} \int_{-T}^T g^\wedge(t) I_\alpha(u-\alpha+it) dt.$$

If we restrict the complex variable z to a vertical line by setting $z = u - \alpha + it$, we get

$$P_0\theta_z(x) = \sum_{\text{Re}(p)+k<0} \frac{b_p(x) x^{p+k}}{k!} (-u+\alpha-it)^k$$

$$= \sum_{k<-\text{Re}(p_0)} c_k(u-\alpha, x)(-it)^k,$$

where the coefficients $c_k(u - \alpha, x)$ depend on the variables $u - \alpha$ and x through the coefficients of t^p for $\operatorname{Re}(p) < 0$. With this, we have

$$I_\alpha(u - \alpha + it)$$
$$= \int_0^\infty \left[\theta(x) e^{-x(u-\alpha+it)} - \sum_{k < -\operatorname{Re}(p_0)} c_k(u - \alpha, x)(-it)^k \right] e^{-\alpha x} dx.$$

Let us define
$$d\mu_\alpha(x) = e^{-\alpha x} dx \quad \text{and} \quad \theta_{u-\alpha}(x) = \theta(x) e^{-(u-\alpha)x}.$$

Therefore, the above calculations yield the equality

$$I_\alpha(u - \alpha + it)$$
$$= \int_0^\infty \left[\theta_{u-\alpha}(x) e^{-itx} - \sum_{k < -\operatorname{Re}(p_0)} c_k(u - \alpha, x)(-it)^k \right] d\mu_\alpha(x),$$

Applying Theorem 4.3 from [JoL 92b] we obtain:

Theorem 5.3. *Assume g satisfies the Fourier conditions to order M. Then, with notation as above, we have*

$$\lim_{T \to \infty} \frac{1}{\sqrt{2\pi}} \int_{-T}^T g^\wedge(t) I_\alpha(u - \alpha + it) dt$$
$$= \int_0^\infty \left[\theta_{u-\alpha}(x) g(-x) - \sum_{k < -\operatorname{Re}(p_0)} c_k(u - \alpha, x) g^{(k)}(0) \right] d\mu_\alpha(x).$$

In summary, we have shown:

Theorem 5.4. *Assume that Φ is of regularized product type or reduced order M. If g is a test function satisfying the four basic Fourier conditions to order M, then the Weil functional $W_{\Phi,u}(g)$ is defined. Further, Theorems 5.1, 5.2 and 5.3 combine to provide an explicit evaluation of this functional.*

For the sake of space, we will explicitly evaluate the Weil functional only for special functions Φ, namely those involving the classical gamma function.

Example: The gamma function. Many known zeta functions, such as those associated to number fields or modular forms, have Euler sums and functional equations with fudge factors which involve the gamma function. For example, the zeta function corresponding to an ideal class in a number field k has a functional equation with $\sigma_0 = 1$ and

$$G(s) = \tilde{G}(s) = A^{s/2}\Gamma(s/2)^{r_1}\Gamma(s)^{r_2},$$

so

$$G'/G(s) = \frac{1}{2}\log A + \frac{r_1}{2}\Gamma'/\Gamma(s/2) + r_2\Gamma'/\Gamma(s).$$

So, the evaluation of the Weil functional in this case reduces to considering the logarithmic derivative of the gamma function. As recalled on page 52 of [JoL 93a] and §1 of Chapter I, the (lesser known) Gauss formula for the gamma function states that for any $z \in \mathbf{C}$ with $\text{Re}(z) > -1$, we have

$$\Gamma'/\Gamma(z+1) = -\int_0^\infty \left[e^{-zt}\theta(t) - \frac{1}{t}\right]e^{-t}dt$$

where

$$\theta(t) = \sum_{n=0}^\infty e^{-nt} = \frac{1}{1-e^{-t}}.$$

From this, and Theorem 4.3 of [JoL 93b], for any f which satisfies the four basic Fourier conditions and $a > -1$, we have

$$-\lim_{T\to\infty} \frac{1}{\sqrt{2\pi}} \int_{-T}^T f^\wedge(t)\Gamma'/\Gamma(a+it+1)dt$$

$$= \int_0^\infty \left[f(-x)e^{-ax}\theta(x) - \frac{f(0)}{x}\right]e^{-x}dx.$$

This is Weil's formula as reworked by Barner (see [Ba 81], page 146).

CHAPTER IV

From Functional Equations to Theta Inversions

The classical Jacobi theta inversion formula for the Riemann theta function of one variable states that for all $t > 0$, we have the equality

$$\frac{1}{2\pi} \sum_{n=-\infty}^{\infty} e^{-n^2 t} = \frac{1}{\sqrt{4\pi t}} \sum_{n=-\infty}^{\infty} e^{-(2\pi n)^2/4t}.$$

If we set

$$\theta(u) = \sum_{n=-\infty}^{\infty} e^{-n^2 \pi u},$$

then the Jacobi inversion formula can be stated as the equality

$$\theta(u) = \frac{1}{\sqrt{u}} \theta(1/u) \quad \text{for } u > 0.$$

Spectrally, the inversion formula can be viewed as expressing a sum over all the eigenvalues of the Laplace operator on the circle (namely the squares of the integers) as equal to another similar sum, with the inversion $t \mapsto 1/t$.

We give the following very simple spectral interpretation of the Jacobi inversion formula. Let $X = 2\pi \mathbf{Z} \backslash \mathbf{R}$ be the circle. The heat kernel for the usual Laplacian on $\mathbf{R}$ is

$$K_{\mathbf{R}}(x, t, y) = \frac{1}{\sqrt{4\pi t}} e^{-(x-y)^2/4t}.$$

The heat kernel on X is the $2\pi \mathbf{Z}$ periodization of the heat kernel $K_{\mathbf{R}}$ on $\mathbf{R}$. On the other hand, the eigenfunction expansion of the heat kernel K_X can be easily computed. When we equate this periodization with the eigenfunction expansion of the heat kernel, we obtain what amounts to a theta inversion formula, namely

$$\frac{1}{\sqrt{4\pi t}} \sum_{n=-\infty}^{\infty} e^{-(x-y+2\pi n)^2/4t} = \frac{1}{2\pi} \sum_{n=-\infty}^{\infty} e^{-n^2 t} e^{inx} e^{-iny}.$$

In Theorem 1.1 below, we show how the above classical theta inversion formula admits a vast extension to much more general theta functions, essentially formed with the sequence of zeros and poles of functions in the fundamental class as defined in Chapter II, §1. Specifically, inversion formulas follow from our general explicit formula when using Gaussian type test functions. In this context, the Jacobi inversion formula comes from the explicit formulas associated to the sine function

$$\sin(\pi i s) = \frac{e^{-\pi s} - e^{\pi s}}{2i} = -\frac{e^{\pi s}}{2i}\left(1 - e^{-2\pi s}\right)$$

which, when written in this form, can be seen to have an Euler sum and functional equation with $\sigma_0 = 0$ and a simple exponential fudge factor. We will prove our general inversion formulas in §1, and give various examples in §2.

We will show conversely in Chapter V how inversion formulas for theta functions satisfying **AS 1**, **AS 2**, and **AS 3** yield Dirichlet series with an additive functional equation.

§1. An application of the explicit formulas.

We shall apply the general explicit formula of Chapter III, to the test function f_t defined for $t > 0$ by

(1) $\quad f_t(u) = \dfrac{1}{\sqrt{4\pi t}} e^{-(\log u)^2/4t} \quad$ so $\quad F_t(x) = \dfrac{1}{\sqrt{4\pi t}} e^{-x^2/4t}.$

Note that F_t is the heat kernel on $\mathbf{R}$. It is immediate that F_t satisfies the four basic Fourier conditions needed in the proof of the explicit formulas. By a direct calculation, we have

(2) $\quad \mathbf{M}_{\sigma_0/2} f_t(s) = e^{(s-\sigma_0/2)^2 t} \quad$ and $\quad F_t^\wedge(r) = \dfrac{1}{\sqrt{2\pi}} e^{-r^2 t}.$

For example, to derive the first formula in (2), write

$$\mathbf{M}_{\sigma_0/2} f_t(s) = \int_0^\infty f_t(u) u^{s-\sigma_0/2} \frac{du}{u}$$

$$= \frac{1}{\sqrt{4\pi t}} \int_{-\infty}^\infty e^{-x^2/4t + x(s-\sigma_0/2)} dx$$

$$= e^{(s-\sigma_0/2)^2 t}.$$

With this, the explicit formula yields the following result, which we call a **theta inversion formula.**

Theorem 1.1. Let $(Z, \tilde{Z}, \Phi)$ be in the fundamental class, and assume that Φ has no zeros or poles on the line $\mathrm{Re}(s) = \sigma_0/2$. Let $\{\rho\}$ and $\{\alpha\}$ be as in Chapter III, §2. Then for all $t > 0$ we have

$$\sum_\rho v(\rho) e^{(\rho-\sigma_0/2)^2 t} + \sum_\alpha v(\alpha) e^{(\alpha-\sigma_0/2)^2 t}$$

$$= \frac{1}{\sqrt{4\pi t}} \left[\sum_{\mathbf{q}} \frac{-c(\mathbf{q}) \log \mathbf{q}}{\mathbf{q}^{\sigma_0/2}} e^{-(\log \mathbf{q})^2/4t} \right]$$

$$+ \frac{1}{\sqrt{4\pi t}} \left[\sum_{\tilde{\mathbf{q}}} \frac{-c(\tilde{\mathbf{q}}) \log \tilde{\mathbf{q}}}{\tilde{\mathbf{q}}^{\sigma_0/2}} e^{-(\log \tilde{\mathbf{q}})^2/4t} \right]$$

$$+ E_\Phi(t),$$

where
$$E_\Phi(t) = \frac{1}{2\pi} \int_{-\infty}^{\infty} e^{-r^2 t} \Phi'/\Phi(\sigma_0/2 + ir) dr.$$

Remark 1. As in Chapter III, sums over ρ, α and the integral for E_Φ are to be understood as limits of sums and integral from $-T_m$ to T_m, taken over a suitably defined sequence $\{T_m\}$, depending on Z and Φ.

Remark 2. If $\Phi(s) = G(s)/\widetilde{G}(\sigma_0 - s)$, then
$$E_\Phi(r) = E_G(r) + E_{\widetilde{G}}(r)$$
where
$$E_G(r) = \frac{1}{2\pi} \int_{-\infty}^{\infty} e^{-r^2 t} G'/G(\sigma_0/2 + ir) dr.$$

In particular, if $G = \widetilde{G}$, then $E_\Phi(r) = 2E_G(r)$.

Let us establish some notation in order to write the formula in Theorem 1.1 in a form fitting **AS 1**, **AS 2** and **AS 3**. Let $L = \{\mu_k\}$ be the set of numbers
$$L = \{\mu_k\} = \{-(\rho - \sigma_0/2)^2, -(\alpha - \sigma_0/2)^2\},$$
ordered as a sequence, with integral multiplicities
$$\{a(\mu_k)\} = \{v(\rho), v(\alpha)\}.$$
Define the associated theta function
$$(3) \quad \theta_L(t) = \sum_\rho v(\rho) e^{(\rho - \sigma_0/2)^2 t} + \sum_\alpha v(\alpha) e^{(\alpha - \sigma_0/2)^2 t}.$$

Similarly, let $L^\vee$ be the family of numbers
$$L^\vee = \{\mu_k^\vee\} = \{(\log \mathbf{q})^2/4, (\log \widetilde{\mathbf{q}})^2/4\}$$
with (not necessarily integral) "multiplicities"
$$\{a(\mu_k^\vee)\} = \left\{ \frac{-c(\mathbf{q})\log \mathbf{q}}{\mathbf{q}^{\sigma_0/2}}, \frac{-c(\widetilde{\mathbf{q}})\log \widetilde{\mathbf{q}}}{\widetilde{\mathbf{q}}^{\sigma_0/2}} \right\},$$

and define the associated theta function

(4)
$$\theta_{L^v}(t) = \sum_{\mathbf{q}} \frac{-c(\mathbf{q})\log \mathbf{q}}{\mathbf{q}^{\sigma_0/2}} e^{-[(\log \mathbf{q})^2/4]t}$$
$$+ \sum_{\tilde{\mathbf{q}}} \frac{-c(\tilde{\mathbf{q}})\log \tilde{\mathbf{q}}}{\tilde{\mathbf{q}}^{\sigma_0/2}} e^{-[(\log \tilde{\mathbf{q}})^2/4]t}.$$

With this notation, we can now write Theorem 1.1 in the following form.

Theorem 1.2. *With notation as above, we have the inversion formula*
$$\theta_L(t) = \frac{1}{\sqrt{4\pi t}} \theta_{L^v}(1/t) + E_\Phi(t).$$

In general, we define a theta inversion formula to be a relation between two theta functions, such as (3) and (4), with an additional term such as $E_\Phi(t)$ which we require to satisfy the asymptotic condition **AS 2**.

Theorem 1.3. *With notation as above, the theta function*
$$\sum_\rho v(\rho) e^{(\rho - \sigma_0/2)^2 t}$$
satisfies the asymptotic conditions **AS 1**, **AS 2**, *and* **AS 3**.

Proof. By the Cramér theorem, we have that the sequences
$$\{\rho/i : Z(\rho) = 0 \text{ with } \operatorname{Im}(\rho) > 0 \text{ and } -a < \operatorname{Re}(\rho) < \sigma_0 + a\}$$
and
$$\{\rho/i : Z(\rho) = 0 \text{ with } \operatorname{Im}(\rho) < 0 \text{ and } -a < \operatorname{Re}(\rho) < \sigma_0 + a\}$$
are such that the associated theta functions satisfy the three basic asymptotic conditions. Therefore, by applying Theorem 7.6 and Theorem 7.7 of [JoL 93a], we conclude that the theta function
$$\sum_\rho v(\rho) e^{(\rho - \sigma_0/2)^2 t}$$

satisfies the asymptotic conditions. □

The theta series taken over $\{\alpha\}$ is incomplete, and it will be more useful in this chapter to deal with the alternate version of the explicit formula given as Theorem 2.3 in §2 of Chapter III.

We let R be of regularized harmonic series type. For each real number u such that R has no pole on the line $\text{Re}(z) = u$, we have the E_u-**transform**

$$E_u R(t) = \lim_{n \to \infty} \frac{1}{2\pi i} \int_{u-iT_n}^{u+iT_n} e^{(z-\sigma_0/2)^2 t} R(z) dz$$

where $\{T_n\}$ is a sequence selected as in Theorem 6.2 of Chapter I.

Directly from Theorem 2.3 of Chapter III, we have the following result.

Theorem 1.4. *Let* $(Z, \tilde{Z}, \Phi)$ *be in the fundamental class. Decompose* Φ *as a product*

$$\Phi = \Phi_{\text{left}} \Phi_{\text{right}}$$

such that Φ_{left} *has all its zeros and poles in a left half plane and* Φ_{right} *has all its zeros and poles in a right half plane. Select* $A > \sigma_0$, $\delta > 0$, *and a such that:*

Φ_{left} *has all its zeros and poles in* $\text{Re}(z) \leq A - \delta$;
Φ_{right} *has all its zeros and poles in* $\text{Re}(z) \geq -a + \delta$.

Let

$$\{\zeta\} = \text{set of zeros and poles of } \Phi_{\text{left}} \text{ with } \text{Re}(\zeta) > -a.$$

Then for all $t > 0$ *we have*

$$\sum_\rho v(\rho) e^{(\rho - \sigma_0/2)^2 t} + \sum_\zeta v(\zeta) e^{(\zeta - \sigma_0/2)^2 t}$$

$$= \frac{1}{\sqrt{4\pi t}} \left[\sum_{\mathbf{q}} \frac{-c(\mathbf{q}) \log \mathbf{q}}{\mathbf{q}^{\sigma_0/2}} e^{-(\log \mathbf{q})^2/4t} \right]$$

$$+ \frac{1}{\sqrt{4\pi t}} \left[\sum_{\tilde{\mathbf{q}}} \frac{-c(\tilde{\mathbf{q}}) \log \tilde{\mathbf{q}}}{\tilde{\mathbf{q}}^{\sigma_0/2}} e^{-(\log \tilde{\mathbf{q}})^2/4t} \right]$$

$$+ E_A(\Phi'_{\text{left}}/\Phi_{\text{left}})(t) + E_{-a}(\Phi'_{\text{right}}/\Phi_{\text{right}})(t).$$

We note that in Theorem 1.4 both theta series satisfy the asymptotic conditions **AS 1**, **AS 2**, and **AS 3**. This assertion is true for the sum over ζ since we assumed Φ was of regularized product type, because we choose $-a$ sufficiently far to the left, and so we can apply Theorem 7.6 and Theorem 7.7 of [JoL 93a]. As for the sum over ρ, one applies Cramér's theorem and Theorem 7.6 and Theorem. 7.7 of [JoL 93a].

Remark 3. In the present application, and the subsequent ones in this chapter and the next, we apply the explicit formula to the simplest types of Gaussian test functions. Even for such test functions, one can give examples where instead of $(4\pi t)^{1/2}$ or $(4\pi t)^{3/2}$ (as in Chapter V, §4) one takes $(4\pi t)^{n/2}$ for an odd integer n, and in addition one also has an arbitrary polynomial as a coefficient. If one uses $(4\pi t)^{n/2}$ with an even integer n, then one gets Bessel series instead of Dirichlet series. All these cases deserve to be treated systematically, since they apply to several important situations of spectral analysis for classical manifolds, and we shall do so elsewhere. Here we selected the simplest cases to serve as examples.

§2. Some examples of theta inversions

To emphasize the significance of the theta inversion formula given in Theorem 1.1, let us now discuss a few specific applications of the theorem.

Example 1: The sine function. Let us write the sine function as

$$\sin(\pi i s) = \frac{e^{-\pi s} - e^{\pi s}}{2i} = -\frac{e^{\pi s}}{2i}\left(1 - e^{-2\pi s}\right) = G(s)Z(s)$$

where $G(s) = -e^{\pi s}/2i$ and $Z(s) = 1 - e^{-2\pi s}$. The fact that the sine function is odd trivially yields the functional equation

$$G(s)Z(s) = -G(-s)Z(-s),$$

so $\sigma_0 = 0$. Further, Z satisfies the Euler sum condition since

$$\log Z(s) = \log\left(1 - e^{-2\pi s}\right) = \sum_{n=1}^{\infty} \frac{1}{n} e^{-2\pi n s},$$

whence $\{q\} = \{e^{2\pi n}\}$ for $n \geq 1$ and $c(e^{2\pi n}) = 1/n$. Since $\sin(\pi i s)$ is zero only when $s \in \mathbf{Z}$, and then with multiplicity one, the inversion formula Theorem 1.1 specializes to the equality

(1) $$\sum_{n=-\infty}^{\infty} e^{-n^2 t} = 2 \cdot \frac{1}{\sqrt{4\pi t}} \sum_{n=1}^{\infty} 2\pi e^{-(2\pi n)^2/4t} + E_\Phi(t),$$

where the factor of 2 appears since the sums over $\{q\}$ and over $\{\tilde{q}\}$ coincide in this example. Also, we have

(2) $$E_\Phi(t) = 2E_G(t) = \frac{1}{\pi} \int_{-\infty}^{\infty} e^{-r^2 t} \pi dr = \frac{\sqrt{\pi}}{\sqrt{t}}.$$

Combining the terms in (1) and (2), we obtain the classical Jacobi inversion formula, which is the relation

$$\frac{1}{2\pi} \sum_{n=-\infty}^{\infty} e^{-n^2 t} = \frac{1}{\sqrt{4\pi t}} \sum_{n=-\infty}^{\infty} e^{-(2\pi n)^2/4t} \quad \text{for } t > 0.$$

Similarly, one can derive other classical formulas using the cosine function and hyperbolic trigonometric functions.

Example 2: Dirichlet polynomials. Recall the definition of a Dirichlet polynomial given in §7 of Chapter I. We saw in Chapter II that any Dirichlet polynomial P can be written as

$$P(s) = a_N b_N^s \left[1 + \sum_{n=1}^{N-1} \frac{a_n}{a_N} \left(\frac{b_n}{b_N} \right)^s \right] = a_N b_N^s \cdot Z(s),$$

where Z has an Euler sum and functional equation with fudge factor

$$\Phi(s) = \frac{a_N}{a_1} \left(\frac{b_N}{b_1} \right)^s.$$

Hence, one can directly evaluate the Weil functional, yielding the formula

$$E_\Phi(t) = \frac{1}{2\pi} \int_{-\infty}^{\infty} e^{-r^2 t} (\log(b_N/b_1)) dr = \frac{\log(b_N/b_1)}{\sqrt{4\pi t}}.$$

Therefore, the associated theta inversion formula is simply

$$\sum_\rho v(\rho) e^{\rho^2 t} = \frac{\log(b_N/b_1)}{\sqrt{4\pi t}} - \frac{1}{\sqrt{4\pi t}} \sum_{\mathbf{q}} c(\mathbf{q})(\log \mathbf{q}) e^{-(\log \mathbf{q})^2/4t}$$
$$- \frac{1}{\sqrt{4\pi t}} \sum_{\tilde{\mathbf{q}}} c(\tilde{\mathbf{q}})(\log \tilde{\mathbf{q}}) e^{-(\log \tilde{\mathbf{q}})^2/4t}.$$

The specific case of $N = 2$ with

$$b_2 = b_1^{-1} = e^{\pi s} \quad \text{and} \quad a_2 = -a_1 = i/2$$

yields the Jacobi inversion formula.

We find this example particularly interesting for the following reason. In Example 1, the Jacobi inversion formula is a relation involving two well-known sets of data, meaning that both theta functions in the inversion formula involves the squares of integers. However, in the case of a Dirichlet polynomial, we have one well-known set of numbers, namely the sets $\{\mathbf{q}\}$ and $\{\tilde{\mathbf{q}}\}$ which are explicitly and simply expressed in terms of the initial set of numbers $\{a_j, b_j\}$, and one unknown set of numbers, namely the set of zeros.

Example 3: Zeta functions of number fields. In §6 of Chapter III we gave an evaluation of the Weil functional associated to the classical gamma function. Hence, there is a theta inversion formula associated to any zeta function of a number field as in Theorem 1.1. The term E_Φ or E_G is given explicitly as follows. If

$$G(s) = A^{s/2} \Gamma(s)^{r_2} \Gamma(s/2)^{r_1},$$

then, by page 146 of [Ba 81], see also §4 of [JoL 93b], we find

$$E_G(t) = \frac{\log A^{1/2}}{2\pi} \int_{-\infty}^{\infty} e^{-r^2 t} dr + \frac{r_2}{2\pi} \int_{-\infty}^{\infty} \Gamma'/\Gamma(1/2 + it) e^{-r^2 t} dr$$

$$+ \frac{r_1/2}{2\pi} \int_{-\infty}^{\infty} \Gamma'/\Gamma(1/4 + it/2) e^{-r^2 t} dr$$

$$= \frac{\log A^{1/2}}{\sqrt{4\pi t}} + \frac{r_2}{\sqrt{4\pi t}} \int_{0}^{\infty} \left[\frac{1}{x} - e^{-x^2/4t} e^{x/2} \theta(x)\right] e^{-x} dx$$

$$+ \frac{r_1/2}{\sqrt{4\pi t}} \int_{0}^{\infty} \left[\frac{1}{x} - 2e^{-x^2/4t} e^{3x/2} \theta(2x)\right] e^{-2x} dx$$

where

$$\theta(x) = \sum_{n=0}^{\infty} e^{-nx} = \frac{1}{1 - e^{-x}}.$$

It is important to note that to zeta functions, we are associating theta functions which admit inversion formulas but are different from the theta functions used in Hecke's proof of the functional equation and meromorphic continuation (see, for example, Chapter XIII of [La 70]).

Example 4: A connection with regularized products. In spectral theory, one meets the situation when a certain operator has the sequence of eigenvalues $L = \{\lambda_k\}$ with integer multiplicities $\{a_k\}$. In such a situation, when the theory of regularized products applies (namely our axioms from [JoL 93a], recalled in Chapter I, §1), one may have the additional relation

$$D_L(s(s - \sigma_0)) = Z(s)G(s).$$

This occurs for instance for the Selberg zeta function and its analogues. Then

$$(\rho_k - \sigma_0/2)^2 = -\lambda_k + \sigma_0^2/4.$$

Thus we have the simple relation

$$\sum_k v(\rho_k) e^{(\rho_k - \sigma_0/2)^2 t} = e^{(\sigma_0^2/4)t} \sum_k a_k e^{-\lambda_k t} = e^{(\sigma_0^2/4)t} \theta_L(t),$$

expressing the theta series formed with the squares $(\rho_k - \sigma_0/2)^2$ in terms of the theta series with the eigenvalues.

As stated above, in the case of the operator d^2/dx^2 on the circle $S^1 = \mathbf{R}/2\pi\mathbf{Z}$, this relation reduces to the Jacobi theta series formed with $\lambda_k = k^2$ and $\sigma_0 = 0$. In both cases we are dealing with the eigenvalues of the positive Laplacian on some Riemannian manifold.

CHAPTER V

From Theta Inversions to Functional Equations

In this chapter we shall carry out the inverse construction of the preceeding sections; that is, from a theta series satisfying an inversion formula, we derive, by means of a Gauss transform, a Dirichlet series satisfying an additive functional equation.

Of course, Riemann's proof of the functional equation of the Riemann zeta function also relied on a theta inversion formula, but in the present situation, our use of theta inversion is different from that of Riemann because we take a Laplace transform, with a quadratic change of variables, of the regularized theta series instead of the Mellin transform of the theta series. Hence we construct new types of zeta functions which are essentially regularized harmonic series. In case the residues are integers, such series are logarithmic derivatives of functions in the fundamental class. Thus, we see that the general theory requires that we consider the additive class rather than the more restrictive multiplicative class.

Let R be of regularized harmonic series type. In §1, we analyze the transform E_u defined by

$$E_u R(t) = \lim_{n \to \infty} \frac{1}{2\pi i} \int_{u-iT_n}^{u+iT_n} e^{(z-\sigma_0/2)^2 t} R(z) dz.$$

In §2 we carry out the properties of the Gauss transform of theta series, inverted theta series, and the transform $E_u R$. These properties are shown to imply the functional equation of the new zeta function in §3. In §4 we work out an example to obtain a zeta function for any compact quotient of the three dimensional hyperbolic space $\mathbf{M_3}$. This zeta function lies in our additive class but, in general, not in our multiplicative class. For extensions of this example, see the remarks at the end of §4.

Throughout the remainder of this chapter, unless otherwise specified, we use the following basic conditions.

The basic conditions

We let σ_0 be a real number ≥ 0.

We let R be a function of regularized harmonic series type.

We let $\{\mu_k\}$ and $\{a_k\}$ satisfy **DIR 1**, **DIR 2**, and **DIR 3**, and we assume that the corresponding theta series

$$\theta(t) = \theta_L(t) = \sum_{k=0}^{\infty} a_k e^{-\mu_k t}$$

satisfies **AS 1**, **AS 2**, and **AS 3**. We let (M_θ, m) be the reduced order of the theta series θ. Note that we have used μ_k instead of λ_k in the notation. This is because in a subsequent application to spectral theory, there will be an operator with eigenvalues λ_k which are translates of μ_k, namely

$$\mu_k = \lambda_k - \sigma_0^2/4.$$

We let R_θ be the regularized harmonic series associated to the theta function θ, as defined in Chapter I, §1; that is,

$$R_\theta(z) = \mathrm{CT}_{s=1} \mathbf{LM}\theta(s, z).$$

We let $\{\mathbf{q}\}$ be a sequence of real numbers > 1 converging to infinity. We let $\{c(\mathbf{q})\}$ be a sequence of complex numbers, and we let

$$\text{Л}(s) = \sum_\mathbf{q} \frac{c(\mathbf{q})}{\mathbf{q}^s}$$

be the associated Dirichlet series, which we assume converges absolutely in some right half plane. Thus,

$$\text{Л}'(s) = \sum_\mathbf{q} \frac{-c(\mathbf{q}) \log \mathbf{q}}{\mathbf{q}^s}.$$

We let

$$\theta_{L^\vee}(t) = \sum_\mathbf{q} \frac{-c(\mathbf{q}) \log \mathbf{q}}{\mathbf{q}^{\sigma_0/2}} e^{-[(\log \mathbf{q})^2/4]t}.$$

§1. The Weil functional of a Gaussian test function.

In this section, we are concerned with a function R of regularized harmonic series type. For each $\zeta \in \mathbf{C}$ we let:
$$a_R(\zeta) = a(\zeta) = \text{ residue of } R \text{ at } \zeta.$$
Observe that if $R = R_\Phi = \Phi'/\Phi$, where Φ is of regularized product type, then $a(\zeta) = v(\zeta)$ is the order of Φ at ζ.

Any function R of regularized series type can be expressed as a sum of two such functions, each of which has poles only in a half plane to the left or a half plane to the right. Let us write such a decomposition as
$$R(z) = R_{\text{left}}(z) + R_{\text{right}}(z).$$
Assume that $A > 0$ is chosen so that all the poles of R_{left} lie in a left half plane of the form $\text{Re}(z) \leq A - \delta$, for some $\delta > 0$. Similarly, assume that $a > 0$ is chosen so that all the poles of R_{right} lie in a right half plane of the form $\text{Re}(z) \geq -a + \delta$, for some $\delta > 0$. Assume that R_{left} has reduced order $(M_{\text{left}}, m_{\text{left}})$ and R_{right} has reduced order $(M_{\text{right}}, m_{\text{right}})$

As a direct application of Theorem 6.2(a) of Chapter I, we obtain the following result.

Lemma 1.1. *With notation as above, we have*
$$E_A R_{\text{left}}(t) = O(t^{-(M_{\text{left}}+2)/2}) \quad \text{for } t \to 0$$
and
$$E_{-a} R_{\text{right}}(t) = O(t^{-(M_{\text{right}}+2)/2}) \quad \text{for } t \to 0.$$

Proof. By Theorem 6.2(a) of Chapter I, we know that on the line $\text{Re}(z) = A$ the function R_{left} has polynomial growth. In fact, the integral giving $E_A R_{\text{left}}$ can be coarsely estimated by
$$O\left(\int_{-\infty}^{\infty} e^{-btu^2} |u|^{M_{\text{left}}+2} \frac{du}{u} \right)$$
with some number $b > 0$. The first estimate asserted then follows from the standard change of variables $y = \sqrt{t}u$. The second estimate is proved similarly. □

The proof of Lemma 1.1 also applies to prove the following asymptotic bounds.

Lemma 1.2. *With notation as above, we have*

$$E_A R_{\text{left}}(t) = O(t^{-(M_{\text{left}}+2)/2}) \quad \text{for } t \to \infty$$

and

$$E_{-a} R_{\text{right}}(t) = O(t^{-(M_{\text{right}}+2)/2}) \quad \text{for } t \to \infty.$$

Note that the bound given in Theorem 6.2(a) of Chapter I is stronger than what is used in the proof of Lemma 1.1 and Lemma 1.2 above. However, the above results are sufficient for our purposes. In fact, one can easily improve the exponent to $-(M+1)/2 + \epsilon$ for any $\epsilon > 0$.

The bounds in Lemma 1.1 and Lemma 1.2 are enough to allow us to deal with the Gauss transforms of $E_A R_{\text{left}}$ and $E_{-a} R_{\text{right}}$ in the next section.

§2. Gauss transforms.

We shall need some analysis concerning Bessel integrals and what we call the Gauss transform, which we carry out in this section. Throughout we let r, t, u, x be real variables with $t, u, x > 0$. We start by recalling the collapse of the Bessel integral $K_s(x, u)$ under certain conditions.

Lemma 2.1. *Let*

$$K_s(x, u) = \int_0^\infty e^{-x^2 t} e^{-u^2/t} t^s \frac{dt}{t}.$$

Then:

a) In the case $s = 1/2$, we have the evaluation

$$K_{1/2}(x, u/2) = \frac{\sqrt{\pi}}{x} e^{-ux}.$$

b) Let $Д_x$ and $Д_u$ be the differential operators

$$Д_x = -\frac{1}{2x}\frac{\partial}{\partial x} \quad \text{and} \quad Д_u = -\frac{1}{2u}\frac{\partial}{\partial u}.$$

Then we have the relations

$$(Д_x)^n K_s(x, u) = K_{s+n}(x, u)$$

and

$$(Д_u)^n K_s(x, u) = K_{s-n}(x, u).$$

For further properties of the Bessel integral, including a proof of Lemma 2.1, the reader is referred to [La 87].

Warning. For our purposes, we use a normalization of the Bessel integral slightly different from the classical one. If, for instance, $K_s^B(c)$ denotes the K-Bessel function which one finds in tables (see Magnus, Oberhettinger, Erdelyi, Whittaker and Watson, etc.), then we have the relation

$$2K_s^B(2c) = K_s(c)$$

where

$$K_s(c) = \int_0^\infty e^{-c(t+1/t)} t^s \frac{dt}{t}.$$

The classical normalization gets rid of some factors in the differential equation, and our normalization gets rid of extraneous factors in the above integral.

For a suitable function f of a single real variable, we recall the Laplace-Mellin transform is defined as

$$\mathbf{LM} f(s,z) = \int_0^\infty f(t) e^{-zt} t^s \frac{dt}{t}.$$

We shall deal with $f = \theta$, where θ is a theta series of the sort considered previously. We shall put $s = N + 1$, where N is an integer ≥ 0, sufficiently large to cancel the singularity at 0, and we shall also make a change of variables with $z = (s - \sigma_0/2)^2$. With this, we obtain an integral operator which we call the **Gauss transform**.

More precisely, let N be a positive integer and let σ_0 be a real number ≥ 0. For a suitable test function f, we define

$$\mathrm{Gauss}^{(N)}_{\sigma_0/2}(f)(s) = (2s - \sigma_0) \int_0^\infty f(t) e^{-(s-\sigma_0/2)^2 t} t^{N+1} \frac{dt}{t}.$$

Then

$$\mathrm{Gauss}^{(N)}_{\sigma_0/2}(f)(s) = \frac{dz}{ds} \cdot \mathbf{LM} f(N+1, (s-\sigma_0/2)^2).$$

Because of the change of variables from z to s, we are led to consider the differential operator

$$\mathcal{D}_{s-\sigma_0/2} = -\frac{d}{ds} \circ \frac{1}{2s - \sigma_0}.$$

Let $x = s - \sigma_0/2$. For every integer $N \geq 0$, we then get the formula

$$2x \cdot \Д_x^N \circ \frac{1}{2x} = \mathcal{D}^N_{s-\sigma_0/2}.$$

Gauss transform of an inverted theta series

We now shall prove inversion formulas for Gauss transforms of certain series. Such series occurred in the previous section in connection with functions in the fundamental class. Here we start with series for which we assume only the basic conditions.

Directly from Lemma 2.1, we have the following theorem.

Theorem 2.2. *Let*

$$G(t) = \frac{1}{\sqrt{4\pi t}} \sum_{\mathbf{q}} \frac{-c(\mathbf{q})\log \mathbf{q}}{\mathbf{q}^{\sigma_0/2}} e^{-(\log \mathbf{q})^2/4t}.$$

Then for s real and sufficiently large, and $N \geq 0$, we have

$$\text{Gauss}_{\sigma_0/2}^{(N)}(G)(s) = \mathcal{D}_{s-\sigma_0/2}^{N} \Pi'(s).$$

Proof. The proof follows immediately from the definitions and Lemma 2.1. For each $\mathbf{q}$, we put $x = s - \sigma_0/2$ and $u = \log \mathbf{q}$ and compute the transform of each individual term

$$g_{\mathbf{q}}(t) = \frac{1}{\sqrt{4\pi t}} e^{-u^2/4t}$$

as follows:

$$\begin{aligned}
\text{Gauss}_{\sigma_0/2}^{(N)}(g_{\mathbf{q}})(s) &= \frac{2x}{\sqrt{4\pi}} K_{N+1/2}(x, u/2) \\
&= \frac{2x}{2\sqrt{\pi}} \mathcal{D}_x^N K_{1/2}(x, u/2) \\
&= \frac{2x}{2\sqrt{\pi}} \mathcal{D}_x^N \left(\frac{\sqrt{\pi}}{x} e^{-ux} \right) \\
&= 2x \mathcal{D}_x^N \left(\frac{1}{2x} e^{-ux} \right) \text{‘} \\
&= \mathcal{D}_x^N e^{-(\log \mathbf{q})x} \\
&= \mathcal{D}_{s-\sigma_0/2}^N \left(\mathbf{q}^{-s} \mathbf{q}^{\sigma_0/2} \right),
\end{aligned}$$

from which the formula in Theorem 2.2 is now clear. $\square$

Gauss transform of a theta series

Next, let us apply the Gauss transform to a theta series itself; that is, let us evaluate

$$\text{Gauss}_{\sigma_0/2}^{(N)}(\theta)(s) = (2s - \sigma_0) \int_0^\infty \theta(t) e^{-(s-\sigma_0/2)^2 t} t^{N+1} \frac{dt}{t}$$

where

$$\theta(t) = \sum_{k=0}^\infty a_k e^{-\mu_k t}.$$

Let ρ_k be complex numbers, all but a finite number of which have positive imaginary part, and such that

$$(\rho_k - \sigma_0/2)^2 = -\mu_k.$$

Theorem 2.3. *For any positive integer N sufficiently large, and s real and sufficiently large, we have*

$$\int_0^\infty \theta(t) e^{-(s-\sigma_0/2)^2 t} t^{N+1} \frac{dt}{t} = \sum_{k=0}^\infty a_k \frac{\Gamma(N+1)}{[(s-\rho_k)(s+\rho_k-\sigma_0)]^{N+1}}$$

$$= \sum_{k=0}^\infty a_k \frac{\Gamma(N+1)}{[(s-\sigma_0/2)^2 - (\rho_k - \sigma_0/2)^2]^{N+1}},$$

hence the series gives the meromorphic continuation of the integral to all $s \in \mathbf{C}$ with poles at the points $s = \rho_k$ and $s = \sigma_0 - \rho_k$.

Proof. This identity follows routinely by integrating term by term, using the change of variables $t \mapsto t/(s - \sigma_0/2)^{1/2}$. There is no problem with the square root since s was assume to be real and sufficiently large. □

Next we obtain the inversion analogous to Theorem 2.2, but for the theta series instead of the **q** series.

Theorem 2.4. *For any positive integer N sufficiently large, and s real and sufficiently large, we have*

$$\text{Gauss}^{(N)}_{\sigma_0/2}(\theta)(s) = \mathcal{D}^{(N)}_{s-\sigma_0/2}\left[(2s-\sigma_0)R_\theta((s-\sigma_0/2)^2)\right].$$

Proof. As a function of s, $(2s - \sigma_0)R_\theta((s - \sigma_0/2)^2)$ is a meromorphic function whose singularities are simple poles at the points $s = \rho_k$ and $s = \sigma_0 - \rho_k$ with residue a_k. For any individual k, we have

$$\left(-\frac{\partial}{\partial z}\right)^N \left(\frac{1}{z+\mu_k}\right) = \frac{\Gamma(N+1)}{(z+\mu_k)^{N+1}}.$$

So,

$$\left(\frac{-1}{2s-\sigma_0}\frac{\partial}{\partial s}\right)^N \left[\frac{1}{(s-\sigma_0/2)^2 - (\rho_k-\sigma_0/2)^2}\right]$$
$$= \frac{\Gamma(N+1)}{[(s-\sigma_0/2)^2 - (\rho_k-\sigma_0/2)^2]^{N+1}},$$

which can be written as

$$= \frac{1}{2s-\sigma_0}\mathcal{D}^N_{s-\sigma_0/2}\left[\frac{2s-\sigma_0}{(s-\sigma_0/2)^2 - (\rho_k-\sigma_0/2)^2}\right].$$

The result now follows from Theorem 2.3 after multiplying by the term $2s - \sigma_0$ and then summing over k, which is valid since for N sufficiently large the series is absolutely convergent. □

Remark 1. From Theorem 7.6 and Theorem 7.7 of [JoL 93a], one can show that it suffices to take $N > (M_\theta + 2)/2$, where θ has reduced order (M_θ, m). In fact, this requirement can be easily improved to $N > (M_\theta + 1)/2 + \epsilon$ for any $\epsilon > 0$.

Remark 2. In the degenerate case when $\rho_k = \sigma_0/2$, the regularized harmonic series with the quadratic change of variables has a double pole at $s = \sigma_0/2$ with zero residue, which is consistent with the above evaluation of residues.

Inverse Gauss transform of a regularized harmonic series type

Finally, we deal with the Gauss transform applied to the terms $E_A R_{\text{left}}$ $E_{-a} R_{\text{right}}$ which arose in Chapter IV. The next lemma evaluates this Gauss transform, and amounts to an inversion formula, showing that up to taking derivatives, the Gauss transform is the inverse of the E_A transform.

Theorem 2.5. *With notation as above, we have, for N sufficiently large and s real and sufficiently large, the equalities*

$$\text{Gauss}^{(N)}_{\sigma_0/2}(E_A R_{\text{left}})(s) = \mathcal{D}^N_{s-\sigma_0/2} R_{\text{left}}(s)$$

and

$$\text{Gauss}^{(N)}_{\sigma_0/2}(E_{-a} R_{\text{right}})(s) = \mathcal{D}^N_{s-\sigma_0/2} R_{\text{right}}(\sigma_0 - s).$$

Proof. Let

$$F(s) = \int_0^\infty E_A R_{\text{left}}(t) e^{-(s-\sigma_0/2)^2 t} t^{N+1} \frac{dt}{t}$$

which is defined for N sufficiently large by Lemma 1.1, and s real and sufficiently large by Lemma 1.2. Then

$$(1) \quad F(s) = \frac{1}{2\pi i} \int_0^\infty \int_{A-i\infty}^{A+i\infty} e^{(z-\sigma_0/2)^2 t} e^{-(s-\sigma_0/2)^2 t} R_{\text{left}}(z) t^{N+1} dz \frac{dt}{t}.$$

By changing the order of integration, which is valid by Theorem 6.2(a) of Chapter I, we may write (1) in the form

$$(2) \quad F(s) = \frac{1}{2\pi i} \int_{A-i\infty}^{A+i\infty} \frac{\Gamma(N+1)}{[(s-\sigma_0/2)^2 - (z-\sigma_0/2)^2]^{N+1}} R_{\text{left}}(z) dz.$$

Let us write

$$\frac{\Gamma(N+1)}{[(s-\sigma_0/2)^2 - (z-\sigma_0/2)^2]^{N+1}}$$
$$= \mathcal{D}^N_{s-\sigma_0/2} \left[\frac{1}{(s-\sigma_0/2)^2 - (z-\sigma_0/2)^2} \right],$$

which allows us to express (2) as

$$F(s)$$
(3)
$$= \frac{1}{2\pi i} \int_{A-i\infty}^{A+i\infty} Д^N_{s-\sigma_0/2} \left[\frac{1}{(s-\sigma_0/2)^2 - (z-\sigma_0/2)^2} \right] R_{\text{left}}(z) dz.$$

Because of the symmetry in s and z, we can write (3), as

$$F(s)$$
$$= \frac{1}{2\pi i} \int_{A-i\infty}^{A+i\infty} (Д_{z-\sigma_0/2})^N \left[\frac{(-1)^N}{(s-\sigma_0/2)^2 - (z-\sigma_0/2)^2} \right] R_{\text{left}}(z) dz$$
$$= \frac{1}{2\pi i} \int_{A-i\infty}^{A+i\infty} (-Д_{z-\sigma_0/2})^N \left[\frac{1}{(s-z)(s+z-\sigma_0)} \right] R_{\text{left}}(z) dz.$$

Integration by parts shows that $-Д_{z-\sigma_0/2}$ and $\mathcal{D}_{z-\sigma_0/2}$ are adjoint operators. Hence, we obtain the formula

$$\text{Gauss}^{(N)}_{\sigma_0/2}(E_A R_{\text{left}})(s) = (2s - \sigma_0) F(s)$$
$$= \frac{1}{2\pi i} \int_{A-i\infty}^{A+i\infty} \left[\frac{1}{s-z} + \frac{1}{s+z-\sigma_0} \right] (\mathcal{D}_{z-\sigma_0/2})^N R_{\text{left}}(z) dz.$$

From the *proof* of Theorem 6.2(a) of Chapter I, one obtains that on the line $\text{Re}(z) = A$ we have

(4) $\quad (\mathcal{D}_{z-\sigma_0/2})^k R_{\text{left}}(A + it) = O(|t|^{M+\epsilon-2k}) \quad$ for $t \to \infty$,

for any $\epsilon > 0$ and positive integer k. So, for $N > (M+1)/2 + \epsilon$, the integration by parts is valid.

Now we can evaluate $F(s)$ using a contour involving a half-circle, opening to the right with diameter along the vertical line $\text{Re}(s) = A$. In this region, $R_{\text{left}}(z)$ is holomorphic, so the only pole of the integrand occurs when $z = s$, and that with residue 1, after correcting the orientation of the contour. The integral along the half-circle will approach zero as the radius approaches infinity since, again by the *proof* of Theorem 6.2(a) of Chapter I, since the function $(\mathcal{D}_{z-\sigma_0/2})^k R_{\text{left}}(z)$ has polynomial growth as in (4). With all this, the first part of Theorem 2.5 is proved.

Concerning the second part of Theorem 2.5, one applies the same argument as above, except using a contour integral opening to the left, together with the identity $\mathcal{D}_z = \mathcal{D}_{-z}$. □

Remark 2. Lemma 1.1 and the estimate in (4) show that it suffices to take $N > (M+1)/2 + \epsilon$ where R has reduced order (M, m).

§3. Theta inversions yield zeta functions

We now put the results of §2 together to show how theta inversion gives rise to an additive functional equation.

Given R of regularized harmonic series type, there exist a and A sufficiently large positive so that we can decompose R into a sum

$$R = R_{\text{left}} + R_{\text{right}}$$

where R_{left} has it poles in a left half plane $\text{Re}(z) \leq A - \delta$, and where R_{right} has it poles in a right half plane $\text{Re}(z) \geq -a + \delta$.

Theorem 3.1. *Assume that the theta series satisfies the inversion formula*

$$\theta_L(t) = \frac{1}{\sqrt{4\pi t}} \theta_{L^\vee}(1/t) + E_A R_{\text{left}}(t) + E_{-a} R_{\text{right}}(t).$$

For N sufficiently large, we have

$$\mathcal{D}^N_{s-\sigma_0/2}\left[(2s-\sigma_0)R_{\theta_L}((s-\sigma_0/2)^2)\right]$$
$$= \mathcal{D}^N_{s-\sigma_0/2}\left[\text{JI}'(s) + R_{\text{left}}(s) + R_{\text{right}}(\sigma_0 - s)\right].$$

Proof. We simply apply Theorem 2.2, Theorem 2.3, and Theorem 2.5. □

Remark 1. It suffices to take

$$N > \frac{1}{2}(\max\{M_{R_{\text{left}}}, M_{R_{\text{right}}}, M_\theta\} + 1) + \epsilon,$$

for any $\epsilon > 0$.

Theorem 3.2. *There is a polynomial P' of degree $\leq 2N-1$ such that for all real s sufficiently large, we have*

$$(2s - \sigma_0)R_{\theta_L}((s - \sigma_0/2)^2)$$
$$= \text{Л}'(s) + R_{\text{left}}(s) + R_{\text{right}}(\sigma_0 - s) + P'(s).$$

Proof. Let

$$F(s) = (2s - \sigma_0)R_{\theta_L}((s - \sigma_0/2)^2) - \text{Л}'(s)$$
$$- R_{\text{left}}(s) - R_{\text{right}}(\sigma_0 - s),$$

and let us write the formula in Theorem 3.1 as

$$\left(-\frac{\partial}{\partial s}\frac{1}{2s - \sigma_0}\right)\mathcal{D}^{N-1}_{s-\sigma_0/2}F(s) = 0.$$

Hence, there exists a polynomial P_1 of degree ≤ 1 such that

$$\mathcal{D}^{N-1}_{s-\sigma_0/2}F(s) = P_1(s).$$

The proof finishes by continuing to unwind the differential operator $\mathcal{D}^{N-1}_{s-\sigma_0/2}$. □

Corollary 3.3. *The Dirichlet series $\text{Л}'(s)$ has a meromorphic continuation such that under that map $s \mapsto \sigma_0 - s$ the function*

$$\text{Л}'(s) + R_{\text{left}}(s) + R_{\text{right}}(\sigma_0 - s) + P'(s)$$

is odd.

This is immediate from Theorem 3.2 and the formula in Theorem 2.3. Upon integrating, the functional equation obtained in the present case is what we call an additive functional equation, to distinguish it from the multiplicative functional equation satisfied by functions in the fundamental class.

We have now concluded the process from which a theta function with an inversion formula leads to a Dirichlet series having an additive functional equation with fudge terms which are of regularized harmonic series type.

Example 1. If we apply the above construction to the Riemann zeta function, then we have

$$R(s) = \frac{1}{2}\Gamma'/\Gamma(s/2) + \frac{1}{2}\Gamma'/\Gamma((1-s)/2) - \log \pi,$$

hence we have

$$R_{\text{left}}(s) = \frac{1}{2}\Gamma'/\Gamma(s/2) - \log \pi,$$

and

$$R_{\text{right}}(s) = \frac{1}{2}\Gamma'/\Gamma((1-s)/2).$$

It is easy to see that $Л'(s) = 2\zeta'_{\mathbf{Q}}/\zeta_{\mathbf{Q}}(s)$ and $M = 0$, hence the above theorems apply with $N = 1$. Therefore, Corollary 3.3 reconstructs the fact that the function

$$2\xi'_{\mathbf{Q}}/\xi_{\mathbf{Q}}(s) = 2\zeta'_{\mathbf{Q}}/\zeta_{\mathbf{Q}}(s) + \Gamma'/\Gamma(s/2) - \log \pi$$

satisfies the functional equation $\xi'_{\mathbf{Q}}/\xi_{\mathbf{Q}}(s) = -\xi'_{\mathbf{Q}}/\xi_{\mathbf{Q}}(1-s)$.

Remark 2. It may seem odd that whereas we started with a triple $(Z, \tilde{Z}, \Phi)$ in our fundamental class with functional equation which is not symmetric, we derived a function in Corollary 3.3 which has a symmetric functional equation up to an additive polynomial factor. The reason for this is the following. Since the test function to which we apply Theorem 2.3 of Chapter III is even, we can not distinguish between elements in the set $\{q\}$ and elements in the set $\{\tilde{q}\}$. Therefore, we actually ended up considering the zeta function $H_1(s) = Z(s)\tilde{Z}(s)$ which satisfies the symmetric functional equation

$$H_1(s)\Phi(s) = H_1(\sigma_0 - s)\Phi(\sigma_0 - s).$$

One could equally well use the odd test function

$$F_t(x) = \frac{-x/2t}{\sqrt{4\pi t}} e^{-x^2/4t},$$

so then

$$\mathbf{M}_{\sigma_0/2} f(s) = -(s - \sigma_0/2) e^{(s-\sigma_0/2)^2 t}.$$

After carrying through the computations as above, we end up with a formula of the form

$$(2s - \sigma_0)R_\theta((s - \sigma_0/2)^2)$$
$$= \Pi'(s) - (s - \sigma_0/2)[R_{\text{left}}(s) - R_{\text{right}}(\sigma_0 - s)] + P'(s).$$

This equation corresponds to the functional equation

$$H_2(s)H_2(\sigma_0 - s) = 1$$

where

$$H_2(s) = \frac{Z(s)}{\tilde{Z}(s)}\Phi(s).$$

Then, combining the two functional equations above, we recover the original functional equation (up to a sign)

$$(Z(s)\Phi(s))^2 = (\tilde{Z}(\sigma_0 - s))^2,$$

in other words

$$Z(s)\Phi(s) = \pm\tilde{Z}(\sigma_0 - s).$$

Example 2. The functional equation for the function H_2 in Remark 2 is like that of the scattering determinant associated to any non-compact hyperbolic Riemann surface of finite volume (see [He 83]).

§4. A new zeta function for compact quotients of M_3.

In this section we will give, what we believe, is a new example of a zeta function. The approach is that of the previous sections, namely to every theta inversion formula there is a corresponding zeta function. As we will see, the "zeta function" has a Dirichlet series in a half-plane, but is a regularized harmonic series whose singularities are simple poles whose residues are not necessarily integers. Hence the function is not the logarithmic derivative of a regularized product type.

Let M_3 be the simply connected three dimensional Riemannian manifold whose metric has constant sectional curvature equal to -1; see page 38 of [Ch 84] for a precise model. On M_3 there is a transitive group of isometries, so the heat kernel relative to the Laplacian acting on smooth functions, which we denote by K_{M_3}, is a function of distance r and t. In fact, let

$$(1) \qquad h_3(r,t) = \frac{1}{(4\pi t)^{3/2}} e^{-r^2/4t} e^{-t} \frac{r}{\sinh r}.$$

Then
$$K_{M_3}(\tilde{x}, t, \tilde{y}) = h_3(d_{M_3}(\tilde{x}, \tilde{y}), t).$$

For details of this computation, see page 150 of [Ch 84] or page 397 of [DGM 76]. The result is basically due to Millson.

Let $X = \Gamma \backslash M_3$ be a compact quotient of M_3. Using the spectral decomposition of the heat kernel K_X on X, one has the expression

$$(2) \qquad K_X(x, t, y) = \sum_{k=0}^{\infty} \phi_k(x) \phi_k(y) e^{-\lambda_k t},$$

where $\{\phi_k\}$ is a complete system of orthonormal eigenfunctions of the Laplacian and $\{\lambda_k\}$ is the corresponding system of eigenvalues. Choose $\tilde{x}$ and $\tilde{y}$ in M_3 lying above x and y in X. Then

$$(3) \qquad K_X(x, t, y) = \sum_{\gamma \in \Gamma} K_{M_3}(\tilde{x}, t, \gamma \tilde{y}).$$

View the pair (x, y) as fixed, and for each $\gamma \in \Gamma$ define

$$\log \mathbf{q}_\gamma(x,y) = \log \mathbf{q}_\gamma = d_{M_3}(\tilde{x}, \gamma\tilde{y}).$$

Thus we obtain a sequence $\{\mathbf{q}_\gamma\}_{\gamma \in \Gamma}$, which we reindex by itself as simply $\{\mathbf{q}\}$. Suppose $x \neq y$. Then by (1), we have

$$K_X(x,t,y) = \frac{2}{(4\pi t)^{3/2}} e^{-t} \sum_{\mathbf{q}} \frac{\log \mathbf{q}}{\mathbf{q} - \mathbf{q}^{-1}} e^{-(\log \mathbf{q})^2/4t}, \tag{4}$$

which we call the **Γ-expansion** or the **q-expansion**. Combining (2) and (4), we immediately obtain the theta inversion formula

$$\sum_{k=0}^{\infty} \phi_k(x) \phi_k(y) e^{-\lambda_k t}$$

$$= \frac{2}{(4\pi t)^{3/2}} e^{-t} \sum_{\mathbf{q}} \frac{\log \mathbf{q}}{\mathbf{q} - \mathbf{q}^{-1}} e^{-(\log \mathbf{q})^2/4t}. \tag{5}$$

Observe that the factor e^{-t} could be brought to the other side. In the notation of §3, $\sigma_0 = 2$ and $\mu_k = \lambda_k - 1$. We note that (5) holds for any points $x \neq y$ on X, and we have not taken a trace of the heat kernel, although we have taken the trace with respect to the infinite Galois group Γ, i.e. the fundamental group of X. Formula (5) simply reflects a combination of the existence and uniqueness of the heat kernel on X together with the spectral expression (2) and the group expression (3) for K_X.

As in the previous sections, any theta inversion formula can be used to obtain a zeta function with additive functional equation. Let us carry through the computations in this case. We are still assuming $x \neq y$. From (5), let us derive two expressions for the function

$$F_{x,y}(s) = (2s - 2) \int_0^\infty K_X(x,t,y) e^{-s(s-2)t} t^2 \, dt. \tag{6}$$

Using the spectral expansion (2), the integral in (6) yields the equality

$$F_{x,y}(s) = (2s - 2) \sum_{k=0}^{\infty} \frac{2\phi_k(x)\phi_k(y)}{(s(s-2) + \lambda_k)^3}, \tag{7}$$

which can be shown to converge uniformly and absolutely on X by combining standard asymptotic formulas for eigenvalues on compact manifolds (see [BGV 92], for example) and Sogge's theorem on sup-norm bounds for eigenfunctions (see page 226 of [St 90]).

On the other hand, we shall also take the Gauss transform of the right side of (5), for which we need the collapse of the Bessel integral

$$K_{3/2}(s-1, \log \mathbf{q}) = \int_0^\infty e^{-(s-1)^2 t} e^{-(\log \mathbf{q})^2/4t} t^{3/2} \frac{dt}{t}$$

$$= \frac{-1}{2s-2} \frac{\partial}{\partial s} \left(\frac{\sqrt{\pi}}{s-1} \mathbf{q}^{-(s-1)} \right),$$

by Lemma 2.1. Therefore, taking the Gauss transform of (3), or the right side of (5), we obtain a Bessel sum for $F_{x,y}$ which collapses to a Dirichlet series, for $\text{Re}[(s-1)^2] > 0$, namely

$$F_{x,y}(s) = \frac{2}{(4\pi)^{3/2}} \sum_{\mathbf{q}} \frac{\log \mathbf{q}}{\mathbf{q} - \mathbf{q}^{-1}} (2s-2) K_{3/2}(s-1, \log \mathbf{q})$$

which becomes

(8) $$F_{x,y}(s) = \frac{1}{2\pi} \frac{\partial}{\partial s} \left(\frac{-1}{2s-2} \sum_{\mathbf{q}} \frac{\log \mathbf{q}}{1 - \mathbf{q}^{-2}} \mathbf{q}^{-s} \right).$$

Using a simple argument involving volumes of fundamental domains, one can show directly that the Dirichlet series (8) converges for $\text{Re}(s) > 2$. Therefore, we have for s in this half-plane, the formula

$$F_{x,y}(s) = \frac{1}{2\pi} \frac{\partial}{\partial s} \left(\frac{-1}{2s-2} \sum_{\mathbf{q}} \frac{\log \mathbf{q}}{1 - \mathbf{q}^{-2}} \mathbf{q}^{-s} \right)$$

$$= (2s-2) \sum_{k=0}^\infty \frac{2\phi_k(x)\phi_k(y)}{(s(s-2) + \lambda_k)^3}.$$

From (7) it follows that the Dirichlet series in (8) has a meromorphic continuation to all $s \in \mathbf{C}$ with the additive functional equation

$$F_{x,y}(s) = -F_{x,y}(2-s).$$

Further, the meromorphic continuation has singularities precisely when $s(s-2) = -\lambda_k$, that is

(9) $$s = 1 \pm i\sqrt{\lambda_k - 1}$$

each singularity being a double pole with zero residue. Upon integrating with respect to s, we obtain the Dirichlet series

$$(10) \qquad Л_{x,y}(s) = \sum_{\mathbf{q}} \frac{1}{1-\mathbf{q}^{-2}} \mathbf{q}^{-s} = \sum_{\mathbf{q}} \sum_{k=0}^{\infty} \mathbf{q}^{-s-2k},$$

which satisfies the relation

$$\mathcal{D}_{s-1} Л'_{x,y}(s) = -2\pi F_{x,y}(s).$$

The function $Л = Л_{x,y}$ plays the role of the logarithm of a zeta function in the fundamental class, and would be such a logarithm if the residues of $Л'$ were integers, but they are not in general.

Theorem 4.1. *The function $Л$ has the following properties:*

  i) *The series for $Л(s)$ converges for $\mathrm{Re}(s) > 2$;*
  ii) *$Л'$ has a meromorphic continuation to all $s \in \mathbf{C}$;*
  iii) *$Л'$ is odd under the map $s \mapsto 2 - s$;*
  iv) *The singularities of the meromorphic continuation of the function $(2s-2)Л'(s)$ are all simple poles at the points (9), with corresponding residue $2\pi \phi_n(x) \phi_n(y)$.*

Property (iv) should be viewed as a type of Riemann hypothesis for the zeta function (10).

We conclude with some remarks in the case $x = y$. We pick $\tilde{x} = \tilde{y}$. Then the sum in (1) must be decomposed into the term with $\gamma = \mathrm{id}$ and the other terms. We have $\mathbf{q}_{\mathrm{id}} = 1$ and $\mathbf{q}_\gamma > 1$ for all $\gamma \neq \mathrm{id}$. In the present case, the term with $\gamma = \mathrm{id}$ is easily computed, and instead of (4) we then obtain
(11)
$$K_X(x,t,x) = \frac{1}{(4\pi t)^{3/2}} e^{-t} + \frac{2}{(4\pi t)^{3/2}} e^{-t} \sum_{\mathbf{q}} \frac{\log \mathbf{q}}{\mathbf{q} - \mathbf{q}^{-1}} e^{-(\log \mathbf{q})^2/4t}$$

where the sum is over $\mathbf{q} = \mathbf{q}_\gamma$ for $\gamma \neq \mathrm{id}$. Thus the identity term must be handled separately. The integral transformation considered in (6) is sufficiently simple so that we have

$$\frac{2s-2}{(4\pi)^{3/2}} \int_0^\infty e^{-(s-1)^2 t} t^{1/2} \frac{dt}{t} = \frac{1}{4\pi}.$$

320

As a result, when considering the series (10) for $x = y$, summing over
$$\mathbf{q}_\gamma(x,x) = d_{\mathbf{M}_3}(\tilde{x}, \gamma\tilde{x})$$
with $\gamma \neq \mathrm{id}$, all properties (i) through (iv) hold with property (iii) being changed to allow an additive factor which is a polynomial of degree three in s. The coefficients of this polynomial can be determined by considering the limit as $s \to \infty$ and using the asymptotic formulas from Chapter I.

Remark 1. Computations similar to the above hold for compact quotients of odd dimensional hyperbolic spaces, since, for these manifolds, one can obtain simple expressions similar to (1) for their heat kernels; see page 151 of [Ch 84]. For non-compact quotients, one must take into account the appearance of the Eisenstein series in the spectral decomposition of the heat kernels, and the subsequent appearance of other terms in the additive functional equation. Examples of such formulas, as well as the more complicated situation of even dimensional hyperbolic manifolds, will be treated in a future publication.

Remark 2. Now that we have constructed a zeta function with additive functional equation and a Dirichlet series, we can apply the methods of [JoL 93c] to study the function

$$(11) \qquad \sum_{n=0}^{\infty} \phi_n(x)\phi_n(y) e^{z \cdot i\sqrt{\lambda_n - 1}},$$

defined for $\mathrm{Im}(z) > 0$. As in [JoL 93c], there is a meromorphic continuation of (11) to complex z with singularities at the points $\{\pm \log \mathbf{q}\}$. This should be viewed as a Duistermaat-Guillemin type theorem, as in [DG 75], since (11) is a wave kernel. We shall deal with this situation at greater length elsewhere, especially since it requires a systematic exposition of the additive fundamental class and its corresponding Cramér theorem and explicit formulas.

Remark 3. The above example shows the necessity of an "additive class" of zeta functions, as discussed in the introduction, by which we mean functions which are meromorphic with simple poles, have Dirichlet series in a half-plane, and have an additive functional equation with additive fudge factors expressible as linear combinations of regularized harmonic series types. The above example is not expressible in the fundamental multiplicative class since the residues, as determined in (iv) above, will certainly not be integers.

Remark 4. Readers will note the distinction between the zeta function defined by Minakshisundaram-Pleijel in [MP 49] by the Dirichlet series

$$\sum_{k=0}^{\infty} \frac{\phi_k(x)\phi_k(y)}{\lambda_k^s}$$

and the new zeta function $Л'_{x,y}$. The Minakshisundaram-Pleijel zeta function is essentially the one obtained as the Mellin transform of the theta series satisfying the basic theta conditions, whereas the new zeta function is obtained as the Gauss transform of the theta series.

CHAPTER VI

A generalization of Fujii's theorem

Let $L_{\mathbf{Q}}^+$ denote the set of zeros of the Riemann zeta function with positive imaginary part, meaning

$$L_{\mathbf{Q}}^+ = \{\rho \in \mathbf{C} : \zeta_{\mathbf{Q}}(\rho) = 0 \text{ and } \operatorname{Im}(\rho) > 0\}.$$

Write $\rho = \beta + i\gamma$ for any $\rho \in L_{\mathbf{Q}}^+$. The zeta function

$$(1) \qquad \sum_{\rho \in L_{\mathbf{Q}}^+} \frac{v(\rho)}{\operatorname{Im}(\rho)^s} = \sum_{\rho \in L_{\mathbf{Q}}^+} \frac{v(\rho)}{\gamma^s},$$

defined for $\operatorname{Re}(s) > 1$, was studied in [Del 66] and [Gu 45], and it was shown that (1) admits a meromorphic continuation to all $s \in \mathbf{C}$ with explicitly computable singularities, including a double pole at $s = 1$. Building on these results, Fujii considered the zeta functions

$$(2) \qquad \sum_{\rho \in L_{\mathbf{Q}}^+} v(\rho) \frac{\sin(\alpha\gamma)}{\gamma^s}$$

and

$$(3) \qquad \sum_{\rho \in L_{\mathbf{Q}}^+} v(\rho) \frac{\cos(\alpha\gamma)}{\gamma^s}$$

for non-zero $\alpha \in \mathbf{R}$ and $\operatorname{Re}(s) > 1$. It was shown in [Fu 83] that (2) admits a holomorphic continuation to all $s \in \mathbf{C}$ for any non-zero α, and (3) admits a mermorphic continuation to all $s \in \mathbf{C}$ with a simple pole at $s = 1$ having residue equal to $-(2\pi)^{-1}\Lambda(e^\alpha)e^{-\alpha/2}$, where

$$(4) \qquad \Lambda(x) = \begin{cases} \log p, & \text{if } x = p^k \text{ where } p \text{ is a prime and } k \in \mathbf{Z}_{>0} \\ 0, & \text{otherwise.} \end{cases}$$

Instead of functions formed separately with sine and cosine, one may as well consider what we call the **Fujii function**

$$F(s; \alpha) = \sum_{L_Q^+} v(\rho) \frac{e^{i\alpha\gamma}}{\gamma^s}$$

and its meromorphic continuation. We should note that Fujii's proof of the meromorphic continuation involves a very detailed study of many integrals arising from a generalization of Delsarte's work [Del 66] involving various integral transforms of the classical Riemann-von Mangoldt formula.

As remarked on page 233 of [Fu 83], one can prove analogous results for the eigenvalues of the Laplace-Beltrami operator on the fundamental domain of the modular group $PSL(2, \mathbf{Z})$. These theorems are given in [Fu 84b] and are as follows.

Let $\{\lambda_k\}$ be the set of eigenvalues of the hyperbolic Laplacian on the space $PSL(2,\mathbf{Z})\backslash \mathbf{h}$ and set $\lambda_j = 1/4 + r_j^2$ with $r_j > 0$. For any $\alpha \in \mathbf{R}^+$, Fujii considered the function

(5)
$$\sum_{r_j} v(r_j) \frac{\sin(\alpha r_j)}{r_j^s},$$

which is defined for $\text{Re}(s) > 2$. Through a rather lengthy and involved application of the Selberg trace formula, it was proved in [Fu 84b] that (5) has an analytic continuation to all $s \in \mathbf{C}$ to a holomorphic function. Again, one could consider the Fujii function

$$\sum_{r_j} v(r_j) \frac{e^{i\alpha r_j}}{r_j^s}$$

for all $\alpha \in \mathbf{R}$ and study its meromorphic continuation, thus capturing many of the results obtained by Fujii in the papers [Fu 84b] and [Fu 88a].

At this point, one could envision a series of articles in which one would define and study a Fujii function associated to every special zeta function, such as zeta functions and L-functions from the theory of modular forms, zeta functions and L-functions of number fields, spectral zeta functions constructed from the eigenvalues associated to the Laplacian acting on any non-compact arithmetic Riemann surface, to name a few examples, with each example yielding a new paper. Such a case-by-case extension of the classical Cramér theorem [Cr 19] has begun to appear in the literature. However, in [JoL 93c], we gave a vast generalization of Cramér's theorem containing all previously known special cases and many more.

Similarly, in this chapter, we obtain a generalization of Fujii's theorem which applies to any zeta function with Euler sum and functional equation whose fudge factors are of regularized product type. This generalization is simply a corollary of the generalized Cramér theorem, and, in particular, applies both to the zeta functions arising from algebraic number theory and to those arising from spectral theory.

In §1 we will state the generalized Fujii theorem, and the proof will be given in §2. To conclude this chapter, we will give various examples of the generalized Fujii theorem in §3.

§1. Statement of the generalized Fujii theorem.

Let us assume the notation of the previous chapters. With this, we can state the following result, which we call the generalized **Fujii theorem**.

Theorem 1.1. *Let $(Z, \tilde{Z}, \Phi)$ be in the fundamental class. Let a be such that $\sigma_0 + a > \sigma_0'$, and let $\{\rho\}$ be the set of zeros and poles of Z in the open infinite rectangle $\mathcal{R}_a$ with vertices at the four points*

$$-a + i\infty, \quad -a, \quad \sigma_0 + a, \quad \sigma_0 + a + i\infty.$$

Let $v(\rho) = \operatorname{ord}_\rho Z$ and set $\{\lambda\} = \{\rho/i\}$. Then:

i) *For any non-zero $\alpha \in \mathbf{R}$, the Fujii function*

$$F_{Z,a}(s;\alpha) = \sum_{\rho \in \mathcal{R}_a} v(\rho) \frac{e^{i\alpha\lambda}}{\lambda^s}$$

has a meromorphic continuation to all $s \in \mathbf{C}$.

ii) *For any non-zero $\alpha \in \mathbf{R}$, the continuation of the Fujii function $F_{Z,a}(s;\alpha)$ is holomorphic for all $s \in \mathbf{C}$ except for simple pole at $s = 1$ with residue*

$$\begin{cases} -(2\pi)^{-1} c(\mathbf{q}) \log \mathbf{q} & \text{if } \alpha = \log \mathbf{q} \\ -(2\pi)^{-1} c(\tilde{\mathbf{q}})(\log \tilde{\mathbf{q}})\tilde{\mathbf{q}}^{-\sigma_0} & \text{if } \alpha = -\log \tilde{\mathbf{q}} \\ 0 & \text{otherwise.} \end{cases}$$

In [Fu 83], Fujii considered the zeta function formed with the imaginary parts of the non-trivial zeros of the Riemann zeta function, assuming the Riemann hypothesis. The following theorem generalizes this result.

Theorem 1.2. *With notation as in Theorem 1.1, assume there is a real constant β_0 such that $\rho = \beta_0 + i\gamma$ for all $\rho \in \mathcal{R}_a$. Then:*

i) *For any non-zero $\alpha \in \mathbf{R}$, the Fujii function*

$$F_{Z,a}^{RH}(s;\alpha) = \sum_{\rho \in \mathcal{R}_a} v(\rho) \frac{e^{i\alpha\gamma}}{\gamma^s}$$

has a meromorphic continuation to all $s \in \mathbf{C}$.

ii) For any non-zero $\alpha \in \mathbf{R}$, the continuation of the Fujii function $F_{Z,a}^{RH}(s;\alpha)$ is holomorphic for all $s \in \mathbf{C}$ except for simple pole at $s = 1$ with residue

$$\begin{cases} -(2\pi)^{-1}c(\mathbf{q})(\log \mathbf{q})\mathbf{q}^{-\beta_0} & \text{if } \alpha = \log \mathbf{q} \\ -(2\pi)^{-1}c(\tilde{\mathbf{q}})(\log \tilde{\mathbf{q}})\tilde{\mathbf{q}}^{-\sigma_0+\beta_0} & \text{if } \alpha = -\log \tilde{\mathbf{q}} \\ 0 & \text{otherwise.} \end{cases}$$

One obtains the Fujii theorem by considering the above series with α and $-\alpha$ since, for example, if $Z = \zeta_\mathbf{Q}$, we have

$$F_{Z,a}(s;\alpha) - F_{Z,a}(s;-\alpha) = 2i \sum_{\rho \in L_\mathbf{Q}^+} v(\rho) \frac{\sin(\alpha\gamma)}{\gamma^s}.$$

Also, as remarked on page 23 of [Fu 84a], we obtain a meromorphic continuation of the series

$$2 \sum_{\rho \in L_\mathbf{Q}^+} v(\gamma) \frac{\cos(\alpha\gamma)}{\gamma^s} = F_{Z,a}(s;\alpha) + F_{Z,a}(s;-\alpha),$$

both with and without a Riemann hypothesis type assumption.

Thus what appeared up to now to be a phenomenom associated to more or less arithmetic situations, for instance the location of poles at the logs of prime powers or their analogues for the Selberg zeta function, is now seen to be quite a general property of our broad class of functions.

Directly from Theorem 1.2, we have the following corollary.

Corollary 1.3. *In addition to the above conditions, assume $Z = \tilde{Z}$, and assume all coefficients $c(\mathbf{q})$ are real. If all zeros of Z in $\mathcal{R}_a$ lie on a vertical line $\mathrm{Re}(s) = \beta_0$, then we necessarily have $\beta_0 = \sigma_0/2$.*

Finally, let us note that the case of $\alpha = 0$ is handled by our Cramér theorem, specifically Corollary 1.3 of [JoL 93c], and our results from [JoL 93a], specifically Theorem 1.8 and Corollary 1.10. For completeness, let us list this theorem and refer to the above mentioned references in our work for a proof.

Theorem 1.4. *With notation as in Theorem 1.1, assume that Φ has reduced order (M, m). Then the zeta function*

$$\sum_{\rho \in \mathcal{R}_a} v(\rho) \frac{1}{\lambda^s},$$

which is defined for $\mathrm{Re}(s) > M + 1$, has a meromorphic continuation to all $s \in \mathbf{C}$. If there is a constant β_0 such that $\rho = \beta_0 + i\gamma$ for all $\rho \in \mathcal{R}_a$, then the zeta function

$$\sum_{\rho \in \mathcal{R}_a} v(\rho) \frac{1}{\gamma^s},$$

which is defined for $\mathrm{Re}(s) > M + 1$, has a meromorphic continuation to all $s \in \mathbf{C}$.

The asymptotic expansion in §5 of [JoL 93c] and Corollary 1.10 of [JoL 93a] combine to give an explicit description of the poles of the zeta functions in Theorem 1.4. We will not state these results here, but will simply remark that the location and order of poles of the zeta functions in Theorem 1.4 are determined by the asymptotic expansion near $t = 0$ of the theta function associated to the fudge factor Φ.

§2. Proof of Fujii's theorem.

Let $z = \alpha + it$ for any non-zero $\alpha \in \mathbf{R}$, and, with notation as above, let $\rho = i\lambda$. In [JoL 93c] we proved various analytic properties of the Cramér function

$$V_Z(z) = \sum_{\rho \in \mathcal{R}_a} v(\rho) e^{\rho z},$$

which is defined when $\text{Im}(z) > 0$. In particular, Theorem 1.1 of [JoL 93c] and subsequent discussion imply that the function

(1) $$2\pi i V_{Z,a}(z) - e^{\sigma_0 z} \int_{\sigma_0 + a - i\infty}^{\sigma_0 + a} e^{-sz} \Phi'/\Phi(s) ds$$

has a meromorphic continuation to all $z \in \mathbf{C}$, whose only singularities are simple poles at the points $\log \mathbf{q}$ and $-\log \tilde{\mathbf{q}}$. The residues of these poles are given on page 390 of [JoL 93c].

Now assume that Φ is of regularized product type of reduced order (M, m). By combining Lemma 3.1, Proposition 3.2, Lemma 3.3, and, quite importantly, Lemma 4.2 of [JoL 93c], we conclude that the integral in (1) has a holomorphic continuation to any non-zero $z \in \mathbf{R}$. Therefore, for $\alpha \neq 0$, the function

$$V_{\alpha, Z, a}(t) = V_{Z, a}(\alpha + it) = \sum_{\rho \in \mathcal{R}_a} v(\rho) e^{i\alpha \lambda} e^{-t\lambda}$$

has an asymptotic behavior of the form

(2) $$V_{\alpha, Z, a}(t) \sim \sum_{n=-1}^{\infty} c_n(\alpha) t^n \quad \text{as } t \to 0,$$

for some constants $c_n(\alpha)$ which depend on α. Further, from the formula in Theorem 1.1 in [JoL 93c], we have

(3) $$c_{-1}(\alpha) = \begin{cases} -(2\pi)^{-1} c(\mathbf{q}) \log \mathbf{q} & \text{if } \alpha = \log \mathbf{q} \\ -(2\pi)^{-1} c(\tilde{\mathbf{q}}) (\log \tilde{\mathbf{q}}) \tilde{\mathbf{q}}^{-\sigma_0} & \text{if } \alpha = -\log \tilde{\mathbf{q}} \\ 0 & \text{otherwise.} \end{cases}$$

By applying the Mellin transform, we conclude from (2) that the function

$$\Gamma(s) \sum_{\rho \in \mathcal{R}_a} v(\rho) \frac{e^{i\alpha\lambda}}{\lambda^s} = \int_0^\infty V_{Z,a}(\alpha + it) t^s \frac{dt}{t} = \Gamma(s) M V_{\alpha,Z,a}(s)$$

has a meromorphic continuation to all $s \in \mathbf{C}$ whose only singularities are simple poles at the points $s \in \mathbf{Z}_{\leq 0}$ and $s = 1$ (see Theorem 1.5 of [JoL 93a]). Since $\Gamma(s)$ has simple poles at these points $s \in \mathbf{Z}_{\leq 0}$, Theorem 1.1 follows.

Let us now assume that any zero or pole ρ of Z in $\mathcal{R}_a$ is such that $\text{Re}(\rho) = \beta_0$, for fixed β_0. Then we can write $\rho = \beta_0 + i\gamma$, so $\lambda = \gamma - i\beta_0$. With $z = \alpha + it$, we have

$$e^{\rho z} = e^{\alpha \beta_0} e^{i\beta_0 t} \cdot e^{i\gamma \alpha} e^{-t\gamma}.$$

Therefore, the function

(4) $$\sum_{\rho \in \mathcal{R}_a} v(\rho) e^{i\gamma\alpha} e^{-\gamma t} = e^{-\alpha\beta_0} e^{-i\beta_0 t} V_{\alpha,Z,a}(t)$$

has asymptotics as in (2). With this, the proof of Theorem 1.2(i) is completed by applying the Mellin transform and the argument given above. Finally, from (3) and (4), one has the asymptotic formula

$$\sum_{\rho \in \mathcal{R}_a} v(\rho) e^{i\gamma\alpha} e^{-\gamma t} \sim e^{-\alpha\beta_0} c_{-1}(\alpha) t^{-1} \quad \text{as } t \to 0,$$

which completes the proof of Theorem 1.2(ii).

The proof of Corollary 1.3 is as follows. Since all numbers γ are real, we have

$$F_{Z,a}^{RH}(s;\alpha) = \sum_{\rho \in \mathcal{R}_a} v(\rho) \frac{\cos(\alpha\gamma)}{\gamma^s} + i \sum_{\rho \in \mathcal{R}_a} v(\rho) \frac{\sin(\alpha\gamma)}{\gamma^s}.$$

If we assume that all numbers $c(\mathbf{q})$ and $\mathbf{q}$ are real, then the residue of the only possible pole of $F_Z^{RH}(s;\alpha)$ is real, hence the series

$$\sum_{\rho \in \mathcal{R}_a} v(\rho) \frac{\sin(\alpha\gamma)}{\gamma^s}$$

has a holomorphic continuation to all $s \in \mathbf{C}$. However, we can also express this series as

$$2i \sum_{\rho \in \mathcal{R}_a} v(\rho) \frac{e^{i\alpha\gamma}}{\gamma^s} = F_{Z,a}^{RH}(s;\alpha) - F_{Z,a}^{RH}(s;-\alpha),$$

which means that the residues at $s = 1$ necessarily cancel, hence $\beta_0 = \sigma_0/2$.

§3. Examples.

As before, specific applications of our general theorems yield several classical theorems, many recent results, and new applications. We list a few examples here.

Example 1: The sine function. Let $Z = \sin(\pi i s)$. In this case, the associated Fujii function is the Dirichlet-like L-function

$$(1) \qquad F_Z(s; \alpha) = \sum_{n=1}^{\infty} \frac{e^{i\alpha n}}{n^s}$$

since $\{\mathbf{q}\} = \{e^{2\pi n}\}$ and $c(e^{2\pi n}) = 1/n$. Theorem 1.2 states that the series (1) has a holomorphic continuation whenever $\alpha \neq 2\pi n$ for some $n \in \mathbf{Z}$, and when $\alpha = 2\pi n$, the series (1) has a simple pole at $s = 1$ with residue 1.

Example 2: The Riemann zeta function. Let $Z = \zeta_\mathbf{Q}$. If we apply Theorem 1.2 to the Riemann zeta function, assuming the Riemann hypothesis, we obtain the Fujii theorem from [Fu 83]. However, without assuming the Riemann hypothesis, we do have the following result. Let S denote the non-trivial zeros of the Riemann zeta function in the upper half plane, and let $c \in \mathbf{C}$. Then the function

$$\sum_{\rho \in S} e^{(\rho+ic)z} = e^{icz} \sum_{\rho \in S} e^{\rho z}$$

satisfies the asymptotic axiom **AS 2**. Therefore, the proof of Theorem 1.1 implies that for non-zero $\alpha \in \mathbf{R}$, the function

$$(2) \qquad \sum_{\rho \in S} \frac{e^{\alpha \rho/i}}{(\rho/i + c)^s}$$

has a meromorphic continuation to all $s \in \mathbf{C}$. The only singularity of the continuation of (2) is a simple pole at $s = 1$ and then only when $\alpha = \pm \log p^n$ where p is a prime.

Without any modification, the above argument applies to Dirichlet L-functions and L-functions of number fields. A list of zeta functions for which the Cramér theorem of [JoL 93c] holds is given in section 7 of [JoL 93c].

Example 3: Selberg zeta functions of compact Riemann surfaces. If Z is the Selberg zeta function associated to a compact Riemann surface, then the theta function coming from the Cramér theorem is

$$\sum e^{(1/2+i\sqrt{\lambda_j - 1/4})z},$$

where λ_j is an eigenvalue of the Laplacian. Hence, this theta function can be viewed as a type of trace of the wave operator.

Example 4: Selberg zeta functions of non-compact Riemann surfaces. As in [Fu 84b], consider the Selberg zeta function associated to $PSL(2, \mathbf{Z})\backslash \mathbf{h}$. It is known that the set of zeros of the Selberg zeta function is the union of two sets: One set being the eigenvalues of the Laplacian, as described above, and the other set associated to the zeros of the Riemann zeta function (see pages 498 and 508 of [He 83]). From the meromorphy of (2), we conclude that the Fujii function formed with the eigenvalues of the Laplacian acting on $PSL(2,\mathbf{Z})\backslash \mathbf{h}$ has a meromorphic continuation to all $\mathbf{C}$. This theorem is the main result of [Fu 84b].

Similarly, since the scattering determinant for any congruence group is expressible in terms of Dirichlet L-series (see [He 83]), the above argument applies to yield the analogue of the Fujii theorem in these cases. The case of a general non-compact hyperbolic Riemann surface, including those associated to non-congruence subgroups, will be considered in [JoL 94].

BIBLIOGRAPHY

[Ba 81] BARNER, K.: On Weil's explicit formula. *J. reine angew. Math.* **323**, 139-152 (1981).

[Ba 90] BARNER, K.: Einführung in die Analytische Zahlentheorie. Preprint (1990).

[BGV 92] BERLINE, N., GETZLER, E., and VERGNE, M.: *Heat Kernels and Dirac Operators,* Grundlehren der mathematicschen Wissenschaften **298**, New York: Springer-Verlag (1992).

[CoG 93] CONREY, J.B., and GHOSH, A.: On the Selberg class of Dirichlet series: small weights. *Duke Math. J.* **72** (1993), 673-693.

[Cr 19] CRAMÉR, H.: Studien über die Nullstellen der Riemannschen Zetafunktion. *Math. Z.* **4** (1919), 104-130.

[DGM 76] DEBIARD, A., GAVEAU, B., and MAZET, E.: Théorèms de Comparison en Géometrie Riemannienne. *Publ. RIMS Kyoto Univ.* **12** (1976) 391-425.

[Del 66] DELSARTE, J.: Formules de Poisson avec reste. *J. Analyse Math.* **17** (1966) 419-431.

[Den 92] DENINGER, C.: Local L-factors of motives and regularized products. *Invent. Math.* **107,** (1992) 135-150.

[Den 93] DENINGER, C.: Lefschetz trace formulas and explicit formulas in analytic number theory. *J. reine angew. Math.* **441** (1993) 1-15.

[DG 75] DUISTERMAAT, J., and GUILLEMIN, V.: The spectrum of positive elliptic operators and periodic bicharacteristics. *Invent. Math.* **29,** (1975) 39-79.

[Fu 83] FUJII, A.: The zeros of the Riemann zeta function and Gibbs's phenomenon. *Comment. Math. Univ. St. Paul* **32** (1983), 99-113.

[Fu 84a] FUJII, A.: Zeros, eigenvalues, and arithmetic. *Proc. Japan Acad.* 60 *Ser. A.* (1984), 22-25.

[Fu 84b] FUJII, A.: A zeta function connected with the eigenvalues of the Laplace-Beltrami operator on the fundamental domain of the modular group. *Nagoya Math. J.* **96** (1984) 167-174.

[Fu 88a] FUJII, A.: Arithmetic of some zeta function connected with the eigenvalues of the Laplace-Beltrami operator. *Adv. Studies in Pure Math.* **13** (1988) 237-260.

[Fu 88b] FUJII, A.: Some generalizations of Chebyshev's conjecture. *Proc. Japan Acad.* 64 *Ser. A.* (1988) 260-263.

[Fu 93] FUJII, A.: Eigenvalues of the Laplace-Beltrami operator and the von-Mangoldt function. *Proc. Japan Acad.* **69** *Ser. A.* (1993), 125-130.

[Gal 84] GALLAGHER, P. X.: Applications of Guinand's formula. pp 135-157, volume 70 of *Progress in Mathematics* Boston: Birkhauser (1984).

[Gu 45] GUINAND, A. P.: A summation formula in the theory of prime numbers. *Proc. London Math. Soç* (2) **50** (1945) 107-119.

[Hej 76] HEJHAL, D.: *The Selberg Trace Formula for $PSL(2, \mathbf{R})$, volume 1*. Lecture Notes in Mathematics vol. **548** New York: Springer-Verlag (1976).

[Hej 83] HEJHAL, D.:The Selberg trace formula for $PSL(2, \mathbf{R})$, vol. 2. Springer Lecture Notes in Mathematics **1001** (1983).

[In 32] INGHAM, A. E.: *The Distribution of Prime Numbers*, Cambridge University Press, Cambridge, (1932).

[JoL 93a] JORGENSON, J., and LANG, S.: Complex analytic properties of regularized products and series. Springer Lecture Notes in Mathematics **1564** (1993), 1-88.

[JoL 93b] JORGENSON, J., and LANG, S.: A Parseval formula for functions with a singular asymptotic expansion at the origin. Springer Lecture Notes in Mathematics **1564** (1993), 1-88.

[JoL 93c] JORGENSON, J., and LANG, S.: On Cramér's theorem for general Euler products with functional equation. *Math. Ann.* **297** (1993), 383-416.

[JoL 94] JORGENSON, J., and LANG, S.: Applications of explicit formulas to scattering determinants of finite volume hyperbolic Riemann surfaces. In preparation.

[Kur 91] KUROKAWA, N.: Multiple sine functions and Selberg zeta functions. *Proc. Japan Acad., Ser A* **67**, (1991) 61-64.

[La 70] LANG, S.: *Algebraic Number Theory*, Menlo Park, CA.: Addison-Wesley (1970), reprinted as Graduate Texts in Mathematics **110**, New York: Springer-Verlag (1986); third edition, Springer-Verlag (1994).

[La 87] LANG, S.: *Elliptic Functions, second edition.* Graduate Texts in Mathematics **112** New York: Springer-Verlag (1987).

[La 93a] LANG, S.: *Complex Analysis,* Graduate Texts in Mathematics **103**, New York: Springer-Verlag (1985), Third Edition (1993).

[La 93b] LANG, S.: *Real and Functional Analysis, Third Edition,* New York: Springer-Verlag (1993).

[Lav 93] LAVRIK, A. F.: Arithmetic equivalents of functional equations of Riemann type. *Proc. Steklov Inst. Math.* **2** (1993) 237-245.

[MiP 49] MINAKSHISUNDARAM, S., and PLEIJEL, A.: Some properties of the eigenfunctions of the Laplace operator on Riemannian manifolds. *Canadian Jour. Math.* **1** (1949) 242-256.

[Mo 76] MORENO, C. J.: Explicit formulas in automorphic forms. *Lecture Notes in Mathematics* **626** Springer-Verlag (1976).

[Se 56] SELBERG, A.: Harmonic analysis and discontinous groups in weakly symmetric Riemannian spaces with applications to Dirichlet series, *J. Indian Math. Soc. B.* **20** (1956) 47-87 (*Collected papers volume I,* Springer-Verlag (1989) 423-463).

[Sel 91] SELBERG, A.: Old and new conjectures and results about a class of Dirichlet series. *Collected papers volume II,* Springer-Verlag: New York (1991) 47-63.

[St 90] STRICHARTZ, R. S.: Book review of *Heat Kernels and Spectral Theory,* by E. B. Davies, *Bull. Amer. Math. Soc.* **23** (1990) 222-227.

[Ti 48] TITCHMARSH, E. C.: *Introduction to the Theory of Fourier-Integrals, 2nd Edition* Oxford University Press, Oxford (1948).

[Ve 78a] VENKOV, A. B.: A formula for the Chebyshev psi function, *Math. Notes of USSR* **23** (1978) 271-274.

[Ve 78b] VENKOV, A. B.: Selberg's trace formula for the Hecke operator generated by an involution, and the eigenvalues of the Laplace-Beltrami operator on the fundamental domain of the modular group $PSL(2,\mathbf{Z})$. *Math. USSR Izv.* **42** (1978) 448-462.

[Ve 81] VENKOV, A. B.: Remainder term in the Weyl-Selberg asymptotic formula. *J. Soviet Math.* **17** (1981) 2083-2097.

[We 52] WEIL, A.: Sur les "formules explicites" de la théorie des nombres premiers, *Comm. Lund* (vol. dédié à Marcel Riesz), 252-265 (1952).

[We 72] WEIL, A.: Sur les formules explicites de la théorie des nombres, *Izv. Mat. Nauk (Ser. Mat.)* **36,** 3-18 (1972).

Math. Ann. 306, 75–124 (1996)

© Springer-Verlag 1996

Extension of analytic number theory and the theory of regularized harmonic series from Dirichlet series to Bessel series

Jay Jorgenson*, Serge Lang

Department of Mathematics, Yale University, 10 Hillhouse Ave., New Haven, CT 06520-8283, USA

Received: 18 May 1995 / Revised version: 4 December 1995

Table of Contents

1 The Bessel fundamental class 77
2 Regularized harmonic series 82
3 Some properties of the K Bessel function 86
4 Cramér's theorem and Fujii's theorem for Bessel series 92
5 Explicit formulas for Bessel series 97
6 Theta inversions 105
7 Example 1: Hyperbolic scattering determinants 111
8 Polynomially split heat kernels 116
9 Example 2: Odd dimensional hyperbolic spaces 118
10 Example 3: The symmetric space of a complex Lie group 120

Mathematics Subject Classification (1991): 11M35, 11M41, 30B50, 30D15, 33A35, 33A40, 35P99, 35S99, 42A99

Introduction

In [JoL 93c] and [JoL 94b] we defined a broad class of zeta functions (Dirichlet series) which behave in a manner similar to that which one would expect if they were built from the eigenvalues of a positive operator, even though no such operator may be present. We have dealt with both a multiplicative theory, involving a normalized Weierstrass product in the theory of regularized products, and with an additive theory of regularized harmonic series, corresponding to a normalization of Mittag-Leffler expansions of certain functions which have only simple poles (called resonances in physics).

---

* The first author acknowledges support from NSF grant DMS-93-07023 and from the Sloan Foundation. The second author thanks the Max-Planck-Institut for its yearly hospitality. We thank the referee for his corrections.

Our zeta functions have been built from theta series, not only via a Mellin transform as is done classically since Riemann, but also via a Gauss transform for the additive theory as in [JoL 94b]. In many important aspects of classical analytic number theory or spectral theory, only the logarithmic derivative of the zeta function occurs. It is both natural, and in some cases essential, to build up such an additive theory directly for functions which may not be logarithmic derivatives. Many aspects are so developed in [JoL 94b], and here we give general definitions along this line.

Furthermore, it turns out that when one takes the above integral transforms one gets Dirichlet series only in special cases, and to cover all the cases we know which arise in number theory and differential geometry, it is essential to deal with more general objects than Dirichlet series. At least two possibilities arise: one is that of Bessel series, and the other is that of certain integrals which follow a similar formalism. The possiblity of an integral will be treated in a subsequent paper. In the present paper, we deal systematically with the Bessel series playing the role of Dirichlet series. We shall prove the analogue of Cramér's theorem for such series, corresponding to the main result of [JoL 93c], and also the main theorem concerning explicit formulas, corresponding to [JoL 94b]. Our formulation of the Cramér theorem essentially states that if a function admits a Dirichlet series representation in a half plane, and now more generally a Bessel series representation, if it satisfies a functional equation, and if the fudge factor is of regularized harmonic series type, then the function itself is of regularized harmonic series type. This formulation, as before, is an inductive one which allows to prove simultaneously that very large classes of functions are of regularized harmonic series type.

There are cases in number theory or spectral theory - differential geometry where series more general than Bessel ones arise with properties similar to those of the present paper. In fact, the referee asked us whether we knew examples where hypergeometric functions occur. We do, see the last paragraph of this introduction. But beyond hypergeometric functions, there are other possiblities, such as solutions of linear differential equations, which classically have Fuchsian singularities similar to the logarithmic fractional singularities we consider in the present paper. We already warned the reader at the end of §2 that there may arise more general cases. Purely from an analyst's point of view, Bochner's paper [Bo 51] led us to trace back some of the literature, and we realized that the thirties saw a systematic concern with summation and inversion formulas, such as those of Doetsch [Do 35], Erdelyi [Er 36a], [Er 36b], [Er 37a], Kober [Ko 35], and Lowry [Lo 32]. These results are about to take their place in the present context of regularized harmonic series, but carrying out this program goes way beyond the limited purpose of the present paper. We regard the situation as open ended, and we do not yet have enough examples or a systematic theoretical insight to axiomatize the situation further at the moment.

We start in §1 by recalling briefly some basic properties of the K-Bessel function, which replaces the exponential function. Then we give the definition of the Bessel fundamental class, and its connection with regularized harmonic series.

In §3 we prove some known results about the K-Bessel function which usually occur in tables but which we develop from scratch from the integral definition we give for the Bessel function. We need both exact formulas and asymptotic results. These are applied in §4 to get the generalized Cramér theorem for the Bessel class, as well as the corresponding generalization of Fujii's theorem, and in §5 to get the "explicit formulas". In §6 we equate the existence of functions with Bessel series expansions and functional equation with the existence of a theta series with a theta inversion formula, as in [JoL 94b].

In the last four sections, we give examples from spectral theory and differential geometry, dealing with the scattering determinant, odd dimensional hyperbolic spaces, and symmetric spaces for complex semi-simple Lie groups. These examples also show that in the Dirichlet or Bessel series expansions one must allow for polynomial coefficients and not just constant coefficients.

We emphasize here that the explicit formulas state that the sum of a function taken over what plays the role of prime powers in number theory and lengths of geodesics in spectral theory, is equal to the sum of the Mellin transform taken over the poles of the regularized harmonic series under consideration. These poles would be the zeros in the critical strip of a corresponding Dirichlet series, if it exists. Thus we obtain in our context results which are analogous to geometric theorems due to Duistermaat-Guillemin [DG 75]. Such results were not known before in some important cases such as the scattering determinant, which does not have an "Euler product" in the classical sense, but nevertheless has a Dirichlet series representation. The generality of our methods thus yields essentially new classes of functions which satisfy the formalism of regularized harmonic series.

Bessel series play at least a dual role. They arise not only as extensions of Dirichlet series, but also as extensions of theta series. The Gauss transforms of such Bessel series involve Legendre functions and, therefore, hypergeometric functions. We postpone to further papers or monographs the systematic analysis of this phenomenon in the context of regularized harmonic series, Mellin transforms and Gauss transforms. These have applications to the theory of Eisenstein series which will then be given as examples among others. For example, we can carry out completely Asai's paper [As 70] for arbitrary number fields. This paper involves K-Bessel theta series, even in several variables. Beyond this, the possibilities are again open ended since in particular they have to include the results of [Bo 51] and the related other papers cited above in the context of Gauss transforms.

1. The Bessel fundamental class

We start by summarizing basic facts about the K-Bessel function. As a function of two real variables (a, b) with $a, b > 0$, and $\alpha \in \mathbf{C}$, we define

$$(1) \qquad K_\alpha(a, b) = \int_0^\infty e^{-(a^2 t + b^2/t)} t^\alpha \frac{dt}{t}.$$

The following properties are classical, and simple proofs directly from the above definition will be found in [La 87], Chapter 20, §3. We have

$$K_\alpha(a,b) = (b/a)^\alpha K_\alpha(ab) \tag{2}$$

where the K-Bessel function in one variable $c > 0$ is defined by

$$K_\alpha(c) = \int_0^\infty e^{-c(u+1/u)} u^\alpha \frac{du}{u}.$$

Warning. Our normalization of the Bessel integral is slightly different from the classical normalization. If we let $K_s^B(c)$ denote the classical normalization, then we have

$$2K_s^B(2c) = K_s(c).$$

Observe that from (2), we get

$$K_\alpha(a,b) = (b/a)^{2\alpha} K_\alpha(b,a) \tag{3}$$

As a function of α, $K_\alpha(a,b)$ is entire, and satisfies the functional equation

$$K_\alpha(c) = K_{-\alpha}(c). \tag{4}$$

For $\alpha = 1/2$, the integral collapses to

$$K_{1/2}(c) = \frac{\sqrt{\pi}}{\sqrt{c}} e^{-2c} \tag{5}$$

and

$$K_{1/2}(a,b) = \frac{\sqrt{\pi}}{a} e^{-2ab}. \tag{6}$$

The K-Bessel function decreases exponentially at infinity. More precisely:

Let $x_0 > 0$ and $\sigma_0 \leq \sigma \leq \sigma_1$. There is a number $C(x_0, \sigma_0, \sigma_1) = C$ such that if $x > x_0$, then

$$K_\sigma(x) \leq C e^{-2x}. \tag{7}$$

This estimate is one of the reasons why series formed with Bessel functions will have convergence properties similar to those of Dirichlet series.

We define the notion of a formal **polynomial Bessel series** to be a series of the form

$$\mathrm{B}(s) = \sum_j P_j(s) K_{\alpha_j}(s, \log \mathbf{q}_j^{1/2})$$

where $\{\mathbf{q}_j\}$ is a sequence of real numbers > 1 tending to infinity; $\{\alpha_j\}$ is a sequence of complex numbers; $\{P_j\}$ is a sequence of polynomials. So far, this notation is purely formal. We have to impose conditions which guarantee appropriate convergence in practice. Thus we impose the following conditions:

PBS 1. The degrees of the polynomials P_j are bounded independently of j.

PBS 2. There exist a sequence of complex numbers $\{c_j\}$ and $\sigma_0' \geq 0$ such that the Dirichlet series $\sum c_j \mathbf{q}_j^{-s}$ converges uniformly and absolutely in the half plane $\text{Re}(s) > \sigma_0'$; and if $\|P\|$ denotes the sup norm of the coefficients of a polynomial P, then

$$\|P_j\| \leq |c_j| \quad \text{for all } j.$$

PBS 3. The sequence $\{\alpha_j\}$ is bounded.

A formal polynomial Bessel series satisfying **PBS 1**, **PBS 2**, and **PBS 3** will simply be said to be a **polynomial Bessel series**. If $d = \max \deg(P_j)$, we call d the **polynomial degree** of $\mathrm{B}(s)$ If d is a bound for the degrees as in **PBS 1**, and $L_c(s) = \sum c_j \mathbf{q}_j^{-s}$, we say that B is **dominated** by the pair (d, L_c). The special case when $\deg P_j \leq 0$ for all j is called the case of the **Bessel series**.

From (2) and (7), we conclude that

$$|K_{\alpha_j}(s, \log \mathbf{q}_j^{1/2})| \leq C \mathbf{q}_j^{-\sigma} (\log \mathbf{q}_j)^{\text{Re}(\alpha_j)} |s|^{-\text{Re}(\alpha_j)} e^{\text{Im}(\alpha_j) \text{Arg}(s)}$$

for all $\text{Re}(s) = \sigma > 0$. Thus the absolute convergence of the Dirichlet series $\sum c_j \mathbf{q}_j^{-s}$ in the half plane $\text{Re}(s) > \sigma_0'$ implies that the Bessel series converges absolutely and uniformly on compact sets in the open right half plane $\text{Re}(s) > \sigma_0'$, because $|P(\sigma + it)|$ may tend polynomially to infinity as $|t| \to \infty$, but $|\mathbf{q}_j^s|$ is independent of t. Note that the possibility of including arbitrary polynomials as coefficients prevents uniform convergence in a closed half plane $\text{Re}(s) \geq \sigma_0' + \varepsilon$, as in the case when the coefficients of a Dirichlet series are constant. This turns out to be the only significant difference in the formalism of polynomial series and ordinary series with constant coefficients.

The notion of a **polynomial Dirichlet series** is defined in the same way, replacing the Bessel function by the exponential $\mathbf{q}_j^{-s}$ (disregarding the parameters α_j). Thus a polynomial Dirichlet series is merely a polynomial Bessel series when $\alpha_j = 1/2$ for all j and P_j is either 0 or a monomial of degree 1. Indeed, (6) yields the relation

$$(8) \qquad \sum_j c_j s K_{1/2}(s, \log \mathbf{q}_j^{1/2}) = \sum_j c_j \sqrt{\pi} \mathbf{q}_j^{-s}.$$

The case when $\alpha_j = 1/2$ for all j is only the simplest manifestation of the more general case when $\alpha_j = n/2$ with an odd integer n. Then again the Bessel series collapses to a (Laurent) Dirichlet series. Indeed, if $\alpha_j = N + 1/2$, then from §6 (1) we find the relation

$$(9) \qquad \sum_j c_j s K_{N+1/2}(s, \log \mathbf{q}_j^{1/2}) = \mathscr{L}_s^N \sum_j c_j \sqrt{\pi} \mathbf{q}_j^{-s}.$$

343

To develop further our additive theory, we have to recall some elementary facts about Mittag-Leffler expansions of certain meromorphic functions. In the theory of regularized products, we constructed certain types of Weierstrass products. In general, as already pointed out in [JoL 93a], we need to deal directly with functions of regularized harmonic series type whose definition will be recalled below. Such series are normalized Mittag-Leffler series, and we also need some facts from elementary complex analysis. For convenience, a meromorphic function with only simple poles will be called a **simple Mittag-Leffler function**.

One constructs canonical simple Mittag-Leffler functions just as one constructs canonical Weierstrass products. Given sequences of complex numbers $\{z_k\}$ and $\{a_k\}$ with $|z_k| \to \infty$, we say that $\{z_k\}$ is of $\{a_k\}$-**strict order** $\leq \rho$ if

$$\sum \frac{|a_k|}{|z_k|^\rho} < \infty.$$

Readers will recognize here condition **DIR 2(a)** from [JoL 93a]. We say that the sequence $\{z_k\}$ is of **strict order** $\leq \rho$ if $\sum 1/|z_k|^\rho < \infty$. This second condition amounts to **DIR 2(b)**. We say that the sequence $\{z_k\}$ has **finite order** if it is strict order $\leq \rho$ for some positive real ρ.

Let d be the smallest integer $\geq \rho$. Let G_d be the (geometric series) polynomial

$$G_d(z) = \sum_{n=0}^{d-1} z^n.$$

Define the **simple Mittag-Leffler term** of order d,

$$F_k(z; z_k) = \frac{a_k}{z - z_k} + \frac{a_k}{z_k} G_d\left(\frac{z}{z_k}\right),$$

and the **canonical Mittag-Leffler series**

$$F(z) = \sum F_k(z; z_k).$$

Then the canonical Mittag-Leffler series converges absolutely and uniformly on every compact set not containing any z_k. Conversely, given an integer d, we say that a meromorphic function F_1 is of **finite Mittag-Leffler order** $\leq d$ if its sequence of poles has finite order, and if F_1 admits a Mittag-Leffler series

$$F_1(z) = \sum [\Pr(F_1; z_k, z) + H_k(z)] + H(z),$$

where $\Pr(F_1; z_k, z)$ is the principal part at the pole z_k, H_k is the Mittag-Leffler polynomial insuring the convergence of the series, H is a polynomial, and

$$\deg(H_k), \deg(H) \leq d - 1 \quad \text{for all } k.$$

Lemma 1.1. *Let F_1 be a simple Mittag-Leffler function of Mittag-Leffler order $\leq d$. Let F be the canonical Mittag-Leffler series formed with the sequence of poles and their residues $\{z_k, a_k\}$. Then there is a polynomial of degree $\leq d$ such that $F = F_1 + G$.*

Proof. Elementary complex analysis, similar to the corresponding proofs for Weierstrass products of finite order.

In the same vein, we shall need two other lemmas for use in the proof of the main result in §4.

Lemma 1.2. *Let $\{z_k\}$ have $\{a_k\}$-strict order $\leq \rho$, and let d be the smallest integer $\geq \rho$.*

(a) *Let U be the complement of the union of all discs $D(z_k, \delta_k)$, centered at z_k, of radius $\delta_k = 1/|z_k|^d$. Then*

$$F(z) = o\left(|z|^{\rho+d}\right) \quad \text{for } z \in U \text{ and } |z| \to \infty.$$

(b) *Let S be a subset of $\mathbf{C}$ at finite non-zero distance from all z_k, that is there exists $c > 0$ such that $|z - z_k| \geq c$ for all $z \in S$ and all k. Then*

$$F_1(z) = O\left(|z|^d\right) \quad \text{for } z \in S \text{ and } |z| \to \infty.$$

Lemma 1.3. *Let $F_1, \ldots, F_r$ be simple Mittag-Leffler functions of finite Mittag-Leffler order. Let S be a vertical strip, say the set of all $z = x + iy$ such that $x \in [x_1, x_2]$ lies in some fixed interval. Then there exists a sequence $\{T_n\}$ with $n \leq T_n \leq n + 1$, and a positive integer N such that for each i,*

$$F_i(z) = O\left(T_n^N\right) \quad \text{for } \operatorname{Im}(z) = \pm T_n, \operatorname{Re}(z) \in [x_1, x_2], \text{ and } n \to \infty.$$

Proof. Both lemmas are proved by routine techniques of elementary complex analysis, and are similar to Lemma 2.1 of [JoL 93c]. We leave the details to the reader.

We say that a triple of functions $(F, \tilde{F}, R)$ satisfies an **additive functional equation** if the relation

$$F(s) + R(s) + \tilde{F}(\sigma_0 - s) = 0$$

holds with some $\sigma_0 \geq 0$.

We shall say that the triple $(\mathrm{B}, \tilde{\mathrm{B}}, R)$ is in the **polynomial Bessel fundamental class** if the following conditions are satisfied.

1. The functions B and $\tilde{\mathrm{B}}$ have polynomial Bessel series satisfying **PBS 1**, **PBS 2**, and **PBS 3**.

2. The functions B and $\tilde{\mathrm{B}}$ are simple Mittag-Leffler functions of finite Mittag-Leffler order, and R is of regularized harmonic series type, whose definition is recalled in §2.

3. The triple $(\mathrm{B}, \tilde{\mathrm{B}}, R)$ satisfies an additive functional equation as above with $0 \leq \sigma_0 \leq \sigma_0'$.

One defines the polynomial Dirichlet fundamental class similarly, replacing the Bessel functions by the usual $\mathbf{q}_j^{-s}$ of Dirichlet series. From (8), we see that the polynomial Dirichlet fundamental class is contained in the polynomial Bessel fundamental class. Examples of the various fundamental classes will be given later.

2. Regularized harmonic series

We now recall some terminology from [JoL 93a] and [JoL 94b]. The referee suggested simply to state that much of the following discussion is taken verbatim from [JoL 93a] and [JoL 94b]; however, for the convenience of the reader, we have decided to include the material.

Let $\{\lambda_k\}$ and $\{a_k\}$ be sequences of complex numbers such that $\text{Re}(\lambda_k) \to \infty$ in a sector contained in the right half plane, and such that the series $\sum a_k \lambda_k^{-s}$ converges uniformly and absolutely for all $\text{Re}(s)$ sufficiently large. We suppose $\lambda_0 = 0$ and $\lambda_k \neq 0$ for $k \geq 1$. We will consider a **theta series**, which is defined by

$$\theta(t) = \sum_{k=0}^{\infty} a_k e^{-\lambda_k t},$$

and, for each integer $N \geq 1$, we define the **asymptotic exponential polynomials** by

$$Q_N(t) = \sum_{k=0}^{N-1} a_k e^{-\lambda_k t}.$$

We are also given a sequence of complex numbers $\{p\} = \{p_j\}$ with

$$\text{Re}(p_0) \leq \text{Re}(p_1) \leq \cdots \leq \text{Re}(p_j) \leq \cdots$$

increasing to infinity, and, to every p in this sequence, we associate a polynomial B_p of degree n_p. We then define the **asymptotic polynomials at** 0 by

$$P_q(t) = \sum_{\text{Re}(p) < \text{Re}(q)} B_p(\log t) t^p.$$

Let $\mathbf{C}\langle T \rangle$ be the algebra of polynomials in T^p with arbitrary complex powers $p \in \mathbf{C}$. Then, with this notation, $P_q(t) \in \mathbf{C}[\log t]\langle t \rangle$.

The function θ on $(0, \infty) = \mathbf{R}_{>0}$ is subject to **asymptotic conditions**:

AS 1. Given a positive number C and $t_0 > 0$, there exists N and $K > 0$ such that

$$|\theta(t) - Q_N(t)| \leq Ke^{-Ct} \text{ for } t \geq t_0.$$

AS 2. For every q, there is an integer $m(q) \geq 0$ such that

$$\theta(t) - P_q(t) = O(t^{\operatorname{Re}(q)}|\log t|^{m(q)}) \text{ for } t \to 0,$$

which shall be written as

$$\theta(t) \sim \sum_p B_p(\log t) t^p.$$

AS 3. Given $\delta > 0$, there exists an $\alpha > 0$ and a constant $C > 0$ such that for all N and $0 < t \leq \delta$ we have

$$|\theta(t) - Q_N(t)| \leq C/t^\alpha.$$

We shall assume throughout that the theta series converges absolutely for $t > 0$. Further, we shall assume that the convergence of the theta series is uniform for $t \geq \delta > 0$ for every δ.

The **Laplace-Mellin transform** of a measurable function f on $(0, \infty)$ is defined by

$$\mathbf{LM}f(s, z) = \int_0^\infty f(t) e^{-zt} t^s \frac{dt}{t}.$$

It is shown in [JoL 93a] that if θ satisfies **AS 1**, **AS 2** and **AS 3**, then $\mathbf{LM}\theta$ has a meromorphic continuation for $s \in \mathbf{C}$ and $z \in \mathbf{U}_L$, where $\mathbf{U}_L$ is the complex plane minus the union of horizontal half lines from $-\infty$ to $-\lambda_k$ and $-\infty$ to 0. Furthermore, for each z, the function $s \mapsto \mathbf{LM}\theta(s, z)$ has poles only at the points $-(p+n)$ with $B_p \neq 0$ in the asymptotic expansion of θ at 0, and the pole at $-(p+n)$ has order at most $n(p')+1$ where $n(p') = \max \deg B_w$ for $\operatorname{Re}(w) \leq \operatorname{Re}(p)$.

The **regularized harmonic series** $R(z)$ is defined to be the meromorphic function such that
$$R(z) = \mathrm{CT}_{s=1} \mathbf{LM}\theta(s, z),$$

where $\mathrm{CT}_{s=1}$ means the constant term at $s = 1$. The regularized harmonic series is a particular meromorphic function that has simple poles at $z = -\lambda_k$ with residue a_k.

A meromorphic function R will be said to be of **regularized harmonic series type** if it is a finite linear combination

$$R(z) = \sum c_j R_j(\alpha_j z + \beta_j) + P'(z) + \sum \frac{c'_k}{z - \beta'_k},$$

where:

(i) each R_j is a regularized harmonic series;
(ii) P' is a polynomial;
(iii) $c_j, c'_k, \alpha_j, \beta_j, \beta'_k \in \mathbf{C}$;

(iv) the α_j and β_j are restricted so that the poles of $R_j(\alpha_j z + \beta_j)$ lie in a finite union of regions of the form:
- a sector opening to the right, meaning
$$\{z \in \mathbf{C} : -\frac{\pi}{2} + \epsilon < \arg(z) < \frac{\pi}{2} - \epsilon\} \text{ for some } \epsilon > 0,$$
- a sector opening to the left, meaning
$$\{z \in \mathbf{C} : \frac{\pi}{2} + \epsilon < \arg(z) < \frac{3\pi}{2} - \epsilon\} \text{ for some } \epsilon > 0,$$
- a vertical strip, meaning
$$\{z \in \mathbf{C} : a < \operatorname{Re}(z) < b\} \quad \text{for some } a, b \in \mathbf{R}.$$

For any regularized harmonic series R, we define the **reduced order** to be the pair of integers (M, m) where, in the notation of the asymptotic condition **AS 2**:

M is the largest integer $< -\operatorname{Re}(p_0)$;

$m = m(p_0) + 1$ if there is a complex index p with
$$\operatorname{Re}(p_0) = \operatorname{Re}(p) \in \mathbf{Z}_{<0} \quad \text{and} \quad B_p \neq 0,$$
otherwise simply set $m = m(p_0)$. A function R of regularized harmonic series type has reduced order (M, m) where $M = \max\{M_j\}$ and m is the maximum over all m_j for which $M = M_j$.

In addition, we shall also find that the asymptotic condition **AS 2** has a complexified version, stronger than that stated in [JoL 93c], and which we now describe. We must consider not only the asymptotics on the positive real axis, tending to 0, but we must consider asymptotics along arbitrary rays, tending to 0, and having as much uniformity as the situation allows. Thus we extend the terminology as follows.

Let $\mathscr{S}$ be a sector, possibly of finite radius > 0, with vertex at the origin to start. It is always assumed that the origin is not part of the sector. We do not exclude the possibility that $\mathscr{S}$ is a single ray, or that $\mathscr{S}$ is an obtuse sector, for instance the complement of a single ray such as the positive real axis. We shall be concerned with the following extension of **AS 2** for a function ψ defined on $\mathscr{S}$

AS 2$\mathscr{S}$. For every $z_0 \in \mathscr{S}$, and so $z_0 \neq 0$ by convention, the function
$$t \mapsto \psi(tz_0) \quad \text{with } t > 0 \text{ and } t \to 0$$
satisfies **AS 2**.

We sometimes express **AS 2**$\mathscr{S}$ by saying that **AS 2** is satisfied for every ray in $\mathscr{S}$ or simply is **satisfied on** $\mathscr{S}$. In some cases, we may wish to deal with a stronger uniform version. So we say that **AS 2**$\mathscr{S}$ holds **uniformly on** $\mathscr{S}$ if the following condition is satisfied.

There exists a sequence $\{p\}$ of complex numbers with $\mathrm{Re}(p) \to \infty$, and for each p a polynomial B_p such that if we put

$$P_q(z) = \sum_{\mathrm{Re}(p) < \mathrm{Re}(q)} B_p(\log z) z^p,$$

then uniformly for $z \in \mathscr{S}$ we have

$$\psi(z) = P_q(z) + O\left(|z|^{\mathrm{Re}(q)} |\log|z||^{m(q)}\right) \quad \text{for } |z| \to 0.$$

Since the complex log is involved, we use this uniformity condition only when the sector is simply connected, and a natural branch of the logarithm is defined on the sector.

We note that if a generalized polynomial P_q exists satisfying the above asymptotic condition, then it is uniquely determined. In particular, if $\mathrm{Re}(q) < \mathrm{Re}(q')$, then the terms of "degree" $< \mathrm{Re}(q)$ in $P_{q'}$ are the same as those in P_q.

By a translation we may define the same asymptotic condition on a sector whose vertex lies at some point z_0, not necessarily the origin. Then the generalized polynomials are of the form $P_q(z - z_0)$, and we say that **AS 2**$\mathscr{S}$ is satisfied uniformly at z_0. This asymptotic condition is local, by taking the radius of the sector sufficiently small.

We shall need to consider singularities not only at arbitrary points z_0, but also of a more general type than just generalized polynomials as above. So we introduce a definition, relevant for our Bessel extension of Cramér's theorem. Let V be an open set in the plane, $z_0 \in V$ and $V^* = V \setminus \{z_0\}$. Let F be a complex valued function on a simply connected, open dense subset of V^*. We say that F has a **logarithmic fractional singularity** at z_0 of **exponent** β if there exists a holomorphic function f on a neighborhood of z_0 in V with $f(z_0) \neq 0$, a complex number β, and a polynomial P such that

$$F(z) - (z - z_0)^\beta P(\log(z - z_0)) f(z)$$

is holomorphic on a neighborhood of z_0 in V, i.e. has a holomorphic extension to such a neighborhood. If $\beta \notin \mathbf{Z}_{\geq 0}$ or if $\deg P = n \geq 1$, we say that F has a logarithmic fractional singularity at z_0 of **order** (β, n). Observe that if F has such a logarithmic fractional singularity, then F has an analytic continuation along every path in some punctured disc centered at z_0. More precisely, if D is a disc of sufficiently small radius centered at z_0, and we delete any ray in the disc with vertex z_0, we obtain a sector $\mathscr{S}$ and F satisfies **AS 2**$\mathscr{S}$ uniformly on this sector.

The above definition fits the precise situation which we shall meet later. More systematically and for other purposes, it is clear that the above definition should be extended to allow for a singularity, say at the origin, of the form

$$P(z) = \sum f_{i,j}(z) z^{p_i} (\log z)^{m_j}$$

where the coefficients $f_{i,j}$ are holomorphic at the origin. One may then translate such expressions to an arbitrary point z_0 to get the possible generalization of the logarithmic fractional singularity at z_0. In this paper, we meet no worse than the special case described above with point exponent β.

In practice, we shall of course not deal with just one possible singularity. Thus suppose V is an open set in the plane and S is a discrete subset of point of V. Let F be holomorphic on a simply connected dense open subset of $V \setminus S$. We say that F has **logarithmic fractional singularities** on S relative to V if F has such singularities at each point of S, and if F can be analytically continued along every path in $V \setminus S$. We then have an explicit formula locally in the neighborhood of each singularity for the analytic continuation of F around this singularity.

3. Some properties of the K Bessel function

In this section, we shall derive basic properties of the one variable K Bessel function which will be needed throughout the article. There are many references for the study of the Bessel function, beginning with the treatise [Wa 44]. First we will obtain an analytic continuation of the one variable K to a larger region than that allowed by the defining integrals. For this, we realize the one variable K Bessel function as a solution of a second order linear differential equation, which can then be expressed as a series.

Lemma 3.1. *For all $\alpha \in \mathbf{C}$ and $x \in \mathbf{R}^+$, we have the differential equation*

$$x^2 K_\alpha''(x) + x K_\alpha'(x) - (4x^2 + \alpha^2) K_\alpha(x) = 0.$$

Proof. If we differentiate under the integral sign, we obtain the relations

(1) $\quad K_\alpha'(x) = -K_{\alpha+1}(x) - K_{\alpha-1}(x) \quad$ and $\quad K_\alpha''(x) = K_{\alpha+2}(x) + 2K_\alpha(x) + K_{\alpha-2}(x)$.

On the other hand, if we integrate by parts, we derive the relation

(2) $\qquad K_{\alpha+1}(x) = K_{\alpha-1}(x) + \dfrac{\alpha}{x} K_\alpha(x).$

By substituting (2) into (1), with α replaced by $\alpha + 1$, we get

(3) $\qquad K_\alpha''(x) = 4 K_\alpha(x) + \dfrac{\alpha+1}{x} K_{\alpha+1}(x) - \dfrac{\alpha-1}{x} K_{\alpha-1}(x).$

Equations (1) and (3) express K_α, K_α' and K_α'' in terms of $K_{\alpha+1}$ and $K_{\alpha-1}$. We can now solve to obtain a single linear relation involving K_α, K_α' and K_α'', which turns out to the statement of the Lemma.

For any $\alpha \in \mathbf{C}$ and $x \in \mathbf{R}^+$ let us define the function

$$I_\alpha(x) = \sum_0^\infty \frac{x^{\alpha+2m}}{m!\Gamma(m+1+\alpha)},$$

with the understanding that for $n \in \mathbf{Z}_{\geq 0}$, we set $1/\Gamma(-n) = 0$. Note that the map $z \mapsto I_\alpha(z)$ is analytic in $\mathbf{C} \setminus P$, where P is any ray in $\mathbf{C}$ with vertex at the origin. Further, $I_\alpha(z)$ has an analytic continuation along any path in $\mathbf{C} \setminus \{0\}$.

Proposition 3.2. *If $\alpha \notin \mathbf{Z}$, then the general solution of the differential equation*

$$x^2 y'' + xy' - (4x^2 + \alpha^2)y = 0$$

is expressible as $c_1 I_\alpha(x) + c_2 I_{-\alpha}(x)$. In particular, the one variable K Bessel function can be written as

$$K_\alpha(x) = \frac{\pi}{\sin(\pi\alpha)}[I_{-\alpha}(x) - I_\alpha(x)].$$

When $\alpha = n \in \mathbf{Z}$, we have $K_n(x) = \lim_{\alpha \to n} K_\alpha(x)$.

Proof. The fact that $I_\alpha(x)$ is a solution of the differential equation follows directly by interchanging sum and differentiation together with elementary manipulation of the series which defines $I_\alpha(x)$. What remains is to express the K Bessel function in terms of the above defined I Bessel functions. Let us write

(4) $$K_\alpha(x) = c_1(\alpha) I_{-\alpha}(x) + c_2(\alpha) I_\alpha(x)$$

for some constants $c_1(\alpha)$ and $c_2(\alpha)$, and assume $\alpha > 0$. To evaluate $c_1(\alpha)$, we examine the asymptotic behavior of both sides of (4) as x approaches zero. Directly from the series, we have

$$I_\alpha(x) = \frac{1}{\Gamma(\alpha+1)} x^\alpha + O(x^{\alpha+1}) \quad \text{as } x \to 0,$$

and

$$I_{-\alpha}(x) = \frac{1}{\Gamma(1-\alpha)} x^{-\alpha} + O(x^{-\alpha+1}) \quad \text{as } x \to 0.$$

Therefore,

$$c_1(\alpha) I_{-\alpha}(x) + c_2(\alpha) I_\alpha(x) = \frac{c_1(\alpha)}{\Gamma(1-\alpha)} x^{-\alpha} + O(x^{-\alpha+1}) \quad \text{as } x \to 0.$$

In addition, we have

$$\begin{aligned} x^\alpha K_\alpha(x) &= x^\alpha \int_0^\infty e^{-x(t+1/t)} t^\alpha \frac{dt}{t} = \int_0^\infty e^{-(u+x^2/u)} u^\alpha \frac{du}{u} \\ &= \Gamma(\alpha) + o(1) \quad \text{as } x \to 0, \end{aligned}$$

from which we conclude

$$\Gamma(\alpha) = c_1(\alpha)/\Gamma(1-\alpha)$$

or $c_1(\alpha) = \Gamma(\alpha)\Gamma(1-\alpha) = \pi/\sin(\pi\alpha)$. The evaluation of $c_2(\alpha)$ follows from the relation $K_\alpha(x) = K_{-\alpha}(x)$. This concludes the proof.

Corollary 3.3. *For any $\alpha \notin \mathbf{Z}$, the functions $K_\alpha(w,b)$, $K_\alpha(b,w)$, and $K_\alpha(w)$ admit analytic continuations to all $w \in \mathbf{C} \setminus \mathbf{R}_{\leq 0}$. Further, for all $\epsilon > 0$, we have*

$$K_\alpha(w) = O(|w|^{-|\mathrm{Re}(\alpha)|})$$

$$K_\alpha(w,b) = O(|w|^{-2|\mathrm{Re}(\alpha)|}) \quad \text{and} \quad K_\alpha(b,w) = O(1)$$

as $|w| \to 0$ with $|\arg(w)| \leq \pi - \epsilon$.

Remark 1. Proposition 3.2 allows for the development of full asymptotic expansions of the K Bessel functions near the origin. However, for our purposes, we only need the lead term asymptotics as stated in Corollary 3.3.

To continue, we shall determine the asymptotic behavior of the K Bessel functions near infinity. For this, we need the following integral formula.

Proposition 3.4. *For all α with $\mathrm{Re}(\alpha) > -1$ and $w \in \mathbf{C} \setminus \mathbf{R}_{\leq 0}$, we have*

$$\Gamma(\alpha+1)K_{\alpha+1/2}(w) = \frac{\sqrt{\pi}}{\sqrt{w}} e^{-2w} \int_0^\infty e^{-u} u^\alpha \left(1 + \frac{u}{4w}\right)^\alpha du$$

$$= 2\sqrt{\pi} w^{\alpha+1/2} \int_1^\infty e^{-2wt}(t^2-1)^\alpha dt.$$

Proof. We shall first prove the stated result with $w = x \in \mathbf{R}_{>0}$, from which the general result follows by analytic continuation. We write down the integrals defining $K_{\alpha+1/2}$ and the gamma function, and we multiply them using Fubini's theorem. Then we make a multiplicative translation $u \mapsto u/t$, followed by the inversion $t \mapsto 1/t$. We then obtain

$$\Gamma(\alpha+1)K_{\alpha+1/2}(x) = \int_0^\infty \int_0^\infty e^{-((x+u)t + x/t)} t^{1/2} \frac{dt}{t} u^{\alpha+1} \frac{du}{u}$$

$$= \sqrt{\pi} \int_0^\infty e^{-2\sqrt{x}\sqrt{x+u}} \frac{u^{\alpha+1}}{\sqrt{x+u}} \frac{du}{u}$$

$$= \frac{\sqrt{\pi}}{\sqrt{x}} \int_0^\infty e^{-2x(1+u)^{1/2}} \frac{x^{\alpha+1} u^{\alpha+1}}{\sqrt{1+u}} \frac{du}{u}.$$

The first equality above uses the integral definitions for the gamma function and the one variable K Bessel function, and the second follows from the transformation $u \mapsto xu$. Now we put

$$(1+u)^{1/2} = 1 + v \quad \text{so} \quad dv = (1+u)^{-1/2} du/2.$$

After this change of variables, the factor e^{-2x} comes out of the integral. Then we let $v = u/2x$. A trivial algebraic manipulation gives the first formula. Substituting $u = 2x(t-1)$ yields the second, thus proving Proposition 3.4.

As an immediate corollary of Proposition 3.4, we have the following asymptotic formula from page 963 of [GR 65].

Corollary 3.5. *For any α with $\mathrm{Re}(\alpha) > -1$ and $\alpha \notin \mathbf{Z}$, and all $\epsilon > 0$, we have the asymptotic expansion*

$$K_{\alpha+1/2}(w)$$
$$= \frac{\sqrt{\pi}}{\sqrt{w}} e^{-2w} \sum_{m=0}^{N-1} \frac{\Gamma(\alpha+m+1)}{\Gamma(m+1)\Gamma(\alpha-m+1)} \frac{1}{(4w)^m} + O\left(e^{-2\mathrm{Re}(w)}|w|^{-N-1/2}\right)$$

as $|w| \to \infty$ with $|\arg(w)| \leq \pi - \epsilon$. If $\alpha = n \in \mathbf{Z}$, then we have the formula

$$K_{n+1/2}(w) = \frac{\sqrt{\pi}}{\sqrt{w}} e^{-2w} \sum_{m=0}^{n} \frac{\Gamma(n+m+1)}{\Gamma(m+1)\Gamma(n-m+1)} \frac{1}{(4w)^m}$$

which holds for all $w \in \mathbf{C} \setminus \mathbf{R}_{\leq 0}$.

Proof. Simply expand the integrand in the formula proved in Proposition 3.4 using the binomial theorem.

Next we shall consider the integral

$$(5) \qquad I(a,b,c;w) = \int_0^1 t^a (1-t^2)^b (1-wt)^c \, dt,$$

which is well defined when $\mathrm{Re}(a) > -1$, $\mathrm{Re}(b) > -1$, all $c \in \mathbf{C}$, provided $w \notin [0,1]$. The following result relates the integral (5) to the Laplace-Mellin transform of the K Bessel function.

Proposition 3.6. *For $z \in \mathbf{C}$ with $\mathrm{Re}(z) > 0$ and $s \in \mathbf{C}$ with $\mathrm{Re}(s) > \mathrm{Re}(\alpha)$ we have*

$$\mathrm{LMK}_\alpha(s,z) = \frac{2\sqrt{\pi}\,\Gamma(s+\alpha)}{2^{s+\alpha}\Gamma(\alpha+1/2)} I(s-\alpha-1, \alpha-1/2, -s-\alpha; -z/2).$$

Therefore, by change of variable, we have for $\beta \in \mathbf{C}$ with $\mathrm{Re}(\beta) > 0$,

$$\text{LM}(K_\alpha \circ \beta)(s,z) = \int_0^\infty K_\alpha(\beta x) e^{-zx} x^s \frac{dx}{x}$$

$$= \frac{2\sqrt{\pi}\,\Gamma(s+\alpha)}{2^{s+\alpha}\beta^s \Gamma(\alpha+1/2)} I(s-\alpha-1, \alpha-1/2, -s-\alpha; -z/2\beta).$$

Proof. Using the second expression in Proposition 3.4, we now integrate the Laplace-Mellin transform by interchanging the order of integration, which yields the formula

$$\int_0^\infty K_\alpha(x) e^{-zx} x^s \frac{dx}{x} = \frac{2\sqrt{\pi}}{\Gamma(\alpha+1/2)} \int_1^\infty \int_0^\infty e^{-2xt}(t^2-1)^{\alpha-1/2} x^\alpha e^{-zx} x^s \frac{dx}{x} dt$$

$$= \frac{2\sqrt{\pi}\,\Gamma(\alpha+s)}{\Gamma(\alpha+1/2)} \int_1^\infty (t^2-1)^{\alpha-1/2}(2t+z)^{-s-\alpha} dt.$$

To finish, one uses the change of variables $t \mapsto 1/t$, and quotes analytic continuation.

Remark 2. One can relate the integral considered in the above proposition to the classical hypergeometric function; see, for example, page 712 of [GR 65]. We shall use Proposition 3.6 in our consideration of the Cramér theorem for the Bessel class, so we are solely interested in a quantitative analysis of the integral considered in Proposition 3.6.

Lemma 3.7. *The integral $I(a,b,c;w)$ has an analytic continuation along every path in $\mathbb{C} \setminus \{1\}$.*

Proof. The holomorphicity for $w \notin [0,1]$ is immediate from the integral. For any $w \in (0,1)$, the holomorphicity follows from changing the contour of integration by integrating over a path connecting $t = 0$ to $t = 1$ by a path not lying on the real line. Near $w = 0$, we can expand $(1-wt)^c$ via the binomial theorem in order to express the integral as a convergent power series which converges for all $|w| < 1$. The lemma follows.

In addition to the Laplace-Mellin transform considered in Proposition 3.6, we shall need to analyze the partial Laplace-Mellin transforms, which are defined by

$$\text{LM}_0^A(K_\alpha \circ \beta)(s,z) = \int_0^A K_\alpha(\beta x) e^{-zx} x^s \frac{dx}{x}$$

and
$$\mathbf{LM}_A^\infty(K_\alpha \circ \beta)(s,z) = \int_A^\infty K_\alpha(\beta x) e^{-zx} x^s \frac{dx}{x}.$$

Proposition 3.8. *Let $\alpha \in \mathbf{C}$, $\beta \in \mathbf{R}$, and $A > 0$.*

(a) *The partial Laplace-Mellin transform $\mathbf{LM}_0^A(K_\alpha \circ \beta)(s,z)$ extends to an entire function of z and s.*
(b) *For any $s \in \mathbf{C}$, the partial Laplace-Mellin transform $\mathbf{LM}_A^\infty(K_\alpha \circ \beta)(s,z)$ extends to a function of $z \in \mathbf{C}$ with logarithmic fractional singularity at -2β. Its fractional order is $(-s+1/2, 0)$ if $-s+1/2 \notin \mathbf{Z}_{\geq 0}$ and $(n, n+1)$ if $-s+1/2 = n \in \mathbf{Z}_{\geq 0}$.*

Proof. For part (a), one can expand the K Bessel function in a series, as given by Proposition 3.2, and the exponential function in a power series to conclude that $\mathbf{LM}_0^A(K_\alpha \circ \beta)(s,z)$ is holomorphic for all z and $\mathrm{Re}(s) > \mathrm{Re}(\alpha)$. To extend to all s, one uses the differential equation
$$\partial_z(\mathbf{LM}_0^A K_\alpha \circ \beta)(s,z) = -(\mathbf{LM}_0^A K_\alpha \circ \beta)(s+1,z).$$
To prove part (b), we first write
$$(\mathbf{LM}_A^\infty K_\alpha \circ \beta)(s,z) = (\mathbf{LM} K_\alpha \circ \beta)(s,z) - (\mathbf{LM}_0^A K_\alpha \circ \beta)(s,z).$$
By Lemma 3.7 and part (a), it follows that $(\mathbf{LM}_A^\infty K_\alpha \circ \beta)(s,z)$ has an analytic continuation along every path in $\mathbf{C} \setminus \{-2\beta\}$. To examine the asymptotic behavior near $z = -2\beta$, we use Corollary 3.5, which shows that it suffices to consider integrals of the form
$$\int_A^\infty e^{-(z+2\beta)x} x^{s-1/2} \frac{dx}{x},$$
from which the proposition follows.

In the language established in the previous section, we obtain:

Corollary 3.9. *The integral $I(a,b,c;w)$ has a logarithmic fractional singularity at $w = 1$. That is, there exist functions $h_1(w)$ and $h_2(w)$ which are holomorphic near $w = 1$ such that near $w = 1$ we have*
$$I(a,b,c;w) = h_1(w) + (w-1)^{b+c+1} h_2(w) \quad \text{if } b+c+1 \notin \mathbf{Z}_{\geq 0}$$
and
$$I(a,b,c;w) = h_1(w) + (w-1)^{b+c+1} \log(w-1) h_2(w) \quad \text{if } b+c+1 \in \mathbf{Z}_{\geq 0}.$$

Proof. Directly from Proposition 3.6 and Proposition 3.8.

4. Cramér's theorem and Fujii's theorem for Bessel series

Let $(\mathrm{B}, \widetilde{\mathrm{B}}, R)$ be in the polynomial Bessel fundamental class. Let $a > 0$ be such that $\sigma_0 + a > \sigma_0'$. We define the following regions in the complex plane:

$\mathscr{R}_a$ = infinite open rectangle bounded by the lines

$$\mathrm{Re}(s) = -a, \quad \mathrm{Re}(s) = \sigma_0 + a.$$

$\mathscr{R}_a^+$ = semi-infinite open rectangle bounded by the lines

$$\mathrm{Re}(s) = -a, \quad \mathrm{Re}(s) = \sigma_0 + a, \quad \mathrm{Im}(s) = 0.$$

$\mathscr{R}_a^+(T)$ = the portion of $\mathscr{R}_a^+$ below the line $\mathrm{Im}(s) = T$.

$\mathscr{R}_a^+(\varepsilon; T)$ = the portion of $\mathscr{R}_a^+(T)$ above the line $\mathrm{Im}(s) = \varepsilon$.

$\mathscr{R}_a^+(\varepsilon)$ = the portion of $\mathscr{R}_a^+$ above the line $\mathrm{Im}(s) = \varepsilon$.

We assume that the functions B, $\widetilde{\mathrm{B}}$, and R have no poles on the vertical boundary components

$$(-a + i\infty, -a] \quad \text{and} \quad [\sigma_0 + a, \sigma_0 + a + i\infty)$$

of $\mathscr{R}_a^+$. In this section, we shall consider the series

$$2\pi i V_{\mathrm{B},a}(z) = 2\pi i \sum_{\rho \in \mathscr{R}_a^+} \mathrm{res}_\rho(\mathrm{B}) e^{\rho z} = \int_{\partial \mathscr{R}_a^+} \mathrm{B}(s) e^{sz} ds$$

which we *formally* write as

$$(1) \quad 2\pi i V_{\mathrm{B},a}(z) = \int_{-a+i\infty}^{-a} e^{sz} \mathrm{B}(s) ds + \int_{-a}^{\sigma_0+a} \mathrm{B}(s) e^{sz} ds + \int_{\sigma_0+a}^{\sigma_0+a+i\infty} \mathrm{B}(s) e^{sz} ds.$$

As in [JoL 93c], we will realize (1) as

$$2\pi i V_{\mathrm{B},a}(z) = \lim_{\substack{\varepsilon \to 0 \\ T \to \infty}} 2\pi i V_{\mathrm{B},a}(z; \varepsilon, T)$$

where

$$(2) \quad 2\pi i V_{\mathrm{B},a}(z; \varepsilon, T) = 2\pi i \sum_{\rho \in \mathscr{R}_a^+(\varepsilon;T)} \mathrm{res}_\rho(\mathrm{B}) e^{\rho z} = \int_{\mathscr{R}_a^+(\varepsilon;T)} \mathrm{B}(s) e^{sz} ds.$$

It is necessary to consider $\varepsilon > 0$ as above since the function B may have poles on the line segment $[-a, \sigma_0 + a]$. With this, the infinite integrals on the vertical lines in (1) are viewed as the limits

Bessel series instead of Dirichlet series 93

$$\int_{\sigma_0+a+i\varepsilon}^{\sigma_0+a+i\infty} = \lim_{m\to\infty} \int_{\sigma_0+a+i\varepsilon}^{\sigma_0+a+iT_m}$$

for a suitable chosen sequence $\{T_m\}$ of real numbers which converge to infinity. The sequence $\{T_m\}$ is determined by the following lemma.

Lemma 4.1. *There is a sequence $\{T_m\}$ of real numbers which converge to infinity such that for any fixed z with $\mathrm{Im}(z) > 0$, we have*

$$\lim_{T_m\to\infty}\left[\int_{-a+iT_m}^{\sigma_0+a+iT_m} \widetilde{B}(s)e^{sz}\,ds\right] = 0.$$

Proof. Follows directly from Lemma 1.3.

Set
$$V_{B,a}(z;\varepsilon) = \lim_{T_m\to\infty} V_{B,a}(z;\varepsilon,T_m),$$
so, upon using the functional equation of $B(s)$, we can write (1) as

$$2\pi i V_{B,a}(z;\varepsilon) = -\int_{\sigma_0+a-i\varepsilon}^{\sigma_0+a-i\infty} e^{z(\sigma_0-s)}\left[\widetilde{B}(s) + R(\sigma_0-s)\right]ds + \int_{-a+i\varepsilon}^{\sigma_0+a+i\varepsilon} B(s)e^{sz}\,ds$$

(3) $$+ \int_{\sigma_0+a+i\varepsilon}^{\sigma_0+a+i\infty} B(s)e^{sz}\,ds.$$

We are interested in a quantative analysis of (3). To begin, we have the following elementary statement.

Lemma 4.2. *The function $V_{B,a}(z;\varepsilon)$ is holomorphic for $z \in \mathbf{C}$ with $\mathrm{Im}(z) > 0$.*

Proof. The integral in (3) over the finite segment from $-a$ to $\sigma_0 + a$ is entire in z. By assumption, the functions $B(s)$ and $\widetilde{B}(s)$ have polynomial growth on vertical lines contained in the half-plane of uniform convergence; hence, the integrals in (3) involving B and $\widetilde{B}$ are holomorphic for $\mathrm{Im}(z) > 0$. Chapter I of [JoL 94b] establishes polynomial growth estimates for regularized harmonic series on vertical lines, which then imply that the integral in (3) containing R as an integrand is holomorphic for z with $\mathrm{Im}(z) > 0$.

Proposition 4.3. *If R is of regularized harmonic series type, then the integral*

$$\int_{\sigma_0+a-i\varepsilon}^{\sigma_0+a-i\infty} e^{z(\sigma_0-s)}R(\sigma_0-s)ds$$

satisfies the asymptotic condition **AS 2.**$\mathscr{S}$ *where $\mathscr{S}$ is any sector of the form $\mathbf{C}\setminus P$ with P being a ray from 0 to ∞ which lies in the closed half plane $\mathrm{Im}(z) \leq 0$.*

357

Proof. The proof is contained in the analysis of §3 and §4 of [JoL 93c].

It remains to study quantitative properties of the integrals

(4) $$\int_{\sigma_0+a+i\varepsilon}^{\sigma_0+a+i\infty} \mathrm{B}(s)e^{sz}\,ds = \int_{\sigma_0+a+i\varepsilon}^{\sigma_0+a+i\infty} \sum_j P_j(s) K_{\alpha_j}(s, \log q_j^{1/2}) e^{sz}\,ds$$

and

(5) $$\int_{\sigma_0+a-i\varepsilon}^{\sigma_0+a-i\infty} \widetilde{\mathrm{B}}(s) e^{z(\sigma_0-s)}\,ds = e^{\sigma_0 z} \int_{\sigma_0+a-i\varepsilon}^{\sigma_0+a-i\infty} \sum_j \tilde{P}_j(s) K_{\tilde{\alpha}_j}(s, \log \tilde{q}_j^{1/2}) e^{-sz}\,ds.$$

We shall consider (4) with the analysis of (5) being identical. For $\mathrm{Re}(s) > \sigma_0'$, we have

$$\lim_{\varepsilon \to 0} \int_{\sigma_0+a+i\varepsilon}^{\sigma_0+a+i\infty} \mathrm{B}(s) e^{sz}\,ds = \int_{\sigma_0+a}^{\sigma_0+a+i\infty} \mathrm{B}(s) e^{sz}\,ds,$$

so, for our purposes, it suffices to consider the setting when $\varepsilon = 0$.

Let us write (4) as a sum of integrals involving a Bessel function of one variable yielding the formula

(6) $$\int_{\sigma_0+a}^{\sigma_0+a+i\infty} \mathrm{B}(s) e^{sz}\,ds = \int_{\sigma_0+a}^{\sigma_0+a+i\infty} \sum_j P_j(s) (\log q_j^{1/2}/s)^{\alpha_j} K_{\alpha_j}(s \log q_j^{1/2}) e^{sz}\,ds.$$

Now set $P_j(s) = \sum_{k=0}^{n_j} c_{j,k} s^k$, so then (6) takes the form

(7) $$= \sum_j \sum_{k=0}^{n_j} c_{j,k} \int_{\sigma_0+a}^{\sigma_0+a+i\infty} s^k \left(\log q_j^{1/2}/s\right)^{\alpha_j} K_{\alpha_j}(s \log q_j^{1/2}) e^{sz}\,ds.$$

By a change of contour along a quarter circle, (7) is equal to

(8) $$= \sum_j \sum_{k=0}^{n_j} c_{j,k} \int_{\sigma_0+a}^{\infty} x^k \left(\log q_j^{1/2}/x\right)^{\alpha_j} K_{\alpha_j}(x \log q_j^{1/2}) e^{xz}\,dx,$$

where it is necessary to assume $\mathrm{Re}(z) < 0$. The interchange of integration and summation which yields (7) from (6) and the change of contour which yields (8) from (7) follows from Lemma 1.2. Observe that (8) provides an analytic continuation of (6) to the half plane $\mathrm{Re}(z) < 0$. By uniform convergence, we can write (8) as

$$= \sum_j \sum_{k=0}^{n_j} c_{j,k} (\log \mathbf{q}_j^{1/2})^{\alpha_j} \int_{\sigma_0+a}^{\infty} K_{\alpha_j}(x \log \mathbf{q}_j^{1/2}) e^{xz} x^{k-\alpha_j+1} \frac{dx}{x}$$

$$(9) \quad = \sum_j \sum_{k=0}^{n_j} c_{j,k} (\log \mathbf{q}_j^{1/2})^{\alpha_j} \mathbf{LM}_{\sigma_0+a}^{\infty}(K_{\alpha_j} \circ \log \mathbf{q}_j^{1/2})(k - \alpha_j + 1, -z).$$

The analysis of Proposition 3.8 now applies to yield the analytic properties of (9), as summarized in the following theorem.

Theorem 4.4. *Let S_B be the discrete set $\{\log \mathbf{q}_j\}$. Then the function of z defined by*

$$\int_{\sigma_0+a}^{\sigma_0+a+i\infty} \mathrm{B}(s) e^{sz} \, ds$$

$$= \sum_j \sum_{k=0}^{n_j} c_{j,k} (\log \mathbf{q}_j^{1/2})^{\alpha_j} \mathbf{LM}_{\sigma_0+a}^{\infty}(K_{\alpha_j} \circ \log \mathbf{q}_j^{1/2})(k - \alpha_j + 1, -z)$$

has logarithmic fractional singularities on $\mathbf{C}$ relative to S_B.

As stated above, the analysis for (5) is identical to the analysis carried out for (4), and the analogue of Theorem 4.4 holds. Combining these results with Proposition 4.3, we have the following generalization of **Cramér's theorem**.

Theorem 4.5. *Choose $\varepsilon > 0$ so that the function $\mathrm{B}(s)$ has no poles in the closed rectangle with vertices at the points*

$$-a, -a + i\varepsilon, \sigma_0 + a, \sigma_0 + a + i\varepsilon.$$

Then the Cramér function $V_{\mathrm{B},a}(z;\varepsilon)$ can be expressed as

$$2\pi i V_{\mathrm{B},a}(z;\varepsilon) = \int_{-a+i\varepsilon}^{\sigma_0+a+i\varepsilon} e^{zs} \mathrm{B}(s) ds - \int_{\sigma_0+a-i\varepsilon}^{\sigma_0+a-i\infty} e^{(\sigma_0-s)z} R(\sigma_0 - s) ds$$

$$+ \sum_j \sum_{k=0}^{n_j} c_{j,k} (\log \mathbf{q}_j^{1/2})^{\alpha_j} \mathbf{LM}_{\sigma_0+a}^{\infty}(K_{\alpha_j} \circ \log \mathbf{q}_j^{1/2})(k - \alpha_j + 1, -z)$$

$$- e^{\sigma_0 z} \sum_j \sum_{k=0}^{\tilde{n}_j} \bar{c}_{j,k} (\log \bar{\mathbf{q}}_j^{1/2})^{\bar{\alpha}_j} \mathbf{LM}_{\sigma_0+a}^{\infty}(K_{\bar{\alpha}_j} \circ \log \bar{\mathbf{q}}_j^{1/2})(k - \bar{\alpha}_j + 1, z)$$

$$+ e^{\sigma_0 z} \int_{\sigma_0+a}^{\sigma_0+a-i\varepsilon} e^{-sz} \tilde{\mathrm{B}}(s) ds - \int_{\sigma_0+a}^{\sigma_0+a+i\varepsilon} e^{sz} \mathrm{B}(s) ds,$$

with the following additional information:

1. *the integral involving R converges for* $\text{Im}(z) > 0$ *and satisfies* **AS 2.$\mathscr{S}$**.
2. *the series over the sequence* $\{\log \mathfrak{q}_j\}$ *converges uniformly and absolutely for all* $z \in \mathbf{C}$ *with* $\text{Re}(z) < 0$ *and extends to a function of* $z \in \mathbf{C}$ *with logarithmic fractional singularities at* S_B;
3. *the series over the sequence* $\{\log \bar{\mathfrak{q}}_j\}$ *converges uniformly and absolutely for all* $z \in \mathbf{C}$ *with* $\text{Re}(z) > 0$ *and extends to a function of* $z \in \mathbf{C}$ *with logarithmic fractional singularites at* $-S_{\widetilde{\mathrm{B}}}$;
4. *the integrals over the finite intervals are entire functions of* z.

Remark 1. The order of the fractional singularities of the extended Cramér function is computed in Proposition 3.8. Specifically, we have

$$\text{ord}_{\log \mathfrak{q}_j}[V_{\mathrm{B},a}(z;\varepsilon)] = -\deg(P_j) + \alpha_j - 1/2,$$

and

$$\text{ord}_{-\log \bar{\mathfrak{q}}_j}[V_{\widetilde{\mathrm{B}},a}(z;\varepsilon)] = -\deg(\tilde{P}_j) + \tilde{\alpha}_j - 1/2.$$

If all polynomials P_j and $\tilde{P}_j$ are monomials of degree 1 and all orders α_j and $\tilde{\alpha}_j$ are equal to $1/2$, then the Bessel series is equal to the Dirichlet series, and the above result generalizes the Cramér theorem from [JoL 93c].

By following the argument in §6 of [JoL 93c] (see also Theorem 4.1 of Chapter II in [JoL 94b]), this time for the additive class of functions defined in §1 above, we have the following corollary of Theorem 4.5 which asserts that the assumption of regularized harmonic series type implies a type of "ladder" phenomenon.

Theorem 4.6. *Let* $(\mathrm{B}, \widetilde{\mathrm{B}}, R)$ *be in the polynomial Bessel fundamental class, and assume that R has reduced order* (M, m). *Then* B *and* $\widetilde{\mathrm{B}}$ *are of regularized harmonic series type and of reduced order* $(M, m+1)$ *if* $-\text{Re}(p_0) \in \mathbf{Z}$, *otherwise the reduced order is* (M, m).

Proof. Let $\{\rho_k\}$ be the sequence of poles of B in $\mathscr{R}_a^+$ with $a_k = -\text{res}_{\rho_k}(\mathrm{B})$. Set $\lambda_k = \rho_k/i$ and let

$$\theta_{\mathrm{B},a}^+(t) = \sum a_k e^{-\lambda_k t}.$$

Similarly construct $\theta_{\mathrm{B},a}^-(t)$. Using Theorem 4.5, let $R_{\mathrm{B},a}^+$ and $R_{\mathrm{B},a}^-$ be the regularized harmonic series associated with the theta series $\theta_{B,a}^+$ and $\theta_{\mathrm{B},a}^-$. By the choice of a, the Bessel series $\mathrm{B}(s)$ has no poles for $\text{Re}(s) \geq \sigma_0 + a$. By the functional equation, the only poles of $\mathrm{B}(s)$ in the half-plane $\text{Re}(s) \leq -a$ coincide in location and with opposite residue as the poles of the fudge factor $R(\sigma_0 - s)$. Since R is of regularized harmonic series type, we can write $R(s) = R_{\text{left},a}(s) + R_{\text{right},a}(s)$, where $R_{\text{left}}(s)$ is of regularized harmonic series type and $\mathrm{B}(s) + R_{\text{left},a}(s)$ is analytic in the half-plane $\text{Re}(s) \leq -a$. Let us now consider the function

$$F(s) = \mathrm{B}(s) - R_{\mathrm{B},a}^+(s) - R_{\mathrm{B},a}^-(s) + R_{\text{right}}(\sigma_0 - s).$$

The above discussion shows that F is entire. The function B is of finite Mittag-Leffler order, by assumptions, and all functions of regularized harmonic series type are of finite Mittag-Leffler order. Therefore, F is an entire function of finite Mittag-Leffler order, hence a polynomial. In order to determine the reduced order of B, one argues as in §6 of [JoL 93c].

Remark 2. As stated on page 117 of [JoL 94b], the Cramér function can be viewed as a type of wave kernel; this point will be discussed further in §8. Therefore, the identification of the fractional singular set of the Cramér function is an analogue of the Duistermaat-Guillemin theorem [DG 75].

With Theorem 4.5, we can state the following result, which we call the generalized **Fujii theorem** for the Bessel class (see Chapter VI of [JoL 94b]).

Corollary 4.7. *Let* $(\mathrm{B}, \widetilde{\mathrm{B}}, R)$ *be in the polynomial Bessel fundamental class. Let* $\{\rho\}$ *be the set of poles of* B *in the open infinite rectangle* $\mathscr{R}_a$, *and then set* $\{\lambda\} = \{\rho/i\}$. *Then for any non-zero* $\alpha \in \mathbf{R}$, *the Fujii function*

$$F_{\mathrm{B},a}(s;\alpha) = \sum_{\rho \in \mathscr{R}_a} \mathrm{res}_\rho(\mathrm{B}) \frac{e^{i\alpha\lambda}}{\lambda^s}$$

has a meromorphic continuation to all $s \in \mathbf{C}$ *with a finite number of simple poles. Furthermore, the function* $F_{\mathrm{B},a}(s;\alpha)$ *has an analytic continuation to all* $\mathbf{C}$ *if* α *is not in* $S_{\mathrm{B}} \cup (-S_{\widetilde{\mathrm{B}}})$.

Proof. Let $z = \alpha + it$ for any non-zero $\alpha \in \mathbf{R}$, and, with notation as above, let $\rho = i\lambda$. For $\alpha \neq 0$, Theorem 4.5 states that the function

$$V_{\mathrm{B},a}(\alpha + it) = \sum_{\rho \in \mathscr{R}_a} \mathrm{res}_\rho(\mathrm{B}) e^{i\alpha\lambda} e^{-t\lambda}$$

satisfies the asymptotic condition **AS 2**. By applying the Mellin transform, we conclude that the function

$$\Gamma(s) \sum_{\rho \in \mathscr{R}_a} \mathrm{res}_\rho(\mathrm{B}) \frac{e^{i\alpha\lambda}}{\lambda^s} = \int_0^\infty V_{\mathrm{B},a}(\alpha + it) t^s \frac{dt}{t}$$

has a meromorphic continuation to all $s \in \mathbf{C}$; see Theorem 2.5 of [JoL 93a] which also describes the location and type of singularities.

5. Explicit formulas for Bessel series

In this section, we shall establish the analogue of the explicit formulas for a triple $(\mathrm{B}, \widetilde{\mathrm{B}}, R)$ in the polynomial Bessel fundamental class. Quite generally, the explicit formulas state that the sum of a function taken over what plays the role of prime powers in number theory and lengths of geodesics in spectral theory is equal to the sum of the Mellin transform taken over the poles of the regularized harmonic series under consideration together with a "term at infinity" coming from the fudge term in the additive functional equation. We leave applications of the explicit formulas to questions such as counting of prime powers or poles for a subsequent paper. Here we will give the aspects of proof in the setting of the polynomial Bessel fundamental class which is new, beyond what was developed in [JoL 94b].

As in [JoL 94b], we let f be a measurable function on $\mathbf{R}^+$ so that under certain convergence conditions we have the **Mellin transform**

$$\mathbf{M}f(s) = \int_0^\infty f(u)u^s \frac{du}{u}.$$

Let $\sigma_0 \in \mathbf{R}_{\geq 0}$. We define $\mathbf{M}_{\sigma_0/2}f$ to be the translate of $\mathbf{M}f$ by $\sigma_0/2$, meaning

$$\mathbf{M}_{\sigma_0/2}f(s) = \mathbf{M}f(s - \sigma_0/2).$$

We put

$$F(x) = f(e^{-x}) \quad \text{so} \quad f(u) = F(-\log u).$$

Then letting $s = \sigma + it$, we find

$$\mathbf{M}_{\sigma_0/2}f(s) = \int_{-\infty}^{\infty} F(x)e^{-(\sigma-\sigma_0/2)x}e^{-itx}\,dx.$$

We denote by $BV(\mathbf{R})$ the space of functions of bounded variation on $\mathbf{R}$. Following Barner [Ba 81], we require the test functions F to satisfy the following basic Fourier conditions.

FOU 1. $F \in BV(\mathbf{R}) \cap L^1(\mathbf{R})$.
FOU 2. F is **normalized**, meaning

$$F(x) = \frac{1}{2}(F(x+) + F(x-)) \quad \text{for all } x \in \mathbf{R}.$$

FOU 3. There exists a constant $a' > 0$ such that the function

$$x \mapsto F(x)e^{(\sigma_0/2+a')|x|}$$

is in $BV(\mathbf{R})$.

If we let N be any integer ≥ 0, then we say that F satisfies the **basic Fourier conditions to order N** if F is N times differentiable and its first N derivatives satisfy the above basic conditions.

Recall that the definition of the Fourier transform $\phi^\wedge$ of an L^1 function ϕ is

$$\phi^\wedge(t) = \frac{1}{\sqrt{2\pi}} \int_{-\infty}^{\infty} \phi(y) e^{-ity} dy.$$

For $\mu \in \mathbf{C}$, one defines the μ-**fractional derivative** of a function ϕ through the inverse Fourier transform formula, namely

$$\phi^{(\mu)}(x) = \frac{1}{\sqrt{2\pi}} \int_{-\infty}^{\infty} \phi^\wedge(t)(it)^\mu e^{itx} dt,$$

provided $\phi^\wedge$ has sufficiently fast decay (exponential decay at infinity suffices). In order to define the μ-fractional derivative, it is necessary to assume that the Fourier transform $\phi^\wedge(t)$ vanishes to high enough order at $t = 0$, which is guaranteed if ϕ has enough derivatives in L^1. If we let $\phi_b(x) = \phi(x) e^{bx}$, then

$$\phi_b(x) = \lim_{T \to \infty} \frac{1}{\sqrt{2\pi}} \int_{-T}^{T} \phi^\wedge(t) e^{(it+b)x} dt$$

so, in particular, abbreviating $\phi_b^{(\mu)} = (\phi_b)^{(\mu)}$, we have

$$\phi_b^{(\mu)}(x) = \lim_{T \to \infty} \frac{1}{\sqrt{2\pi}} \int_{-T}^{T} \phi^\wedge(t)(it+b)^\mu e^{(it+b)x} dt$$

and

(1) $$\phi_b^{(\mu)}(0) = \lim_{T \to \infty} \frac{1}{2\pi} \int_{-T}^{T} \int_{-\infty}^{\infty} \phi(y)(it+b)^\mu e^{-ity} dy\, dt.$$

We let:

$\mathscr{R}_a$ be the infinite rectangle bounded by the vertical lines

$$\mathrm{Re}(s) = -a \quad \text{and} \quad \mathrm{Re}(s) = \sigma_0 + a.$$

$\mathscr{R}_a(T)$ be the finite rectangle bounded by the above vertical lines and the lines

$$\mathrm{Im}(s) = -T \quad \text{and} \quad \mathrm{Im}(s) = T.$$

We assume that R has no poles on $\partial \mathscr{R}_a$. The explicit formulas follows from a consideration of the integral

(2) $$\frac{1}{2\pi i} \int_{\partial \mathscr{R}_a} \mathrm{B}(s) \mathbf{M}_{\sigma_0/2} f(s) ds = \lim_{n \to \infty} \frac{1}{2\pi i} \int_{\partial \mathscr{R}_a(T_n)} \mathrm{B}(s) \mathbf{M}_{\sigma_0/2} f(s) ds,$$

for a suitably chosen sequence of real numbers $T_n \to \infty$. The only aspect of the analysis from Chapter III of [JoL 94b] which changes concerns the integrals considered in §3 of [JoL 94b]. The purpose of this section is to evaluate these integrals. First, we recall necessary notation from [JoL 94b], and then we state the explicit formulas for the polynomial Bessel fundamental class. After this, we give the analysis needed in order to prove the explicit formulas.

We let:

$\{\rho\}$ = the set of poles of B in the *full* strip $-a \leq \sigma \leq \sigma_0 + a$;
$\{\alpha\}$ = the set of poles of R in the *half* strip $-a \leq \sigma \leq \sigma_0/2$;

With this, we are interested in the infinite sum

$$V_{\mathrm{B},a}(f) = \sum_{\rho \in \mathcal{R}_a} \mathrm{res}_\rho(\mathrm{B}) \mathbf{M}_{\sigma_0/2} f(\rho),$$

which is understood in the limiting sense

$$\sum_{\rho \in \mathcal{R}_a} \mathrm{res}_\rho(\mathrm{B}) \mathbf{M}_{\sigma_0/2} f(\rho) = \lim_{n \to \infty} \sum_{\rho \in \mathcal{R}_a(T_n)} \mathrm{res}_\rho(\mathrm{B}) \mathbf{M}_{\sigma_0/2} f(\rho).$$

Since the similar sum over the family $\{\alpha\}$ is taken on the left half interval, we use the notation

$$\begin{aligned} V_{R,a,\sigma_0/2}(f) &= \sum_{\mathrm{Re}(\alpha) \leq \sigma_0/2} \mathrm{res}_\alpha(R) \mathbf{M}_{\sigma_0/2} f(\alpha) \\ &= \lim_{n \to \infty} \sum_{\substack{\alpha \in \mathcal{R}_a(T_n) \\ \mathrm{Re}(\alpha) \leq \sigma_0/2}} \mathrm{res}_\alpha(R) \mathbf{M}_{\sigma_0/2} f(\alpha). \end{aligned}$$

For any u such that R has no pole on the line $\mathrm{Re}(s) = u$, we define

$$W^{\#}_{R,u,\sigma_0}(f) = \lim_{n \to \infty} \frac{1}{2\pi i} \int_{u-iT_n}^{u+iT_n} \mathbf{M}_{\sigma_0/2}(f)(s) R(\sigma_0 - s) ds.$$

Under suitable conditions on the test function F, which will be expressed in terms of the above Fourier conditions, we shall prove:

Bessel series instead of Dirichlet series   101

The Explicit Formula.

$$V_{\mathrm{B},a}(f) + V_{R,a,\sigma_0/2}(f) - W^{\#}_{R,-a,\sigma_0}(f)$$

$$= \sum_j \sum_{k=0}^{n_j} c_{j,k} (\log \mathbf{q}_j^{1/2})^{\alpha_j} \int_0^\infty F^{(k-\alpha_j)}_{\sigma_0/2}\left(-\log \mathbf{q}_j^{1/2}(u+1/u)\right) u^{\alpha_j} \frac{du}{u}$$

$$+ \sum_j \sum_{k=0}^{\tilde{n}_j} \tilde{c}_{j,k} (\log \tilde{\mathbf{q}}_j^{1/2})^{\tilde{\alpha}_j} \int_0^\infty F^{(k-\tilde{\alpha}_j)}_{-\sigma_0/2}\left(\log \tilde{\mathbf{q}}_j^{1/2}(u+1/u)\right) u^{\tilde{\alpha}_j} \frac{du}{u}.$$

Note that the two sums on the right can be written more compactly as follows using D for ordinary differentiation:

$$\sum_j (\log \mathbf{q}_j^{1/2})^{\alpha_j} \int_0^\infty P_j(D) F^{(-\alpha_j)}_{\sigma_0/2}\left(-\log \mathbf{q}_j^{1/2}(u+1/u)\right) u^{\alpha_j} \frac{du}{u}$$

and

$$\sum_j (\log \tilde{\mathbf{q}}_j^{1/2})^{\tilde{\alpha}_j} \int_0^\infty \tilde{P}_j(D) F^{(-\tilde{\alpha}_j)}_{-\sigma_0/2}\left(\log \tilde{\mathbf{q}}_j^{1/2}(u+1/u)\right) u^{\tilde{\alpha}_j} \frac{du}{u}.$$

The main result of this section is the following theorem.

Theorem 5.1. *Let* $(\mathrm{B}, \tilde{\mathrm{B}}, R)$ *be in the polynomial Bessel fundamental class, with no poles on the line* $\mathrm{Re}(s) = -a$. *Then for any function F which satisfies the Fourier conditions to sufficiently large order, the explicit formula holds.*

As in our formulation of Cramér's theorem, the above theorem is an inductive one, expressing the sum $V_{\mathrm{B},a}(f)$ in terms of a similar sum concerning $V_{R,a,\sigma_0/2}$, the Weil functional, and terms involving the sequences $\mathbf{q}_j$ and $\tilde{\mathbf{q}}_j$.

Remark 1. Observe that the sum $V_{\mathrm{B},a} + V_{R,a,\sigma_0/2} - W^{\#}_{R,-a,\sigma_0}(f)$ is independent of a even though no individual term is independent of a.

Remark 2. The analysis of the Weil functional $W^{\#}_{R,-a,\sigma_0}(f)$ is quite important since, in Weil's original formulation of the explicit formulas for Euler products with functional equations whose fudge factors are gamma functions, the Weil functional was left in a form which required complicated arguments to identify it with the classical forms in the classical special cases. Barner reformulated the Weil functional in more practical terms in [Ba 81] in the case considered by Weil. In [JoL 93b], we developed a Parseval type formula for a general regularized harmonic series which then applies to evaluate the Weil functional in the general setting considered here. Since all applications of the explicit formulas for the polynomial Bessel fundamental class are left for a subsequent paper, the reader is referred to that forthcoming article as well as [JoL 93b] for further analysis of the Weil functional.

Remark 3. In the case when $k = 1$ and $\alpha = 1/2$, we can write

$$F^{(1/2)}_{\sigma_0/2}(c(u+1/u))$$
$$= \lim_{T \to \infty} \frac{1}{2\pi} \int_{-T}^{T} \int_{-\infty}^{\infty} F(x) e^{-itx} e^{-(it+\sigma_0/2)c(u+1/u)} (it+\sigma_0/2)^{1/2} dx dt.$$

If we use the collapse of the Bessel function at $\alpha = 1/2$, namely the formula

$$K_{1/2}((it+\sigma_0/2)c) = \frac{\sqrt{\pi}}{\sqrt{c}} \frac{e^{-2c(it+\sigma_0/2)}}{\sqrt{it+\sigma_0/2}},$$

we then obtain the formula

$$\int_0^\infty F^{(1/2)}_{\sigma_0/2}(c(u+1/u)) u^{1/2} \frac{du}{u} = \frac{\sqrt{\pi}}{\sqrt{c}} e^{-\sigma_0 c} F(-2c).$$

If $c = \log \mathbf{q}_j^{1/2}$, then

$$\frac{\sqrt{\pi}}{\sqrt{c}} e^{-\sigma_0 c} F(-2c) = \frac{\sqrt{2\pi}}{\sqrt{\log \mathbf{q}_j}} \mathbf{q}_j^{-\sigma_0/2} F(-\log \mathbf{q}_j).$$

If we take

$$c_j = (-c(\mathbf{q}_j) \log \mathbf{q}_j)/\sqrt{\pi}, \quad k = 1 \quad \text{and} \quad \alpha_j = 1/2,$$

we then have

$$c_j (\log \mathbf{q}_j^{1/2})^{1/2} \int_0^\infty F^{(1/2)}_{\sigma_0/2}\left(\log \mathbf{q}_j^{1/2}(u+1/u)\right) u^{1/2} \frac{du}{u}$$
$$= \frac{-c(\mathbf{q}_j) \log \mathbf{q}_j}{\mathbf{q}_j^{\sigma_0/2}} F(-\log \mathbf{q}_j),$$

see formula (8) of §1. Using these calculations, Theorem 5.1 specializes to the results in §3, Chapter III of [JoL 94b].

Next we give the detailed analysis proving Theorem 5.1.

Lemma 5.2. *Let* $(\mathrm{B}, \widetilde{\mathrm{B}}, R)$ *be in the polynomial Bessel fundamental class, and assume that R is of regularized harmonic series type. Assume f satisfies the basic Fourier conditions to sufficiently large order. Then there is a sequence $\{T_n\}$ of real numbers which converge to infinity such that*

$$\lim_{T_n \to \infty} \left[\int_{-a \pm iT_n}^{\sigma_0 + a \pm iT_n} \mathrm{B}(s) \mathbf{M}_{\sigma_0/2} f(s) ds \right] = 0.$$

Proof. By Theorem 4.6, the Bessel series Б is of regularized harmonic series type. By Lemma 1.3, there is an integer N and a sequence of real numbers $T_n \to \infty$ such that for all $x \in [-a, \sigma_0 + a]$, we have the asymptotic relation

$$\text{Б}(x \pm iT_n) = O(T_n^N) \quad \text{as } T_n \to \infty.$$

Lemma 1.1, Chapter III of [JoL 94b] shows that if f satisfies the basic Fourier conditions to order $N + 1$, then

$$\mathbf{M}_{\sigma_0/2}f(s) = O(1/|s|^{N+2}) \quad \text{for } |s| \to \infty.$$

With this, the lemma follows.

Lemma 5.3. *Let* $(\text{Б}, \widetilde{\text{Б}}, R)$ *be in the polynomial Bessel fundamental class, and assume that R is of regularized harmonic series type. Write* $P_j(s) = \sum_{k=0}^{n_j} c_{j,k} s^k$. *Assume f satisfies the basic Fourier conditions to sufficiently large order, and set* $F_b(x) = F(x)e^{xb}$. *Then*

$$\lim_{T_n \to \infty} \left[\frac{1}{2\pi i} \int_{\sigma_0+a-iT_n}^{\sigma_0+a+iT_n} \text{Б}(s)\mathbf{M}_{\sigma_0/2}f(s)ds \right]$$

$$= \sum_j \sum_{k=0}^{n_j} c_{j,k} (\log \mathbf{q}_j^{1/2})^{\alpha_j} \int_0^\infty F_{\sigma_0/2}^{(k-\alpha_j)}\left(-\log \mathbf{q}_j^{1/2}(u + 1/u)\right) u^{\alpha_j} \frac{du}{u}.$$

Proof. We first write

$$\lim_{T \to \infty} \frac{1}{2\pi i} \int_{\sigma_0+a-iT}^{\sigma_0+a+iT} \text{Б}(s)\mathbf{M}_{\sigma_0/2}f(s)ds$$

$$= \lim_{T \to \infty} \frac{1}{2\pi i} \int_{\sigma_0+a-iT}^{\sigma_0+a+iT} \sum_j P_j(s) K_{\alpha_j}(s, \log \mathbf{q}_j^{1/2}) \mathbf{M}_{\sigma_0/2}f(s)ds$$

$$= \lim_{T \to \infty} \sum_j \sum_{k=0}^{n_j} \frac{c_{j,k}(\log \mathbf{q}_j^{1/2})^{\alpha_j}}{2\pi i} \int_{\sigma_0+a-iT}^{\sigma_0+a+iT} s^{k-\alpha_j} K_{\alpha_j}(s \log \mathbf{q}_j^{1/2})\mathbf{M}_{\sigma_0/2}f(s)ds.$$

Let us first consider integrals of the form

$$(3) \qquad \frac{1}{2\pi i} \int_{\sigma_0+a-iT}^{\sigma_0+a+iT} s^\mu K_\alpha(sc) \mathbf{M}_{\sigma_0/2}f(s)ds.$$

If we let $s = \sigma_0 + a + it$, we can write (3) as

$$(4) = \frac{1}{2\pi} \int_{-T}^{T} \int_{-\infty}^{\infty} \int_{0}^{\infty} F(x) e^{-(\sigma_0/2 + a + it)x} (\sigma_0 + a + it)^\mu e^{-c(\sigma_0 + a + it)(u + 1/u)} u^\alpha \frac{du}{u} dxdt.$$

By interchanging the order of integration, we are left with understanding the inner integrals, that is

$$(5) \quad \frac{1}{2\pi} \int_{-\infty}^{\infty} \int_{-T}^{T} F(x) e^{x\sigma_0/2} e^{-(\sigma_0 + a + it)x} (\sigma_0 + a + it)^\mu e^{-c(\sigma_0 + a + it)(u + 1/u)} dt dx.$$

Let $x = y - c(u + 1/u)$, then (5) can be written as

$$(6) \quad \frac{1}{2\pi} \int_{-\infty}^{\infty} \int_{-T}^{T} F(y - c(u + 1/u)) e^{[y - c(u+1/u)]\sigma_0/2} e^{-(\sigma_0 + a + it)y} (\sigma_0 + a + it)^\mu dt dy.$$

Define the function

$$H_\beta(y; c) = F(y - c(u + 1/u)) e^{[y - c(u+1/u)]\sigma_0/2} e^{-\beta y},$$

so then (6) is simply

$$(7) \quad \frac{1}{2\pi} \int_{-\infty}^{\infty} \int_{-T}^{T} H_{\sigma_0 + a}(y; c)(\sigma_0 + a + it)^\mu e^{-ity} dt dy$$

and (4) takes the form

$$\int_{0}^{\infty} \left(\frac{1}{2\pi} \int_{-\infty}^{\infty} \int_{-T}^{T} H_{\sigma_0 + a}(y; c)(\sigma_0 + a + it)^\mu e^{-ity} dt dy \right) u^\alpha \frac{du}{u}.$$

With all this, the integral we are considering is then expressible as

$$(8) \quad \frac{1}{2\pi i} \int_{\sigma_0 + a - iT}^{\sigma_0 + a + iT} \mathrm{B}(s) \mathbf{M}_{\sigma_0/2} f(s) ds = \int_{0}^{\infty} \left(\int_{-\infty}^{\infty} \int_{-T}^{T} H_{\sigma_0 + a}(y) e^{-ity} dt dy \right) \frac{du}{u}$$

where

$$(9) \quad H_{\sigma_0 + a}(y) = \sum_{j} \sum_{k=0}^{n_j} \frac{c_{j,k}(\log \mathbf{q}_j^{1/2})^{\alpha_j}}{2\pi} H_{\sigma_0 + a}(y; \log \mathbf{q}_j^{1/2})(\sigma_0 + a + it)^{k - \alpha_j} u^{\alpha_j}.$$

We can now recognize that (8) is precisely of the form (1). For an individual integral in (8), we obtain the formula

$$\left(\lim_{T \to \infty} \frac{1}{2\pi} \int_{-\infty}^{\infty} \int_{-T}^{T} H_{\sigma_0 + a}(y; c)(\sigma_0 + a + it)^\mu e^{-ity} dy dt \right) = F_{\sigma_0/2}^{(\mu)}(-c(u + 1/u)),$$

where we recall that
$$F_b^{(\mu)} = (F_b)^{(\mu)}.$$

The formula asserted in the statement of the lemma follows by including the integral with respect to u and then summing all terms.

Let us now discuss briefly the convergence questions associated to the above calculations. In the case Б is a Dirichlet series, complete details are given on pages 73 to 77 of [JoL 94b]. The basic assumptions stated in §1 imply that the Bessel series Б is dominated by a convergent Dirichlet series, as defined in §1. Also recall from [JoL 94b] and Cramér's theorem (Theorem 4.6) that the asymptotic behavior of the Bessel series is determined by the order of the additive factor R, which in turn established the requirement that the test function F in the explicit formulas satisfies the Fourier conditions to sufficiently large order. Therefore, even though the Bessel series has polynomial growth on vertical lines in the half plane of convergence, as opposed to being bounded in the case of a Dirichlet series, the integral in Lemma 5.3 necessarily converges. Combining this with **FOU 1** and **FOU 3**, implies that the function (9) is in L^1. By arguing similarly, one shows that (9) is of bounded variation.

This concludes the proof of Lemma 5.3. Applying a tilde and following the same calculations, we obtaing the following result.

Lemma 5.4. *Let $(Б, \widetilde{Б}, R)$ be in the polynomial Bessel fundamental class, and assume that R is of regularized harmonic series type. Write $\tilde{P}_j(s) = \sum_{k=0}^{n_j} \tilde{c}_{j,k} s^k$. Assume f satisfies the basic Fourier conditions to sufficnetly large order, and set $F_b(x) = F(x)e^{xb}$. Then*

$$\lim_{T_n \to \infty} \left[\frac{1}{2\pi i} \int_{\sigma_0+a-iT_n}^{\sigma_0+a+iT_n} \widetilde{Б}(s) \mathbf{M}_{\sigma_0/2} f(\sigma_0 - s) ds \right]$$
$$= \sum_j \sum_{k=0}^{\tilde{n}_j} \tilde{c}_{j,k} (\log \tilde{\mathbf{q}}_j^{1/2})^{\tilde{\alpha}_j} \int_0^\infty F_{-\sigma_0/2}^{(k-\tilde{\alpha}_j)} \left(\log \tilde{\mathbf{q}}_j^{1/2}(u + 1/u) \right) u^{\tilde{\alpha}_j} \frac{du}{u}.$$

Putting all the lemmas together proves Theorem 5.1.

6. Theta inversions

Let N be a positive integer. For a suitable test function f, we define

$$\text{Gauss}^{(N)}(f)(s) = 2s \int_0^\infty f(t) e^{-s^2 t} t^{N+1} \frac{dt}{t}.$$

Then

$$\text{Gauss}^{(N)}(f)(s) = \frac{dz}{ds} \cdot \mathbf{LM}f(N+1, z) \quad \text{where } z = s^2.$$

Because of the change of variables from z to s, we are led to consider the differential operator

$$\mathscr{D}_s = -\frac{\partial}{\partial s} \circ \frac{1}{2s}.$$

Then

(1) $$\mathscr{D}_x^N [2xK_\alpha(x, u)] = 2xK_{\alpha+N}(x, u).$$

Although we shall not need it here, we note for the record the other operator

$$D_s = -\frac{1}{2s}\frac{\partial}{\partial s}$$

for which we have trivially the formulas:

(2) $$D_s \text{Gauss}^{(N)}(f)(s) = \text{Gauss}^{(N+1)}(f)(s),$$

(3) $$2xD_x^N \circ \frac{1}{2x} = \mathscr{D}_x^N \quad \text{and} \quad D_x^N K_\alpha(x, u) = K_{\alpha+N}(x, u).$$

When we have to make a translation by a real number $\sigma_0 > 0$, we write

$$\text{Gauss}^{(N)}_{\sigma_0/2}(f)(s) = \text{Gauss}^{(N)}(f)(s - \sigma_0/2).$$

We can define $\mathscr{D}_s^{-1}$ in the obvious way by

$$\mathscr{D}_s^{-1} = -2sI_s = \mathscr{I}_s,$$

where I_s is the integral taken with zero constant of integration applied to powers $\mathbf{q}^{-s}$, so that

$$I_s(\mathbf{q}^{-s}) = -\mathbf{q}^{-s}/\log \mathbf{q}.$$

One applies I_s to products $Q(s)\mathbf{q}^{-s}$, where Q is a polynomial, by using the rule for integration by parts lowering the degree of the polynomial coefficient of $\mathbf{q}^{-s}$ step by step. Then we define

$$\mathscr{I}_s^k = \mathscr{D}_s^{-k} = (\mathscr{D}_s^{-1})^k$$

for positive integers k.

We let $\{\mathbf{q}_j\}$ be a sequence of real numbers > 1 tending to infinity, and we let $\{b_j\}$ be complex numbers such that $\sum b_j \mathbf{q}_j^{-s}$ converges absolutely in some half plane. Let $\{Q_j\}$ be a sequence of polynomials of degrees $\leq d$ for some integer d. We suppose that

$$\|Q_j\| \leq |b_j| \quad \text{for all } j.$$

Given a polynomial $Q(X)$ of degree $\leq d$, we define the **dual polynomial**

$$Q^\vee(X) = X^d Q(1/X),$$

so $Q \mapsto Q^\vee$ is an involution on polynomials which depends on the choice of integer d. Let $\{\alpha_j\}$ be a sequence of complex numbers which are bounded in absolute value. We define the **(generalized polynomial) inverted theta series** by

$$\bar{\theta}^\vee(t) = \sum Q_j(1/4\pi t)(4\pi t)^{-\alpha_j} e^{-(\log q_j)^2/4t}.$$

Remark 1. We are dealing here with a slightly more general notion than in [JoL 94b] because of the powers $t^{-\alpha_j}$ which come in and are not integral or half-integral powers. The assumptions on the degrees of the polynomials Q_j and on their coefficients guarantees that the Gauss transform is a Bessel series which converges absolutely in some half plane.

For the next theorems, we make the following assumptions:

Basic Assumptions.

- The theta series $\theta(t) = \sum a_k e^{-\mu_k t}$ satisfies **AS 1**, **AS 2**, and **AS 3**.
- The inverted theta series $\bar{\theta}^\vee(t)$ is constructed as above and converges absolutely for all $t > 0$.
- The function R is of regularized harmonic series type, and there exist a and A sufficiently large positive and a decomposition

$$R = R_{\text{left}} + R_{\text{right}}$$

where R_{left} has its poles in a left half plane $\text{Re}(z) \leq A - \delta - \sigma_0/2$ and where R_{right} has its poles in a right half plane $\text{Re}(z) > -a + \delta + \sigma_0/2$.
- These functions satisfy the inversion formula

$$\theta(t) = \bar{\theta}^\vee(t) + E_A R_{\text{left}}^{(\sigma_0/2)}(t) + E_{-a} R_{\text{right}}^{(\sigma_0/2)}(t).$$

where $R^{(\sigma_0/2)}(s) = R(s + \sigma_0/2)$, and $E_u R$ is the transform of R defined by

$$E_u R(t) = \lim_{n \to \infty} \frac{1}{2\pi i} \int_{u-iT_n}^{u+iT_n} e^{z^2 t} R(z) dz.$$

The following two propositions are proved in [JoL 94b], Chapter V, Theorems 2.4 and 2.5.

Proposition 6.1. *For any positive integer N sufficiently large, and s real and sufficiently large, we have*

$$\text{Gauss}^{(N)}(\theta)(s) = \mathscr{D}_s^N \left[2s R_\theta(s^2) \right].$$

371

Proposition 6.2. *With notation as above, we have, for N sufficiently large and s real and sufficiently large, the equalities*

$$\text{Gauss}^{(N)}(E_A R_{\text{left}}^{(\sigma_0/2)})(s) = \mathscr{D}_s^N R_{\text{left}}^{(\sigma_0/2)}(s)$$

and

$$\text{Gauss}^{(N)}(E_{-a} R_{\text{right}}^{(\sigma_0/2)})(s) = (-\mathscr{D}_s)^N R_{\text{right}}^{(\sigma_0/2)}(-s).$$

We now evaluate the Gauss transform of the inverted theta series by computing the Gauss transform of each individual term.

Proposition 6.3. *Let N be a positive integer. Then*

$$\text{Gauss}^{(N)}(\bar{\theta}^\vee)(s) = \frac{1}{(4\pi)^d} \mathscr{D}_s^{N-d} \sum_j Q_j^\vee (4\pi \mathscr{D}_s) \left[(4\pi)^{-\alpha_j} 2s K_{-\alpha_j+1}(s, \log \mathbf{q}_j^{1/2}) \right].$$

Proof. Put $x = s$ and for each $\mathbf{q}$, let $u = \log \mathbf{q}^{1/2}$. Let $0 \le k \le d$, $\alpha = \alpha_j$, and

$$g(t) = \frac{(4\pi t)^k}{(4\pi t)^{d+\alpha}} e^{-u^2/t}.$$

Then by (1)

$$\begin{aligned}
\text{Gauss}^{(N)}(g)(x) &= \frac{(4\pi)^k 2x}{(4\pi)^{d+\alpha}} K_{N-d+k-\alpha+1}(x, u) \\
&= \frac{(4\pi)^k}{(4\pi)^{d+\alpha}} \mathscr{D}_x^{N-d} \mathscr{D}_x^k \left[2x K_{-\alpha+1}(x, u) \right].
\end{aligned}$$

This proves Proposition 6.3.

Immediately from the lemma and the definitions, we obtain:

Corollary 6.4. *Assume the basic conditions. Suppose $\alpha_j = 1/2$ for all j. Let $N \ge d$ be sufficiently large. Then for s real and sufficiently large, we have*

$$\text{Gauss}^{(N)}(\bar{\theta}^\vee)(s) = \frac{1}{(4\pi)^d} \mathscr{D}_s^{N-d} \sum_j Q_j^\vee (4\pi \mathscr{D}_s) \mathbf{q}_j^{-s}.$$

Proof. For $\alpha = 1/2$, we recall that

$$K_{1/2}(a, b) = \frac{\sqrt{\pi}}{a} e^{-2ab}.$$

The asserted formula in terms of the exponentials $\mathbf{q}_j^{-s}$ then drops out.

For the next theorem, we define:

the polynomial Bessel series

$$\mathrm{B}(s) = \sum_j Q_j(4\pi \mathcal{Z}_s)[(4\pi)^{-\alpha_j} 2sK_{-\alpha_j+1}(s, \log \mathbf{q}_j^{1/2})],$$

the polynomial Dirichlet series

$$L(s) = \sum_j Q_j(4\pi \mathcal{Z}_s)\mathbf{q}_j^{-s}.$$

Theorem 6.5. *There exists a polynomial Q' of degree $\leq 2N+2d-1$, annihilated by $\mathscr{D}_s^{N+d}$, such that*

$$2sR_\theta(s^2) = \mathrm{B}(s) + R_{\text{left}}^{(\sigma_0/2)}(s) + R_{\text{right}}^{(\sigma_0/2)}(-s) + Q'(s).$$

Therefore the Bessel series has a meromorphic continuation, and satisfies a functional equation, namely both sides are odd with respect to the map $s \mapsto -s$.

Proof. First we apply Propositions 6.1, 6.2, and 6.3 to get the relation

$$\begin{aligned}\mathscr{D}_s^N[2sR_\theta(s^2)] &= \frac{1}{(4\pi)^d}\mathscr{D}_s^{N-d}\sum Q_j^\vee(4\pi\mathscr{D}_s)[(4\pi)^{-\alpha_j}2sK_{-\alpha_j+1}(s, \log\mathbf{q}_j^{1/2})]\\ &\quad+\mathscr{D}_s^N[R_{\text{left}}^{(\sigma_0/2)}(s) + R_{\text{right}}^{(\sigma_0/2)}(-s)].\end{aligned}$$

Then we integrate back as follows. Let F be a meromorphic function such that $\mathscr{D}_s F(s) = 0$. Then there is a polynomial P_1 of degree ≤ 1 such that

$$F(s) = P_1(s).$$

We may then peal off inductively the powers of $\mathscr{D}_s$, or more precisely, we apply $\mathcal{Z}_s^{N-d}$ first, and then $\mathcal{Z}_s^d = \mathscr{D}_s^{-d}$. Using the definition of the dual polynomial, we find the asserted expression. This concludes the proof of Theorem 6.5.

Corollary 6.6. *If $\alpha_j = 1/2$ for all j, then the Bessel series is the polynomial Dirichlet series $L(s)$, and thus*

$$2sR_\theta(s^2) = L(s) + R_{\text{left}}(s + \sigma_0/2) + R_{\text{right}}(-s + \sigma_0/2) + Q'(s).$$

Note that by formula (9) of §1, the Bessel series in Corollary 6.6 also reduces to a polynomial Dirichlet series when $\alpha_j = n/2$ and n is an odd integer. We shall see another example of this in §9.

Remark 2. We may of course make a translation $s \mapsto s - \sigma_0/2$ in the above formula, in which case it reads

$$(2s - \sigma_0)R_\theta((s - \sigma_0/2)^2)$$
$$= L(s - \sigma_0/2) + R_{\text{left}}(s) + R_{\text{right}}(\sigma_0 - s) + Q'(s - \sigma_0/2),$$

and the right side is then odd with respect to $s \mapsto \sigma_0 - s$.

Remark 3. We emphasize that the functional equation comes from a new feature built into the symmetry $s \mapsto -s$ of the Gauss transform.

Remark 4. We have now concluded the process from which a theta function with an inversion formula leads to a polynomial Bessel series having an additive functional equation with fudge terms which are of regularized harmonic series type. The converse is also interesting, and stems from a study of the explicit formulas for certain test functions. However, we already said that we will leave for later the analysis of the transform which we encountered in the statement of the explicit formulas in §5.

Special Case 1. The polynomials Q_j are constants and all $\alpha_j = 1/2$. In this case, we are dealing with the Dirichlet series

$$L(s) = \sum b_j \mathbf{q}_j^{-s}.$$

The inverted theta series is

$$\bar{\theta}^\vee(t) = \frac{1}{(4\pi t)^{1/2}} \sum b_j e^{-(\log \mathbf{q}_j)^2/4t},$$

and the theta inversion reads as before

$$\theta(t) = \bar{\theta}^\vee(t) + E_A R_{\text{left}}^{(\sigma_0/2)}(t) + E_{-a} R_{\text{right}}^{(\sigma_0/2)}(t).$$

There is a polynomial P' of degree $\leq 2N - 1$ such that we have the functional equation

$$(2s - \sigma_0)R_\theta((s - \sigma_0/2)^2) = \quad L(s - \sigma_0/2) +$$
$$R_{\text{left}}(s) + R_{\text{right}}(\sigma_0 - s) + P'(s - \sigma_0/2).$$

Special Case 2. The polynomials Q_j are monomials with $d \geq 1$, and all $\alpha_j = 1/2$. In this case, $L(s)$ is the same as in Case 1 but

$$Q_j^\vee(t) = b_j t^d.$$

The inverted theta series has the form

$$\bar{\theta}^\vee(t) = \frac{1}{(4\pi t)^{n/2}} \sum b_j e^{-(\log \mathbf{q}_j)^2/4t},$$

with $n = 2d$ or $2d + 1$. The theta inversion and functional equation read the same as before.

7. Example 1: Hyperbolic scattering determinants

In this section we shall recall known background material concerning scattering determinants and Eisenstein series associated to hyperbolic Riemann surfaces of finite volume, and we will show how the scattering determinant is an example of the type of function to which we can apply the analysis of the previous sections. The functions we consider in this section do not admit classical Euler product expansions but admit Euler sum expansions as considered in [JoL 94b], and so they are in the Dirichlet polynomial fundamental class. Hence, there is no need of the Bessel series here. However, we include this example because we had not included it in [JoL 94b], and it shows the power of the techniques we use. The reader is referred to [Sel 56], [He 83], and references therein for details and proofs.

Let X be a non-compact hyperbolic Riemann surface of finite volume with h cusps, realized as the quotient manifold $X = \Gamma\backslash\mathbf{h}$ where $\mathbf{h}$ is the hyperbolic upper half plane and $\Gamma \subset PSL(2,\mathbf{R})$ is a discrete group of isometries acting on $\mathbf{h}$. Associated to each cusp P_j on X is an Eisenstein series $E_j(s,z)$. Let $\varphi_{i,j}(s)$ be the constant term of $E_j(s,z)$ in the Fourier series expansion at the cusp P_i, and set

$$\phi(s) = \det[\varphi_{i,j}(s)].$$

In [He 83], following Selberg [Sel 56], it is shown that the **scattering determinant** ϕ and the matrix $[\varphi_{i,j}(s)]$ extend to meromorphic functions of $s \in \mathbf{C}$ and satisfy the functional equations

(1) $\qquad [\varphi_{i,j}(s)] \cdot {}^t[\varphi_{i,j}(1-s)] = I_h \quad \text{and} \quad \phi(s)\phi(1-s) = 1,$

where I_h denotes the $h \times h$ identity matrix, and the vector of Eisenstein series $\mathscr{E} = [E_j]$ satisfies the functional equation

$$\mathscr{E}(s,z) = [\varphi_{i,j}(s)]\mathscr{E}(1-s,z).$$

It is shown in [Sel 56] that there exist constants $c_1, c_2 \in \mathbf{R}$ and sequences $\{\mathbf{r}\}$ and $a(\mathbf{r})$, with $\mathbf{r} > 1$ tending to infinity, such that

$$\phi(s) = G(s)H(s)$$

where

$$G(s) = \left[\frac{\pi^{-(s-1/2)}\Gamma_1(s-1/2)}{\pi^{-s}\Gamma_1(s)}\right]^h e^{c_1 s + c_2}$$

and, for $\operatorname{Re}(s) > 1$, $H(s)$ has a Dirichlet series of the form

$$H(s) = 1 + \sum_{\mathbf{r}} \frac{a(\mathbf{r})}{\mathbf{r}^s}.$$

Remark 1. If $\Gamma = PSL(2,\mathbf{Z})$, one can show the associated scattering determinant is equal to

$$\phi(s) = \frac{\pi^{-(s-1/2)}\Gamma(s-1/2)\zeta_Q(2s-1)}{\pi^{-s}\Gamma(s)\zeta_Q(2s)}$$

(see the above references or page 46 of [Kub 73]). For general congruence subgroups of level N, the scattering determinant ϕ can be expressed in terms of the gamma function and Dirichlet L-functions with character which divides N (see [Hu 84] as well as [He 83]).

In general, $H(s)$ has a Dirichlet series for $\mathrm{Re}(s)$ sufficiently large, which is most conveniently written as a Dirichlet series for $H'/H(s)$, that is

(2) $$H'/H(s) = \sum_{\mathbf{q}} \frac{b(\mathbf{q})}{\mathbf{q}^s}.$$

It is immediate that the set $\mathbf{q}$ is simply the set of finite products of elements from $\mathbf{r}$, and the coefficients $\{b(\mathbf{q})\}$ can be expressed in terms of the coefficients $\{a(\mathbf{r})\}$ and various binomial coefficients. The functional equation (1) for $\phi(s)$ can be written as

(3) $$H(s)G(s) = \tilde{H}(1-s)\tilde{G}(1-s)$$

where $\tilde{H} = 1/H$ and $\tilde{G} = 1/G$. In the notation of §1, we have $\sigma_0 = 1$. Let

$$\Phi(s) = G(s)/\tilde{G}(1-s) = G(s)G(1-s) = \left[\frac{\Gamma(s-1/2)\Gamma(1/2-s)}{\Gamma(s)\Gamma(1-s)}\right]^h \pi^h e^{c_1+2c_2},$$

so then $\Phi(s) = \Phi(1-s)$. Therefore, we can write (3) as

(4) $$H'/H(s) + \Phi'/\Phi(s) + \tilde{H}'/\tilde{H}(1-s) = 0.$$

Notice that Φ has a pole of order $2h$ at $s = 1/2$. From the functional equation (1) one has that $\phi(1/2)^2 = 1$, hence H itself has a zero of order h at $s = 1/2$.

Remark 2. As an immediate consequence of the functional equation (4) and Dirichlet series for H and $\tilde{H}$, it is shown in [JoL 93c] that the scattering determinant ϕ is of regularized product type of reduced order $(0, 1)$. The functional equation of the Selberg zeta function associated to X is such that we can now conclude that the Selberg zeta function associated to a non-compact hyperbolic Riemann surface of finite volume is of regularized product type of reduced order $(1, 1)$.

Since ϕ is of regularized product type, the explicit formulas of [JoL 94b] apply. We shall now study the expression given by the explicit formulas, and we will evaluate the Weil functional for certain test functions.

Consider a test function F which, in addition to satisfying the basic conditions from §5, is odd, so $F(x) = -F(-x)$. As we will see, this will allow us to obtain an explicit formula involving only the zeros of the scattering determinant.

For the remainder of this section, we will assume F is odd.

By direct calculation, we have

$$\mathbf{M}_{1/2}f(s) = -\mathbf{M}_{1/2}f(1-s).$$

The proof of this formula is quite elementary, namely

$$-\mathbf{M}_{1/2}f(1-s) = -\int_{-\infty}^{\infty} F(x)e^{x/2}e^{-(1-s)x}dx = \int_{\infty}^{-\infty} F(-u)e^{u/2}e^{-us}du = \mathbf{M}_{1/2}f(s).$$

Since $H = 1/\bar{H}$, the sets $\{\mathbf{q}\}$ and $\{\bar{\mathbf{q}}\}$ are equal, but $b(\mathbf{q}) = -b(\bar{\mathbf{q}})$. Therefore, the two terms in the explicit formula involving $\{\mathbf{q}\}$ and $\{\bar{\mathbf{q}}\}$ are equal for an odd test function. That is,

$$\sum_{\mathbf{q}} \frac{b(\mathbf{q})}{\mathbf{q}^{1/2}} F(-\log \mathbf{q}) = \sum_{\mathbf{q}} \frac{-b(\mathbf{q})}{\mathbf{q}^{1/2}} F(\log \mathbf{q})$$

and

$$\sum_{\bar{\mathbf{q}}} \frac{b(\bar{\mathbf{q}})}{\bar{\mathbf{q}}^{1/2}} F(\log \bar{\mathbf{q}}) = \sum_{\mathbf{q}} \frac{-b(\mathbf{q})}{\mathbf{q}^{1/2}} F(\log \mathbf{q}),$$

so the two terms in the explicit formulas involving sums over $\mathbf{q}$ and $\bar{\mathbf{q}}$ combine to give

$$2\sum_{\mathbf{q}} \frac{-b(\mathbf{q})}{\mathbf{q}^{1/2}} F(\log \mathbf{q}).$$

Let $Z_a(\phi)$ denote the set of zeros $\{\rho\}$ of ϕ in $\mathscr{R}_a$, and let $P_a(\phi)$ denote the set of poles of ϕ in $\mathscr{R}_a$. Define

$$v(z) = \mathrm{ord}_z(\phi).$$

By the functional equation (1), one has that ρ is a zero of ϕ precisely when $1-\rho$ is a pole of ϕ, and $v(\rho) = -v(1-\rho)$. Let $Z_{a,1/2}(\Phi)$ denote the zeros of Φ in the strip $-a < \mathrm{Re}(s) \leq 1/2$, and let $P_{a,1/2}(\Phi)$ denote the poles of Φ in the strip $-a < \mathrm{Re}(s) \leq 1/2$. Observe that

$$P_a(\phi) = P_a(H) \cup Z_{a,1/2}(\Phi) \quad \text{and} \quad Z_a(\phi) = Z_a(H) \cup P_{a,1/2}(\Phi).$$

If we use (1), we then have

$$\begin{aligned} V_{H'/H,a}(f) + V_{\Phi'/\Phi,a,1/2}(f) &= \sum_{\rho \in Z_a(\phi)} v(\rho)\mathbf{M}_{1/2}f(\rho) + \sum_{\rho \in P_a(\phi)} v(\rho)\mathbf{M}_{1/2}f(\rho) \\ &= \sum_{\rho \in Z_a(\phi)} \left[v(\rho)\mathbf{M}_{1/2}f(\rho) + v(1-\rho)\mathbf{M}_{1/2}f(1-\rho)\right] \\ &= 2\sum_{\rho \in Z_a(H)} v(\rho)\mathbf{M}_{1/2}f(\rho). \end{aligned}$$

Remark 3. The set of poles of Φ'/Φ in the half strip $-a \leq \mathrm{Re}(s) \leq 1/2$ is:

$s = 1/2$ with multiplicity $2h$ and residue -1;

$s = -n$ with $-a \leq -n \leq 0$, multiplicity h and residue 1;

$s = -n + 1/2$ with $-a - 1/2 \leq -n < 0$, multiplicity h and residue -1;

Since the pole at $s = 1/2$ lies on the boundary of the critical strip, one can either integrate by including a "bump" in the contour to include or exclude the point, or one can (formally) integrate over the point with $1/2$ of the residue, which is equal to h.

Finally, it remains to compute the term coming from the Weil functional. By moving the line of integration, we obtain

$$W_{\Phi'/\Phi}(F) = \lim_{m \to \infty} \frac{1}{2\pi i} \int_{1+a-iT_m}^{1+a+iT_m} \mathbf{M}_{1/2}f(s)\Phi'/\Phi(s)ds$$

(7)
$$= \lim_{m \to \infty} \frac{1}{\sqrt{2\pi}} \int_{-T_m}^{T_m} F^{\wedge}(t)\Phi'/\Phi(1/2 + it)dt.$$

Since F is odd, so is the Fourier transform $F^{\wedge}$, so then

(8) $$W_{\Phi'/\Phi}(F) = \lim_{m \to \infty} \frac{2h}{\sqrt{2\pi}} \int_{-T_m}^{T_m} F^{\wedge}(t)\left[\Gamma'/\Gamma(it) - \Gamma'/\Gamma(1/2 + it)\right]dt.$$

Observe that the integral in (8) is bounded near $t = 0$ since the zero of the Fourier transform will cancel with the pole of $\Gamma'/\Gamma(it)$. The vanishing of the Fourier transform at $t = 0$ allows one to justify the change of contour in (7). These calculations which evaluate the integrals in (8) are given in §2 of [JoL 93b] (see also [Ba 81] as well as page 116 of [JoL 94b]). Observe that since F is assumed to be odd, $F(0) = 0$ which provides a further simplification, yielding the formula

(9) $$\lim_{m \to \infty} \frac{1}{\sqrt{2\pi}} \int_{-T_m}^{T_m} F^{\wedge}(t)\Gamma'/\Gamma(c + 1 + it)dt = \int_0^{\infty} F(x)e^{-(1+c)x}\theta_{\mathbf{Z}}(x)dx$$

where

$$\theta_{\mathbf{Z}}(x) = \sum_{k=0}^{\infty} e^{-kx} = \frac{1}{1 - e^{-x}}.$$

Since F is odd, (9) is valid for all $c \geq -1$. Therefore, we have

$$W_{\Phi'/\Phi}(F) = 2h \int_0^{\infty} F(x)\theta_{\mathbf{Z}}(x)(1 - e^{-x/2})dx = 2h \int_0^{\infty} F(x)(1 + e^{-x/2})^{-1}dx.$$

Putting all this together, we have the following explicit formula for the scattering determinant.

Theorem 7.1. *Assume that F satisfies the three basic conditions to sufficiently large order and F is odd. Then*

$$\sum_{\rho \in Z_a(H)} v(\rho) \mathbf{M}_{1/2} f(\rho) = \sum_{\mathbf{q}} \frac{-b(\mathbf{q})}{\mathbf{q}^{1/2}} F(\log \mathbf{q}) + h \int_0^\infty F(x)(1 + e^{-x/2})^{-1} dx.$$

Theorem 7.1 is a new example of an explicit formula which results from our generalization of the class of functions under consideration. Because of the vast geometric information contained in the scattering determinant, a complete study of Theorem 7.1 and its applications is warranted. To conclude this section, we shall remark on a few of the aspects of Theorem 7.1 which we find interesting.

Remark 4. Theorem 7.1 gives a precise relation between the sequence $\mathbf{q}$ and the set of zeros (without poles) of the scattering determinant. Furthermore the sequence $\{\mathbf{q}\}$ can be expressed in terms of various realizations of the uniformizing Fuchsian group (see, for example, [Sel 56]). Hence, Theorem 7.1 relates the zeros of the scattering determinant to information concerning the Fuchsian group in a manner similar to the classical Riemann-von Mangolt formula.

Remark 5. For a general non-compact hyperbolic Riemann surface of finite volume, one can study the corresponding Fujii function. Let $Z_a^+(\phi)$ denote the zeros of the scattering determinant in $\mathcal{R}_a$ with positive imaginary part, and let $P_a^+(\phi)$ denote the poles of the scattering determinant in $\mathcal{R}_a$ with positive imaginary part. Then the Fujii function

(10) $$\sum_{Z_a^+(\phi) \cup P_a^+(\phi)} v(\rho) \frac{e^{i\alpha\lambda}}{\lambda^s}$$

has a meromorphic continuation to a holomorphic function for all $s \in \mathbf{C}$. We do not see how to use the trace formula to prove the meromorphic continuation of (10). Similarly, the proof that the Selberg zeta function associated to X is of regularized product type follows from our analysis, whereas attempts to prove such a result via the trace formula have thus far not been successful.

Remark 6. It would be interesting to see if one can use Theorem 7.1 in order to obtain expressions for integrals of the form

$$\int_{-\infty}^\infty h(r) \phi' / \phi(1/2 + ir) dr,$$

which appear in the trace formula for non-compact surfaces. Such expressions would yield formulas relating eigenvalues of the Laplacian on X to the length spectrum and series as in Theorem 7.1. In the case $X = PSL(2, \mathbf{Z}) \backslash \mathbf{h}$, one would obtain a formulas involving eigenvalues of the Laplacian on X, the length spectrum on X and the ordinary primes. Such a result has been studied by Venkov and further studied by Fujii.

Remark 7. It is important to note that we obtained an explicit formula for the scattering determinant when, as far as is known, the scattering determinant for a general surface X does not have a classical Euler product and does not satisfy the Riemann hypothesis. By varying the surface X, one obtains a family of explicit formulas which, in some cases, contain certain classical explicit formulas, as indicated by Remark 1.

Remark 8. One can use Theorem 7.1 in order to obtain results on the distribution of the zeros of a scattering determinant, continuing a study initiated by Selberg in [Sel 90].

8. Polynomially split heat kernels

Let X be a Riemannian manifold, with positive Laplacian Δ. A **heat kernel** for X is a smooth function $\mathbf{K}_X$ of variables (t,x,y) with $x, y \in X$ and $t \in \mathbf{R}^+$, such that $\mathbf{K}_X$ is annihilated by the **heat operator** $\Delta_y + \partial/\partial t$, and $\mathbf{K}_X$ satisfies the Dirac property

$$\lim_{t \to 0}(\mathbf{K}_X * \varphi)(t,x) = \varphi(x) \quad \text{for } \varphi \in C_c^\infty(X).$$

One can define similarly the heat operator for other types of differential or pseudo differential operators, elliptic, more or less positive, for instance lower order perturbations of positive elliptic operators, but let us stick to the Laplacian for concreteness, to avoid lesser known terminology. When X is compact, the heat kernel is uniquely determined. We shall also deal with the non-compact case. Then there is a uniquely determined "minimal" heat kernel which satisfies the properties stated below, cf. [Ch 84].

If $\tilde{X}$ is the simply connected universal covering of X, then the heat kernel on X is given as the "trace" with respect to Γ, namely

HK 1. $\mathbf{K}_X(t,x,y) = \sum_{\gamma \in \Gamma} \mathbf{K}_{\tilde{X}}(t, \gamma \tilde{x}, \tilde{y}),$

where $(\tilde{x}, \tilde{y})$ lies above the point $(x,y) \in X \times X$. If X is compact, then the heat kernel has the eigenfunction expansion

HK 2. $\mathbf{K}_X(t,x,y) = \sum_{k=0}^\infty \varphi_k(x)\overline{\varphi_k(y)} e^{-\lambda_k t},$

where $\{\varphi_k\}$ is a complete orthonormal system of L^2-eigenfunctions of Δ, and $\{\lambda_k\}$ is the corresponding sequence of eigenvalues. Equating the expressions in **HK 1** and **HK 2** gives rise to theta inversion formulas.

For the euclidean space $\mathbf{R}^n$ the heat kernel has the form

$$\mathbf{K}_{\mathbf{R}^n}(t,x,y) = \mathbf{E}_n(t, \text{dist}(x,y))$$

where

$$\mathbf{E}_n(t,r) = \frac{1}{(4\pi t)^{n/2}} e^{-r^2/4t},$$

so we substitute $r = r(x,y) = \text{dist}(x,y) = |x - y|$. In general, the heat kernel is some sort of perturbation of the euclidean heat kernel, but such perturbations can be of several types. The present section applies to one type of perturbation, when the heat kernel of the universal covering space **splits**, by which we mean that

$$\mathbf{K}_{\tilde{X}}(t,\tilde{x},\tilde{y}) = \mathbf{E}_n(t,r) e^{-(\sigma_0^2/4)t} G(\tilde{x},\tilde{y}),$$

with some real number $\sigma_0 \geq 0$, and some function G. Of course in this case,

$$r = r(\tilde{x},\tilde{y}) = \text{dist}_g(\tilde{x},\tilde{y})$$

is the Riemannian distance wth respect to the metric g lifted to the universal cover $\tilde{X}$. Often, $G(\tilde{x},\tilde{y}) = \mathbf{j}(\tilde{y}^{-1}\tilde{x})$, where $\mathbf{j}$ is a normalized Jacobian of an exponential map, in various contexts of differential geometry. Even when the heat kernel is split, the function G may depend only on the distance, or it may not.

It is also necessary to consider the analogous case when there is an extra polynomial, just as for the theta functions. Thus we say that the heat kernel is **polynomially split** if there exists a function $G(T,\tilde{x},\tilde{y})$, polynomial in T, such that

$$\mathbf{K}_{\tilde{X}}(t,\tilde{x},\tilde{y}) = \mathbf{E}_n(t,r) e^{-(\sigma_0^2/4)t} G(t,\tilde{x};\tilde{y}).$$

We shall see examples of both situations in the following two sections.

Remark 1. In general, the heat kernel does not split, even polynomially, but is only expressible as a spectral integral, which gives rise to a similar but somewhat adjusted formalism, when polynomial Bessel and Dirichlet series have to be adjusted with extra integrals, which we treat elsewhere.

When the heat kernel of the universal covering space is split, or is polynomially split, then for a compact quotient X, we obtain an example of theta inversion when $x \neq y$, as follows. Let $n = \dim_{\mathbf{R}} X$, and let:

$$\log \mathbf{q}_\gamma = \log \mathbf{q}_\gamma(\tilde{x},\tilde{y}) = \text{dist}(\gamma\tilde{x},\tilde{y}),$$
$$\mu_k = \lambda_k - \sigma_0^2/4,$$
$$\theta(t) = \theta_{x,y}(t) = \sum \varphi_k(x)\overline{\varphi_k(y)} e^{-\mu_k t} = e^{(\sigma_0^2/4)t} \mathbf{K}_X(t,x,y),$$
$$\tilde{\theta}^\vee(t) = \frac{1}{(4\pi t)^{n/2}} \sum_{\gamma \in \Gamma} G(t,\gamma\tilde{x},\tilde{y}) e^{-(\log \mathbf{q}_\gamma)^2/4t}.$$

The theta inversion formula has the shape

$$\theta(t) = \tilde{\theta}^\vee(t).$$

When $x = y$, one has to separate the term with γ = id. A similar formalism holds in the non-compact case, but is more complicated because of other regularizing terms which have to be introduced, such as Eisenstein series.

The next two sections give the specific, concrete formulas which come out in two important special cases for which the heat kernel is polynomially split: the hyperbolic odd dimensional manifolds, and the complex semisimple Lie groups symmetric spaces. It remains to characterize by differential geometric conditions all cases when the heat kernel is split.

When one has the theta inversion for each pair (x,y), one can do many things, such as putting $x = y$ and integrate with respect to the volume form over the manifold to get an inversion relation for the trace of the heat kernel. These operations may be viewed as homomorphic images of heat kernel relations, as in [JoL 94a].

Finally, we can connect the above with the wave operator. We define the **wave operator** $L_W = \Delta + (\partial/\partial t)^2$. Let ω_k be the positive square root of μ_k for almost all k. We can then define the **wave theta series** to be

$$\theta_W(\tau, x, y) = \theta_W(\tau) = \sum \phi_k(x)\overline{\phi_k(y)}e^{i\omega_k \tau}$$

for complex τ with $\text{Im}(\tau) > 0$. The wave operator L_W acts, say, on the x variable. Then

$$L_W \theta_W = (\sigma_0^2/4)\theta_W.$$

One views θ_W as a wave kernel. From this point of view, the theory of the wave equation is reduced to the theory of the heat equation via the above procedure, especially the analysis carried through in the study of the Cramér theorem. This point was already made in §4. Cf. also the analysis of [JoL 93a], §7.

9. Example 2: Odd dimensional hyperbolic spaces

The case of hyperbolic 3-space $\mathbf{h}_3$ was described in detail in [JoL 94b]. The heat kernel in this case is split. However, in [JoL 94b] we made a translation by $\sigma_0/2$ (= 1 in this case), and there are also some misprints, as well as one garbled formula, so we make corrections here. We define the function

$$F_{x,y}(s) = \frac{1}{(4\pi)^{3/2}} \sum_{\mathbf{q}} \frac{2\log \mathbf{q}}{\mathbf{q} - \mathbf{q}^{-1}} 2sK_{3/2}(s, \log \mathbf{q}^{1/2})$$

$$= \frac{1}{2\pi} \mathscr{D}_s \sum_{\mathbf{q}} \frac{\log \mathbf{q}}{1 - \mathbf{q}^{-2}} \mathbf{q}^{-s-1}$$

(1) $$= 2s \sum_{\mathbf{q}} \frac{2\varphi_k(x)\bar{\varphi}_k(y)}{(s^2 + \mu_k)^3}.$$

We define the polynomial Dirichlet series

(2) $$L_{x,y}(s) = \frac{2s}{2\pi} \sum_q \frac{1}{1-q^{-1}} q^{-s-1},$$

converging for $\text{Re}(s) > 1$. Then

$$\mathscr{D}_s^2 L_{x,y} = \mathscr{D}_s^2(2sR_\theta(s^2)),$$

and there is a polynomial Q' annihilated by $\mathscr{D}_s^2$ such that

(3) $$2sR_\theta(s^2) = L_{x,y}(s) + Q'(s).$$

This gives the continuation of the Dirichlet series to a meromorphic function (essentially a regularized harmonic series), with only simple poles at $s = \pm\sqrt{-\mu_k}$, and residues $\varphi_k(x)\bar{\varphi}_k(y)$. The polynomial Q' is odd, and $L_{x,y}(s)$ satisfies the functional equation of being odd under the map $s \mapsto -s$.

The higher dimensional case of $\mathbf{h}_n$ with odd integers n is entirely similar, but the heat kernel is only polynomially split. For these manifolds, there is a recursive formula for their heat kernels, due to Millson, as follows (see page 151 of [Ch 84] as well as [Da 89], 5.7). We define a function $h_n(t,r)$ of two variables (t,r), recursively for any integer $n \geq 3$ by

$$h_n(t,r) = e^{-(n-2)t} \partial_r h_{n-2}(t,r) \frac{1}{2\pi \sinh r}.$$

Then the heat kernel is, as before,

$$\mathbf{K}_n(t,x,y) = h_n(t, d_{\mathbf{h}_n}(x,y)).$$

In particular, for odd $n = 2d+1$, since $1 + 3 + \cdots + (2d-1) = d^2$, there exists a polynomial P_n in four variables such that

$$h_n(t,r) = \frac{1}{(4\pi t)^{n/2}} e^{-r^2/4t} e^{-d^2 t} \frac{P_n(t, r, \cosh r, \sinh r)}{(\sinh r)^{n-2}}.$$

The same formalism with the fundamental group Γ and a compact quotient $X = \Gamma \backslash \mathbf{h}_n$ applies, as for the case of dimension 3 which was carried out in Chapter V of [JoL 94b]. We put

$$\sigma_0 = 2d = n - 1 \quad \text{so} \quad \sigma_0^2/4 = d^2.$$

In the first place, we have for $x \neq y$:

$$\theta_{x,y}(t) = e^{d^2 t} \mathbf{K}_X(t,x,y) = \sum_k \varphi_k(x) \overline{\varphi_k(y)} e^{-\mu_k t}$$

where $\mu_k = \lambda_k - \sigma_0^2/4 = \lambda_k - d^2$ in the present case.

Secondly, we have

$$\bar{\theta}^\vee_{x,y}(t) = \frac{1}{(4\pi t)^{n/2}} \sum_\gamma G(t, \gamma\tilde{x}, \tilde{y}) e^{-(\log q_\gamma)^2/4t},$$

383

where

$$\log \mathbf{q}_\gamma = d_{\mathbf{h}_n}(\gamma \bar{x}, \bar{y}) \quad \text{and} \quad G(t, \gamma \bar{x}, \bar{y}) = \frac{P_n(t, \log \mathbf{q}_\gamma, C(\mathbf{q}_\gamma), S(\mathbf{q}_\gamma))}{S(\mathbf{q}_\gamma)^{n-2}}$$

putting

$$C(\mathbf{q}) = (\mathbf{q} + \mathbf{q}^{-1})/2 \quad \text{and} \quad S(\mathbf{q}) = (\mathbf{q} - \mathbf{q}^{-1})/2.$$

Having assumed $x \neq y$ we have the theta inversion

$$\theta(t) = \bar{\theta}^\vee(t).$$

Thus, we are exactly under the conditions of Theorem 6.5.

Remark 1. For non-compact quotients, one must take into account the appearance of the Eisenstein series in the spectral decomposition of the heat kernels, and the subsequent appearance of other terms in the additive functional equation. Examples of such formulas, as well as the more complicated situation of even dimensional hyperbolic manifolds, will be treated elsewhere.

Remark 2. The space $\mathbf{h}_3$ is actually a homogeneous space for a complex Lie group, that is

$$\mathbf{h}_3 \approx SL_2(\mathbf{C})/K$$

where K is a maximal compact subgroup. Thus the case of $\mathbf{h}_3$ is subsumed under the examples of the next section. However, a similar result is not true for $\mathbf{h}_n$ with n odd > 3, although as we have seen, the heat kernel still has the general shape which allows for a polynomial Dirichlet zeta function. The higher dimensional $\mathbf{h}_n$ are homogeneous spaces only for real semisimple Lie groups, specifically the connected component of $O(n, 1)$. There are also several models for $\mathbf{h}_n$. Cf. for instance Helgason [Hel 78], pp. 227 and 564; [Hel 84], p. 226; and for the general uniqueness proof up to isometry, see for instance [La 95], p. 248.

Remark 3. The zeta function studied by Shintani and Millson [Mil 78] for compact manifolds of dimension $4n - 1$ may be viewed as a special case. In retrospect, we may view Millson's Theorem 3.1 as expressing the logarithmic derivative of their zeta function as what we call a Gauss transform, after he takes the trace of the operator (setting $x = y$ and integrating over the manifold).

10. Example 3: The symmetric space of a complex Lie group

Gangolli [Ga 68] has already shown that the heat kernel gives rise to a "theta relation" on complex semisimple Lie groups. We shall describe here how Gangolli's Proposition 4.6 fits in our formalism.

Gangolli [Ga 68] determined the heat kernel on semisimple Lie groups as an integral transform by means of spherical functions. Basically, this followed from Harish-Chandra's determination of the expansion in spherical functions, which is an analogue of the determination of the spectral measure on semisimple Lie groups. In the case of complex semisimple Lie groups, the integral giving the heat kernel collapses to an expression similar to the simple expression for the odd dimensional simply connected hyperbolic spaces. Specifically, let G be a non-compact semisimple Lie group. (We do not recall all the definitions, and intend this example either for those who know the terminology, or those who are willing to accept it axiomatically. The article [Ga 68] contains an excellent summary.) There is a decomposition $G = KA_pK$, where K is a maximal compact subgroup, and A_p is certain abelian subgroup isomorphic to a euclidean space via the exponential map. An element x in G can thus be written in the form $x = k_1ak_2$ with $k_1, k_2 \in K$ and $a \in A_p$, the element a being uniquely determined by x.

Since G is complex, the heat kernel $\mathbf{K} = \mathbf{K}(t, x, y)$ is given in terms of a function g_t ($t > 0$) of one variable in G, namely

$$\mathbf{K}(t, x, y) = g_t(y^{-1}x),$$

and this function g_t is K-bi-invariant, that is $g_t(k_1ak_2) = g_t(a)$. For $x \in G$ we write $g_t(x) = g_t(a)$ if $x = k_1ak_2$ with $k_1, k_2 \in K$.

Proposition 3.2 of [Ga 68] shows that if $n = \dim_{\mathbf{R}}(G/K)$, then

$$g_t(a) = \frac{1}{(4\pi t)^{n/2}} e^{-|\log a|^2/4t} e^{-t\sigma_0^2/4} \mathbf{j}(a)$$

where:
- For $a \in A_p$, $\log a$ is the corresponding element in the Lie algebra, and $|\log a|$ is the norm for the natural scalar product on the Lie algebra, the Killing form.
- We define $\sigma_0 = |\sigma_*|$, where σ_* is the sum of the positive roots.
- Let Q be the set of positive roots. For each $\alpha \in Q$ we let

$$C_\alpha = 2\pi \langle \alpha, \sigma_* \rangle^{-1}.$$

- The factor $\mathbf{j}(a)$ is given by

$$\mathbf{j}(a) = \prod_{\alpha \in Q} \mathbf{j}_\alpha(a) \quad \text{where} \quad \mathbf{j}_\alpha(a) = C_\alpha \frac{\alpha(\log a)}{\sinh(\alpha(\log a))},$$

which is a quotient of Jacobian determinants.

We see that for complex semisimple Lie groups, the heat kernel is split, in the terminology introduced in §8. In general the factor $\mathbf{j}$ does not depend only on the Riemannian distance. If it happens that G/K is 2-point homogeneous (that is, given any two pairs of equidistant points, there is an element of G which carries one pair to the other), then indeed, the factor $\mathbf{j}$ depends only on the distance. However, a simply connected G/K is 2-point homogeneous if and only if it is euclidean space, or a globally symmetric space of rank one, so the occurrence is relatively rare, even though significant. Cf. Wolf [Wo 67/72], Theorem 8.12.2, and Helgason [Hel 78], p. 535 for a discussion of this point and references to the literature on Lie groups.

That **j** depends on more than the distance does not affect the formalism of the applicability of theta inversion and the Gauss transform, which we now make explicit.

Let Γ be a discrete subgroup of G acting without fixed point on G/K, and co-compact, so $X = \Gamma \backslash G/K$ is a compact Riemannian manifold. Let $\tilde{X} = G/K$. Suppose $\tilde{X}$ is simply connected. Then Γ is the fundamental group, and we are facing an example of §8, with the heat kernel relation giving a theta inversion of the type we have considered. As in §8 we put

$$(1) \qquad \theta(t) = e^{t\sigma_0^2/4} \mathbf{K}_X(t, \tilde{x}, \tilde{y}) = \sum \varphi_k(x)\overline{\varphi_k(y)} e^{-\mu_k t}.$$

Furthermore, by *assuming* $x \neq y$, we also obtain the dual theta function. We let

$$a(\gamma \tilde{x}, \tilde{y}) = a(\tilde{y}^{-1} \gamma \tilde{x}) \quad \text{and} \quad \log \mathbf{q}_\gamma = |\log a(\gamma \tilde{x}, \tilde{y})|.$$

We define

$$G(\gamma \tilde{x}, \tilde{y}) = \mathbf{j}(a(\tilde{y}^{-1} \gamma \tilde{x})),$$

and we let

$$\tilde{\theta}^\vee(t) = \frac{1}{(4\pi t)^{n/2}} \sum_\gamma G(\gamma \tilde{x}, \tilde{y}) e^{-(\log \mathbf{q}_\gamma)^2/4t}.$$

Then the two heat kernel expressions **HK 1** and **HK 2** of §8 yield the theta inversion formula

$$\theta(t) = \tilde{\theta}^\vee(t),$$

or

$$\sum \varphi_k(x)\overline{\varphi_k(y)} e^{-\mu_k t} = \frac{1}{(4\pi t)^{n/2}} \sum_\gamma G(\gamma \tilde{x}, \tilde{y}) e^{-(\log \mathbf{q}_\gamma)^2/4t}.$$

Thus when $x \neq y$ we are in a situation when Theorem 6.5 applies, and Corollary 6.6 applies if n is odd, after using the appropriate differential operator.

Remark 1. As noted in §8, if $x = y$, then there is an extra term corresponding to γ equal the identity, and this term has to be handled separately. This term will then give rise to the extra fudge terms corresponding to R_{left} and R_{right} in the general formula.

Remark 2. In [Ga 68], Gangolli takes the trace of the heat kernel, and as a result obtains the "theta relation" of his Proposition 4.6 without the pair of elements (x, y) which have been eliminated by setting $x = y$ and integrating over X. On the other hand, his formula is also valid when Γ has fixed points, but that is another story, concerning the behavior of the heat kernel under such groups, when X may have singularities. The condition **HK 1** must be worked out carefully in this case.

Remark 3. The example of this section is of polynomial type, although the polynomial still appears in a fairly simple way, as bt^d with some constant b. Nevertheless, the corresponding Dirichlet or Bessel series is of polynomial type.

Remark 4. Given the theta inversion formula above, we can take arbitrary powers $n \in \mathbf{Z}_{\geq 0}$ yielding theta inversion formulas of the form

$$(\theta(t))^n = (\bar{\theta}^\vee(t))^n.$$

The regularized harmonic series associated to the theta series $(\theta(t))^n$ has poles at points of the form $\mu_{k_1} + \cdots + \mu_{k_n}$.

References

[As 70] ASAI, T.: On a certain function analogous to $\log|\eta(z)|$. *Nagoya Math. J.* **40** (1970) 193-211.
[Ba 81] BARNER, K.: On Weil's explicit formula. *J. reine angew. Math.* **323,** 139-152 (1981).
[Bo 51] BOCHNER, B.: Some properties of modular relations. *Ann. Math.* **53** (1951) 332-363.
[Ch 84] CHAVEL, I.: *Eigenvalues in Riemannian Geometry.* New York: Academic Press (1984).
[Cr 19] CRAMÉR, H.: Studien über die Nullstellen der Riemannschen Zetafunktion. *Math. Z.* **4,** (1919) 104-130.
[Da 89] DAVIES, E. B.: *Heat Kernels and Spectral Theory.* Cambridge: Cambridge University Press (1989).
[Do 35] DOETSCH, G.: Summatorische Eigenschaften der Besselschen Funktionen und andere Funktionalrelationen, die mit der linearen Transformationsformel des Thetafunktionals äquivalent sind. *Comp. Math.* **1** (1934) 85-97.
[DG 75] DUISTERMAAT, J., and GUILLEMIN, V.: The spectrum of positive elliptic operators and periodic bicharacteristics. *Invent. Math.* **29,** (1975) 39-79.
[Er 36a] ERDELYI, A.: Über eine Methode zur Gewinnung von Funktionalbeziehungen zwischen konfluenten hypergeometrischen Funktionen. *Monat. f. Math. u. Phys.* **45** (1936) 31-52.
[Er 36b] ERDELYI, A.: Funktionalrelationen mit konfluenten hypergeometrischen Funktionen, Erste Mitteilung: Additions- und Multiplikationstheoreme. *Math. Zeitschr.* **42** (1936) 125-143.
[Er 37a] ERDELYI, A.: Über gewisse Funktionalbeziehungen. *Monat. f. Math. u. Phys.* **45** (1937) 251-279.
[Ga 68] GANGOLLI, R.: Asymptotic behavior of spectra of compact quotients of certain symmetric spaces. *Acta Math.* **121** (1968) 151-192.
[Ga 77] GANGOLLI, R.: Zeta functions of Selberg's type for compact space forms of symmetric space of rank one. *Ill. Math. J.* **21** (1977) 1-42.
[GW 80] GANGOLLI, R., and WARNER, G.: Zeta functions of Selberg's type for some noncompact quotients of symmetric spaces of rank one. *Nagoya Math. J.* **78** (1980) 1-44.
[GR 65] GRADSHTEYN, I. S., and RYZHIK, I. M.: *Tables of Integrals, Series, and Products.* New York: Academic Press (1965).
[He 83] HEJHAL, D. A.: The Selberg Trace Formula for *PSL*(2, **R**), volume 2. Lecture Notes in Mathematics **1001** Berlin-Heidelberg: Springer-Verlag (1983).
[Hel 64] HELGASON, S.: Fundamental solutions of invariant differential operators on symmetric spaces. *Am. J. Math.* **86** (1964) 565 - 601.
[Hel 78] HELGASON, S.: *Differential Geometry, Lie Groups and Symmetric Spaces.* New York: Academic Press (1978).
[Hel 84] HELGASON, S.: *Groups and Geometric Analysis.* New York: Academic Press (1978).

[Hel 94] HELGASON, S.: *Geometric analysis on symmetric spaces.* Math. Surveys and Monographs, AMS, 1994.
[Hu 84] HUXLEY, M. N.: "Scattering matrices for congruence subgroups", in Modular Forms, R. A. Rankin ed., John Wiley and Sons: New York (1984) 157-196.
[JoL 93a] JORGENSON, J., and LANG, S.: Complex analytic properties of regularized products and series. Lecture Notes in Mathematics **1564** Berlin-Heidelberg: Springer-Verlag (1993), 1-88.
[JoL 93b] JORGENSON, J., and LANG, S.: A Parseval formula for functions with a singular asymptotic expansion at the origin. Lecture Notes in Mathematics **1564** Berlin-Heidelberg: Springer-Verlag (1993), 89-122.
[JoL 93c] JORGENSON, J., and LANG, S.: On Cramér's theorem for general Euler products with functional equation. *Math. Ann.* **297** (1993), 383-416.
[JoL 94a] JORGENSON, J., and LANG, S.: Artin formalism and heat kernels. *J. reine angew. Math.* **447** (1994) 165-200.
[JoL 94b] JORGENSON, J., and LANG, S.: Explicit formulas for regularized products and series. Lecture Notes in Mathematics **1593** Berlin-Heidelberg: Springer-Verlag (1994), 1-134.
[Ko 35] KOBER, H.: Transformationalformeln gewisser Besselscher Reihen, Beziehungen zu Zeta-Funktionen. *Math. Zeitschr.* **39** (1935) 609-624.
[Kub 73] KUBOTA, T.: *Elementary theory of Eisenstein series,* New York: John Wiley and Sons (1973).
[La 70] LANG, S.: *Algebraic Number Theory,* Menlo Park, CA.: Addison-Wesley (1970); Graduate Texts in Mathematics **110,** New York: Springer-Verlag (1986); third edition, Springer-Verlag (1994).
[La 87] LANG, S.: *Elliptic Functions, second edition.* Graduate Texts in Mathematics **112** New York: Springer-Verlag (1987).
[La 93a] LANG, S.: *Complex Analysis,* Graduate Texts in Mathematics **103,** New York: Springer-Verlag (1985), Third Edition (1993).
[La 93b] LANG, S.: *Real and Functional Analysis, 3rd Edition,* New York: Springer-Verlag (1993).
[La 95] LANG, S.: *Differential and Riemannian Manifolds,* New York: Springer-Verlag (1993).
[Lo 32] LOWRY, H.: Operational calculus II, The values of certain integrals and the relationships between various polynomials and series obtained by operational methods. *Phil. Mag.* (7) **13** (1932) 1144-1163.
[Mil 78] MILLSON, J.: Closed geodesics and the η-invariant. *Ann. Math.* **108** (1978), 1-39.
[Sel 56] SELBERG, A.: Harmonic Analysis and discontinuous groups in weakly symmetric Riemannian spaces with applications to Dirichlet series, *J. Indian Math. Soc. B.* **20** (1956) 47-87. (*Collected papers volume I,* Berlin-Heidelberg: Springer-Verlag (1989) 423-463).
[Sel 90] SELBERG, A.: Remarks on the distribution of poles of Eisenstein series. *Israel Mathematical Conference Proceedings* **3** (1990) 251-278. (*Collected papers volume II,* Berlin-Heidelberg: Springer-Verlag (1991) 16-45.)
[Wa 44] WATSON, G. N.: *A Treatise on the Theory of Bessel Functions, 2nd edition.* Cambridge University Press: Cambridge (1944).
[We 52] WEIL, A.: Sur les "formules explicites" de la théorie des nombres premiers, *Comm. Lund* (vol. dédié à Marcel Riesz), 252-265 (1952).
[We 72] WEIL, A.: Sur les formules explicites de la théorie des nombres, *Izv. Mat. Nauk (Ser. Mat.)* **36,** 3-18 (1972).
[Wo 67/72] WOLF, J.: *Spaces of Constant Curvature,* Berkeley (1967), second edition (1972).

J. Jorgenson and S. Lang
Nagoya Math. J.
Vol. 153 (1999), 155–188

HILBERT-ASAI EISENSTEIN SERIES, REGULARIZED PRODUCTS, AND HEAT KERNELS

JAY JORGENSON AND SERGE LANG

Abstract. In a famous paper, Asai indicated how to develop a theory of Eisenstein series for arbitrary number fields, using hyperbolic 3-space to take care of the complex places. Unfortunately he limited himself to class number 1. The present paper gives a detailed exposition of the general case, to be used for many applications. First, it is shown that the Eisenstein series satisfy the authors' definition of regularized products satisfying the generalized Lerch formula, and the basic axioms which allow the systematic development of the authors' theory, including the Cramér theorem. It is indicated how previous results of Efrat and Zograf for the strict Hilbert modular case extend to arbitrary number fields, for instance a spectral decomposition of the heat kernel periodized with respect to SL_2 of the integers of the number field. This gives rise to a theta inversion formula, to which the authors' Gauss transform can be applied. In addition, the Eisenstein series can be twisted with the heat kernel, thus encoding an infinite amount of spectral information in one item coming from heat Eisenstein series. The main expected spectral formula is stated, but a complete exposition would require a substantial amount of space, and is currently under consideration.

The Hilbert modular case for totally real number fields has been well understood for many decades. Asai gave a beautiful treatment showing how to deal with the general case [As 70], but unfortunately he limited himself to number fields of class number one. As far as a general exposition is concerned, matters were not much improved in [EGM 85], which limited itself to imaginary quadratic fields. Of course, [EGM 87] then pushed matters in a deeper way in the direction of special values of Eisenstein series.

As shown below, a general exposition of the Eisenstein series for arbitrary number fields turns out not to be more difficult without any restriction and doing so actually forces a clarification of the terminology and the notation. We shall deal with applications in a different direction from that

---

Received May, 19, 1996.
Jorgenson acknowledges support from NSF Grants, from the Institute for Advanced Study, and from a Sloan Fellowship. Lang thanks the Max-Planck-Institut for productive yearly visits.

of [EGM 85]. Asai ran across theta-type series formed with the K-Bessel function, aside from the Eisenstein series. Bessel functions play a dual role. On the one hand, they give rise to series playing the role of Dirichlet series. In [JoL 96], we showed that taking the Gauss transform of ordinary theta series yields Bessel series in lieu of Dirichlet series, and we showed how these Bessel series fit in our general framework of regularized products. Here we shall first take the usual Mellin transform of Bessel theta series to get the corresponding Dirichlet series. As mentioned in [JoL 96], we leave to another paper the fuller theory which arises from taking the Gauss transform of the Bessel series in general, yielding Legendre series. The tabulations of the present paper will be useful in this subsequent work.

In §1 and §2, we carry out the Riemann-Hecke arguments for a functional equation in a fairly broad context, applying to Bessel theta series. Independently, in §3, §4 and §5 we formulate the general properties of Hilbert-Asai Eisenstein series. In §6, we show how the Eisenstein series fit in the theory of regularized products as developed beginning in the articles [JoL 93] and [JoL 94]. In §7, we show how the heat kernel on the symmetric space associated to the number field can be defined in terms of the Eisenstein series, so that we can apply the general theory and especially [JoL 96]. Finally, in §8, we reformulate the results of §7 in terms of a new Eisenstein series which we call the heat Eisenstein series, thus naturally leading to work in progress concerning spectral expansion on other symmetric spaces.

Thus we can take the Gauss transform of the heat kernel inversion formula to obtain new zeta functions with functional equations. In the case of Riemann surfaces, these functions correspond to logarithmic derivatives of Selberg zeta functions.

Among other things, the present paper provides significant examples for the general theory of regularized products and series. We note that these examples also serve to emphasize the sufficiency of a Dirichlet series representation, as opposed to the more classical emphasis on the existence of Euler products as such. So far, Euler products have not played a role in the general theory we are developing.

§1. Functional equation and Mellin transform on a product of G_m^+

Let $S = \{v\}$ be a finite set and let

$$Y = \prod_{v \in S} Y_v \quad \text{where} \quad Y_v = \mathbf{R}^+.$$

Let

$$d^* y_v = \frac{dy_v}{y_v} \quad \text{and} \quad d^* y = \prod_{v \in S} \frac{dy_v}{y_v}.$$

For each v we let N_v be a positive real number and we let $N = \sum N_v$. We define the **norm**

$$\mathbf{N} y = \prod_{v \in S} y_v^{N_v}.$$

Thus the norm $\mathbf{N}: Y \to \mathbf{R}^+$ is a continuous homomorphism. We let

$$Y^0 = \operatorname{Ker} \mathbf{N} = \{y \in Y \text{ such that } \prod_{v \in S} y_v^{N_v} = 1\}.$$

Let U be an abelian group with a fixed homomorphism onto a discrete subgroup V of Y^0 such that Y^0/V is compact. Since there is an isomorphism

$$\log: Y \to \mathbf{R}^{\#(S)},$$

our assumption on U amounts to saying that $\log Y^0$ is a hyperplane in $\log Y$ and the image of U in $\log Y^0$ is a lattice. Thus, Y^0/V is isomorphic to a real torus of dimension $\#(S) - 1$.

We note that $\mathbf{R}^+$ acts on Y and can even be embedded as a subgroup of Y, namely for $a \in \mathbf{R}^+$ we let

$$ay = (\ldots, a^{1/N} y_v, \ldots).$$

so a is embedded as $(\ldots, a^{1/N}, \ldots)$ in Y. Then $\mathbf{N} a = a$. This embedding splits the sequence

$$0 \to Y^0 \to Y \to \mathbf{R}^+ \to 0,$$

so an element $y \in Y$ can be written uniquely in the form $y = ty^0$ with $t \in \mathbf{R}^+$, $y^0 \in Y^0$.

The measure $d^* y$ determines a measure on Y/V. It can also be written as a product measure

$$d^* y = d^* y^0 \frac{dt}{t}$$

where y^0 is the variable in Y^0, and d^*y^0 is a Haar measure on Y^0, uniquely determined by d^*y.

Before carrying out a general theorem, we recall two distinct special cases arising in classical situations. Let us call a pair of functions h, h_0 on Y **admissible** if they satisfy the conditions:

ADM 1: h, h_0 are invariant under the action of V, so are defined on Y/V;
ADM 2: h_0 extends continuously to $Y^0/V \times [0, \infty)$;
ADM 3: $h_0(y)$ decreases exponentially as $\mathbf{N}y \to \infty$.

For algebraic number fields: The function h satisfies the inversion conditions:
INV 1: (Number Fields) $h(y) = h_0(y) + c_0$ for some constant c_0;
INV 2: (Number Fields) $\mathbf{N}y^{-\frac{1}{2}} h(y^{-1}) = h(y)$.

For Eisenstein series as in [As 70], see also Theorem 4.4 below.
INV 1: (Eisenstein) $h(y) = h_0(y) + c_0 \mathbf{N}y$;
INV 2: (Eisenstein) $h(y^{-1}) + \log \mathbf{N}y = h(y)$.

We define the **Mellin transform** on Y/V, with respect to $\mathbf{N}$:

$$\mathbf{M}_{Y/V} h_0(s) = \int_{Y/V} h_0(y) \mathbf{N}y^s d^*y.$$

The next theorem covers the special case considered in Asai [As 70].

THEOREM 1.1. *Under the Eisenstein conditions* **INV 1**, **INV 2** *above, the function* $\mathbf{M}_{Y/V} h_0$ *has a meromorphic continuation and satisfies the functional equation*

$$\mathbf{M}_{Y/V} h_0(-s) = \mathbf{M}_{Y/V} h_0(s).$$

The above theorem extends easily to the case when we include a character as follows. By a **Hecke character** χ on Y, with respect to the given action of V and $\mathbf{R}^+$, we mean a continuous homomorphism

$$\chi: Y \longrightarrow \mathbf{C}^1$$

into the unit circle, such that χ is $\mathbf{R}^+$- and V-invariant, that is,

$$\chi(ay) = \chi(uy) = \chi(y)$$

for all $y \in Y$, $a \in \mathbf{R}^+$ and $u \in V$. Then χ induces a character on Y^0/V, and since Y^0/V is a real torus, it follows that the group of Hecke characters is isomorphic to $\mathbf{Z}^r$, where $r = \#(S) - 1$.

For a Hecke character χ, we define
$$\mathbf{M}_{Y/V} h_0(s, \chi) = \int_{Y/V} h_0(y) \chi(y) \mathbf{N} y^s d^* y.$$

Then one obtains the functional equation with Hecke characters:

THEOREM 1.2. *Under the Eisenstein conditions, assume that χ is non-trivial. Then*
$$\mathbf{M}_{Y/V} h_0(s, \chi) = \int_{\mathbf{N}y \geq 1} h_0(y) [\chi(y) \mathbf{N} y^s + \overline{\chi}(y) \mathbf{N} y^{-s}] d^* y.$$

In particular,
$$\mathbf{M}_{Y/V} h_0(s, \chi) = \mathbf{M}_{Y/V} h_0(-s, \overline{\chi}).$$

Recall that a **quasi character**
$$Y/V \longrightarrow \mathbf{C}^*$$

is simply a continuous homomorphism. It is therefore uniquely determined by a vector $\beta = (\ldots, \beta_v, \ldots)$ of complex numbers, not necessarily of absolute value 1, satisfying the orthogonality relations

OR 1. $\sum_{v \in S} \beta_v N_v \log u_v \equiv 0 \mod 2\pi i \mathbf{Z}$ for all $u \in V$.

The value of the quasi character $[\beta]$ corresponding to β is given by
$$[\beta](y) = \prod_{v \in S} y_v^{N_v \beta_v}.$$

If we let χ be the Hecke character induced on Y^0/V, and such that $\chi(\mathbf{R}^+) = 1$, then there exists a unique complex number s such that
$$[\beta](ty^0) = \chi(y^0) t^s, \text{ that is } [\beta] = \chi \cdot \mathbf{N}^s.$$

The quasi character is equal to the Hecke character if and only if we also have the orthogonality relation

OR 2. $\sum_{v \in S} \beta_v N_v = 0$.

Thus we say $\chi = [\beta]$ if and only if **OR 1** and **OR 2** are satisfied.

Under these two conditions, it follows that $\text{Re}(\beta_v) = 0$ *for all* v, *that is β_v is pure imaginary for all v, so evidently $[\beta]$ has values on the unit circle.*

The above situation can be further generalized in a manner necessary for certain applications. For one thing, one need not deal only with the pair (h, h_0), but two pairs can intervene. Furthermore, the powers of $\mathbf{N}y$ or $\log \mathbf{N}y$ need not be 1. So we consider the following more general formulations of **INV 1**, **INV 2**. For this, we need the notion of generalized polynomials as they have appeared systematically in the [JoL] series, namely a finite sum

$$P(T) = \sum c_{p,m}(\log T)^m T^p = \sum B_p(\log T) T^p.$$

with $p \in \mathbf{C}$, $c_{p,m} \in \mathbf{C}$, $m \in \mathbf{Z}_{\geq 0}$, and a polynomial B_p for each p. The conditions then read:

INV 1. *There is a generalized polynomial $P(T)$ such that*

$$h(y) = h_0(y) + P(\mathbf{N}y),$$

and there are admissible functions $\tilde{h}$, $\tilde{h}_0$ such that

$$\tilde{h}(y) = \tilde{h}_0(y) + \tilde{P}(\mathbf{N}y)$$

for some generalized polynomial $\tilde{P}$.

INV 2. *There is a generalized polynomial Q and $s_0 \in \mathbf{C}$ such that*

$$\mathbf{N}y^{-s_0/2}\tilde{h}(y^{-1}) + Q(\mathbf{N}y) = h(y).$$

Define the **truncated Mellin transform** of a generalized polynomial $P(T)$ as written above on the interval $[0, 1]$ to be

$$\mathbf{M}_0^1 P(s) = \sum c_{p,m} \left(\frac{d}{ds}\right)^m \frac{1}{s+p} = \sum c_{p,m} \frac{(-1)^m m!}{(s+p)^{m+1}}.$$

LEMMA 1.3. *Let* $\mathrm{Re}(s+p) > 0$ *for all p such that $c_{p,m} \neq 0$. Then*

$$\int_{\mathbf{N}y \leq 1} P(\mathbf{N}y)\chi(y)\mathbf{N}y^s d^*y = \delta_\chi \mu^*(Y^0/V)\mathbf{M}_0^1 P(s).$$

where $\delta_\chi = 1$ if χ is trivial and 0 otherwise

Proof. Immediate.

We define
$$P^-(T) = P(T^{-1}).$$

THEOREM 1.4. *We have the meromorphic continuation*

$$\mathbf{M}_{Y/V} h_0(s, \chi) = \int_{\mathbf{N}y \geq 1} [h_0(y)\chi(y)\mathbf{N}y^s + \tilde{h}_0(y)\overline{\chi}(y)\mathbf{N}y^{-s+s_0/2}] d^*y$$
$$+ \delta_\chi \mu^*(Y^0/V)\mathbf{M}_0^1(Q + \tilde{P}^- - P)(s).$$

The integral is entire in s and in particular

$$\mathbf{M}_{Y/V} h_0(s, \chi) = M_{Y/V}\tilde{h}_0(-s, \overline{\chi}) + \mathrm{Rat}(s),$$

where $\mathrm{Rat}(s)$ is a rational function in s.

Proof. The proof follows a classical pattern and runs as follows:

$\mathbf{M}_{Y/V} h_0(s, \chi)$

$$= \int_{\mathbf{N}y \leq 1} + \int_{\mathbf{N}y \geq 1} h_0(y)\chi(y)\mathbf{N}y^s d^*y$$

$$= \int_{\mathbf{N}y \leq 1} h(y)\chi(y)\mathbf{N}y^s d^*y - \int_{\mathbf{N}y \leq 1} P(\mathbf{N}y)\chi(y)\mathbf{N}y^s d^*y$$

(1)
$$+ \int_{\mathbf{N}y \geq 1} h_0(y)\chi(y)\mathbf{N}y^s d^*y$$

$$= \int_{\mathbf{N}y \leq 1} \mathbf{N}y^{-s_0/2}\tilde{h}(y^{-1})\chi(y)\mathbf{N}y^s d^*y + \int_{\mathbf{N}y \leq 1} Q(y)\chi(y)\mathbf{N}y^s d^*y$$

$$- \int_{\mathbf{N}y \leq 1} P(\mathbf{N}y)\chi(y)\mathbf{N}y^s d^*y + \int_{\mathbf{N}y \geq 1} h_0(y)\chi(y)\mathbf{N}y^s d^*y.$$

First integral on right

(2) $$\begin{aligned}&= \int_{\mathbf{N}y\geq 1} \tilde{h}(y)\overline{\chi(y)}\mathbf{N}y^{-s+s_0/2}d^*y \\ &= \int_{\mathbf{N}y\geq 1} \tilde{h}_0(y)\overline{\chi(y)}\mathbf{N}y^{-s+s_0/2}d^*y + \int_{\mathbf{N}y\geq 1} \tilde{P}(\mathbf{N}y)\overline{\chi(y)}\mathbf{N}y^{-s}\,d^*y \\ &= \int_{\mathbf{N}y\geq 1} \tilde{h}_0(y)\overline{\chi(y)}\mathbf{N}y^{-s+s_0/2}d^*y + \int_{\mathbf{N}y\leq 1} \tilde{P}^-(\mathbf{N}y)\chi(y)\mathbf{N}y^s\,d^*y.\end{aligned}$$

Putting (1) and (2) together yields the asserted result.

Note that Theorem 1.4 covers both Theorem 1.1 and Theorem 1.2, except that the formulations in the previous theorems are simpler because the rational function disappears due to the simpler conditions on h and h_0.

§2. Bessel theta series and Mellin transforms

In this section we see how one can form Bessel series which play the role of theta series.

We let K_α be the K-Bessel function, with $\alpha \in \mathbf{C}$, which is normalized as in [La 73/87] and [JoL 96], that is, for $c > 0$:

$$K_\alpha(c) = \int_0^\infty e^{-c(t+1/t)} t^\alpha \frac{dt}{t}.$$

If K_α^B denotes the one found in classical tables, then

$$2K_\alpha^B(2c) = K_\alpha(c).$$

We suppose given a real number $\sigma_1 > 0$. We suppose that h_0 is a function which can be expressed as a Bessel theta series, meaning the following. There are constants $a_k \in \mathbf{C}$, $c_{k,v} \in \mathbf{R}^+$, $\alpha_v \in \mathbf{C}$ such that

(1) $$h_0(y) = \sum_{k=1}^\infty a_k \sum_{v \in V} \prod_v K_{\alpha_v}(c_{k,v} y_v) y_v^{N_v \sigma_1}.$$

satisfies the **Bessel-theta convergence condition:**

B-TH. *There is a number $\sigma_0' \geq 0$ such that for $\mathrm{Re}(s) > \sigma_0'$ the series*

$$\sum_{k=1}^\infty |a_k| \mathbf{N} c_k^{-(s+\sigma_1)}$$

converges absolutely and thus uniformly in any half plane $\mathrm{Re}(s) \geq \sigma_0' + \varepsilon$.

Of course, we put

$$c_k = (\ldots, c_{k,v}, \ldots) \quad \text{and} \quad \mathbf{N}c_k = \Pi_v c_{k,v}^{N_v}.$$

Note that the condition $\sigma_1 > 0$ implies that h_0 extends continuously to the space $Y^0 \times [0, \infty)$, and h_0 is invariant under the action of V. Furthermore, from **K7** of [La 73/87], Chapter 20, §3, it follows that $h_0(y)$ is exponentially decreasing for $\mathbf{N}y \to \infty$, so h_0 is admissible.

By a standard Bessel indentity, we know that for $\mathrm{Re}(s) > |\mathrm{Re}(\alpha)|$,

$$(2) \qquad \int_0^\infty K_\alpha(ct) t^s \frac{dt}{t} = \frac{1}{2} c^{-s} \Gamma\left(\frac{s-\alpha}{2}\right) \Gamma\left(\frac{s+\alpha}{2}\right).$$

A proof follows by the same technique as in [La 73], Chapter 20, §3.

Since

$$\int_Y = \int_{Y/V} \sum_V \quad \text{and} \quad d^*y = d^*y^0 \frac{dt}{t}$$

the above Bessel identity immediately allows us to compute the Mellin transform of the Bessel series on Y/V. We assume that d^*y is normalized the usual way,

$$d^*y = \prod \frac{dy_v}{y_v}.$$

Then for a Hecke character $\chi = [\beta]$, remembering that **OR 1** and **OR 2** are satisfied, we get

$$\mathbf{M}_{Y/V} h_0((s, \chi)) = \sum_{k=1}^\infty a_k \prod_v \int_{Y_v} K_{\alpha_v}(c_{k,v} y_v) y_v^{N_v \sigma_1} y_v^{N_v s} y_v^{N_v \beta_v} \frac{dy_v}{y_v}.$$

We have

$$\chi(c_k) = \prod_v c_{k,v}^{n_v \beta_v}.$$

We let

$$s_v = N_v(s + \sigma_1 + \beta_v).$$

We then obtain:

THEOREM 2.1. *Let $Z(s)$ be the Dirichlet series*

$$Z(s) = \sum_{k=1}^{\infty} \frac{2^{-\#(S)} \chi(c_k^{-1}) a_k}{\mathbf{N} c_k^{s+\sigma_1}}$$

and let

$$G_v(s) = \Gamma\left(\frac{1}{2}(s_v - \alpha_v)\right) \Gamma\left(\frac{1}{2}(s_v + \alpha_v)\right)$$

$$G(s) = \prod_v G_v(s_v).$$

Then for $\mathrm{Re}(s) > \sigma_0'$.

$$\mathbf{M}_{Y/V} \cdot h_0(s, \chi) = G(s) Z(s).$$

Remark. When $\alpha = 1/2$ then $K_{1/2}$ collapses to the exponential function and the Gauss duplication formula shows that the product of the two gamma factors actually collapses to one gamma factor. This is precisely what happens in the most classical case of the Dedekind zeta function of number fields.

Theorems 1.1 and 1.2 apply to the Bessel series which will be defined below, but the rest of this paper is logically independent of what precedes.

For the convenience of the reader, we now recall Hecke's functional equation for the Dedekind zeta function of a number field F, to be used below. We let $\mathbf{D}_F$ denote the absolute value of the discriminant and we let $\mathfrak{K}$ denote an ideal class. We let $\mathfrak{d}$ be the different. As usual, r_1 and r_2 denote the number of real resp. complex conjugate embeddings of F. We recall the classical zeta function associated with a fractional ideal $\mathfrak{a} \neq (0)$, namely

$$\zeta(s, \mathfrak{a}) = \mathbf{N}\mathfrak{a}^s \sum_{(\mu)} \mathbf{N}(\mu)^{-s}.$$

where $\mathbf{N}$ is the absolute norm and the sum is taken over all principal ideals $(\mu) \subset \mathfrak{a}$ with $\mu \neq 0$. Because of the factor $\mathbf{N}\mathfrak{a}^s$, one sees that $\zeta(s, \mathfrak{a})$ depends only on the ideal class of $\mathfrak{a}$. If $\mathfrak{a} \in \mathfrak{K}^{-1}$ with an ideal class $\mathfrak{K}$, we set

$$\zeta(s, \mathfrak{a}) = \zeta(s, \mathfrak{K}).$$

THEOREM 2.2 (Hecke functional equation). *Let G be the function*

$$G(s) = G_F(s) = \mathbf{D}_F^{s/2}\left(\pi^{-s/2}\Gamma(s/2)\right)^{r_1}\left((2\pi)^{-s}\Gamma(s)\right)^{r_2}$$
$$= A^{s/2}\Gamma(s/2)^{r_1}\Gamma(s)^{r_2}$$

where $A = \mathbf{D}_F \pi^{-r_1}(2\pi)^{-2r_2}$. Let

$$\xi(s, \mathfrak{K}) = G(s)\zeta(s, \mathfrak{K}).$$

Let $\mathfrak{K}'$ be the dual class, i.e. the ideal class $\mathfrak{d}^{-1}\mathfrak{K}^{-1}$. Then $\xi(s, \mathfrak{K})$ is holomorphic in s except for simple poles at $s = 0$ and $s = 1$ and

$$\xi(s, \mathfrak{K}) = \xi(1-s, \mathfrak{K}').$$

§3. Hyperbolic spaces, number fields and Eisenstein series

Let F be a number field of degree N over $\mathbf{Q}$. We let $S = S_\infty$ be the set of absolute values of absolute values at infinity and to each $v \in S_\infty$ we suppose chosen a fixed embedding

$$F \hookrightarrow \mathbf{R} \quad \text{or} \quad F \hookrightarrow \mathbf{C}$$

according as v is real or complex. We let $\mathfrak{o} = \mathfrak{o}_v$ be the ring of algebraic integers in F at a finite place v and we let $\mathfrak{d} = \mathfrak{d}_{\mathfrak{o}/\mathbf{Z}}$ be the different as above.

We need certain spaces from differential geometry. We let:

$\mathbf{h}_2$ = upper half plane = $\mathbf{R} \times \mathbf{R}^+$ with its usual Poincaré metric.

$\mathbf{h}_3$ = hyperbolic 3-space = $\mathbf{C} \times \mathbf{R}^+$, on which we make more comments.

The unique simply connected Riemannian manifold of dimension 3, with constant negative curvature -1, up to isometry, has many models, of which the following is the relevant one, as described in Kubota [Ku 68]. We let $\mathbf{h}_3$ be the space of matrices

$$z = \begin{pmatrix} x & -y \\ y & \bar{x} \end{pmatrix} \quad \text{with } x \in \mathbf{C} \text{ and } y \in \mathbf{R}, \ y > 0.$$

We then write $y = y(z)$.

The group $\mathrm{SL}_2(\mathbf{C})$ operates on $\mathbf{h}_3$ in a natural way as follows. Let

$$\sigma = \begin{pmatrix} \alpha & \beta \\ \gamma & \delta \end{pmatrix} \in \mathrm{SL}_2(\mathbf{C}).$$

Let α be a complex number. We write
$$\tilde{\alpha} = \begin{pmatrix} \alpha & 0 \\ 0 & \bar{\alpha} \end{pmatrix}$$

We define
$$\sigma\langle z\rangle = \left(\tilde{\alpha}z + \tilde{\beta}\right)\left(\tilde{\gamma}z + \tilde{\delta}\right)^{-1}.$$

Quite generally, let $\gamma, \delta \in \mathbf{C}$ not both 0. For $z \in \mathbf{h}_3$ define
$$y(\gamma, \delta; z) = \frac{y(z)}{|\gamma x + \delta|^2 + |\gamma|^2 y^2}.$$

Then a straightforward calculation shows that
$$y(\sigma\langle z\rangle) = y(\gamma, \delta; z).$$

This is the analogue of the standard formula for the imaginary part of the image of a complex number under an element of $\mathrm{SL}_2(\mathbf{R})$.

Having the above simple notions we relate them to the number field F as follows. We let:

$\mathbf{h}_v = \mathbf{h}_2$ if v is real, and $= \mathbf{h}_3$ if v is complex.
$$\mathbf{h}_F = \prod_{v \in S_\infty} \mathbf{h}_v.$$

An element z of $\mathbf{h}_F$ is thus a vector, $z = (\ldots, z_v, \ldots)$, and
$$z_v = (x_v, y_v).$$

For v real, $x_v \in \mathbf{R}$ and for v complex, $x_v \in \mathbf{C}$. In both cases, $y_v \in \mathbf{R}^+$.

We define $y(z)$ to be the vector
$$y(z) = (\ldots, y(z_v), \ldots)_{v \in S_\infty}.$$

We define the **Norm**
$$\mathbf{N}y(z) = \prod_{v \in S_\infty} y(z_v)^{\mathbf{N}_v},$$

where $\mathbf{N}_v = 1$ or 2 according as v is real or complex.

If $\mu, \nu \in F$ are not both 0 and $z \in \mathbf{h}_F$, then we put
$$y(\mu, \nu; z) = \left(\ldots, \frac{y_v}{|\mu_v x_v + \nu_v|^2 + |\mu_v|^2 y_v^2}, \ldots\right)$$

where $y_v = y(z_v)$.

Given a pair of elements (μ, ν) in $F \times F$ *not both* 0 we define the **equivalence class**

$$\{\mu, \nu\}$$

to consist of all pairs (μ_1, ν_1) such that there exists a unit ϵ for which

$$(\mu_1, \nu_1) = \epsilon(\mu, \nu) = (\epsilon\mu, \epsilon\nu).$$

Let $\mathfrak{a}$ be a fractional ideal $\neq 0$. We define:

$\text{Equ}(\mathfrak{a})$ = equivalence classes of pairs $\{\mu, \nu\}$ with $\mu, \nu \in \mathfrak{a}$;

$\text{Equ}^*(\mathfrak{a})$ = equivalence classes of pairs $\{\mu, \nu\}$ with $(\mu, \nu) = \mathfrak{a}$.

We define the **Eisenstein series**

(1) $$E(z, s, \mathfrak{a}) = \sum_{\{\mu, \nu\} \in \text{Equ}(\mathfrak{a})} \mathbf{N}y(\mu, \nu; z)^s \mathbf{N}\mathfrak{a}^{2s}.$$

Thus the Eisenstein series is a higher dimensional version of the zeta series $\zeta(s, \mathfrak{a})$. Like the zeta series, the Eisenstein series converges absolutely for $\text{Re}(s) > 1$.

Let $\mathfrak{K}$ be an ideal class of F. Let $\mathfrak{a} \in \mathfrak{K}^{-1}$ be a fractional ideal. We define the $\mathfrak{K}$-**Eisenstein series**

$$E(z, s, \mathfrak{K}) = E(z, s, \mathfrak{a}).$$

The fact that we put a factor $\mathbf{N}\mathfrak{a}^{2s}$ in the definition of the Eisenstein series shows immediately that the series depend only on the ideal class of $\mathfrak{a}$, or that of $\mathfrak{a}^{-1}$. We now define the **primitive Eisenstein series**

(2) $$E^*(z, s, \mathfrak{K}) = \sum_{\{\mu, \nu\} = \mathfrak{a}} \mathbf{N}y(\mu, \nu; z)^s \mathbf{N}\mathfrak{a}^{2s}$$

where the sum is taken over all equivalence classes of pairs

$$\{\mu, \nu\} \in \text{Equ}^*(\mathfrak{a}),$$

namely such that (μ, ν) generates precisely the ideal $\mathfrak{a}$. Thus $E^*(z, s, \mathfrak{K})$ is a partial sum of the complete Eisenstein series $E(z, s, \mathfrak{K})$. Immediately from the definition. since a fractional ideal (μ, ν) in $\mathfrak{a}$ can be written in the form $\mathfrak{ab}$ with some ideal $\mathfrak{b} \subset \mathfrak{o}$. we obtain

(3) $$E(z, s, \mathfrak{K}) = \sum_{\mathfrak{b}} \frac{\mathfrak{E}^*(\mathfrak{z}, \mathfrak{s}, \mathfrak{ab})}{\mathbf{N}\mathfrak{b}^{2s}}$$

where the sum is taken over all ideals $\mathfrak{b} \neq 0$ in $\mathfrak{o}$. Since $E^*(z, s, \mathfrak{ab})$ depends only on the ideal class of $\mathfrak{ab}$. which is $\mathfrak{b}\mathfrak{K}^{-1}$, we therefore obtain the relation

(4) $$E(z, s, \mathfrak{K}) = \sum_{\mathfrak{L}} \sum_{\mathfrak{b} \in \mathfrak{L}} \mathbf{N}\mathfrak{b}^{-2s} E^*(z, s, \mathfrak{L}^{-1}\mathfrak{K})$$
$$= \sum_{\mathfrak{L}} \zeta(2s, \mathfrak{L}) E^*(z, s, \mathfrak{L}^{-1}\mathfrak{K}).$$

Summing over $\mathfrak{K}$, let us define the **total Eisenstein series**, independent of the class, by

(5) $$E_F(z, s) = \sum_{\mathfrak{K}} E(z, s; \mathfrak{K}) \quad \text{and} \quad E_F^*(z, s) = \sum_{\mathfrak{K}} E^*(z, s; \mathfrak{K}).$$

Since $\sum_{\mathfrak{L}} \zeta(s, \mathfrak{L}) = \zeta_F(s)$, we obtain

(6) $$E_F(z, s) = \zeta_F(2s) E_F^*(z, s).$$

Thus the zeta function of F appears as a natural factor of the Eisenstein series.

We shall derive a meromorphic continuation for $E(z, s, \mathfrak{a})$, by a certain inversion of theta series. We shall thus be led to a dual notion of equivalence of pairs. as follows. For $\mu, \nu' \in F$ and $\mu\nu' \neq 0$, we define $[\mu, \nu'] =$ equivalence class of pairs under the equivalence relation

$$(\mu, \nu') \sim (\epsilon\mu, \epsilon^{-1}\nu')$$

for all units ϵ. Observe that the product $\mu\nu'$ depends only on the class.

We shall obtain a Bessel series for $E(z, s, \mathfrak{a})$. For this purpose, with $\mu, \nu' \in F$ and $\mu\nu' \neq 0$ we let the v-**Bessel factor** be

$$B_v(\mu, \nu', y, s) = y_v^{N_v/2} \int_0^\infty \exp(-N_v \pi y_v(|\mu_v|^2 t_v + |\nu_v'|^2/t_v)) t_v^{N_v s} \frac{dt_v}{t_v}$$

$$= y_v^{N_v/2} \left|\frac{\nu_v'}{\mu_v}\right|^{N_v s} K_{N_v s}(N_v \pi y_v |\mu_v \nu_v'|).$$

Note that the product

$$\prod_v B_v(\mu, \nu', y, s) = \mathbf{N} y^{1/2} \left(\frac{\mathbf{N}(\nu')}{\mathbf{N}(\mu)}\right)^s \prod_v K_{N_v s}(N_v \pi y_v |\mu_v \nu'_v|)$$

depends only on the equivalence class $[\mu, \nu']$, because the absolute value of the norm of a unit is equal to 1.

We let $R(\mathfrak{a})$ be a set of representatives of elements $\neq 0$ in $\mathfrak{a}$ for the equivalence

$$\mu_1 \sim \mu \quad \text{if and only if } \mu_1 = \epsilon\mu \text{ for some unit } \epsilon \in \mathfrak{o}.$$

The sum over (μ) in the definition of $\zeta(s, \mathfrak{a})$ could also be taken for $\mu \in R(\mathfrak{a})$.

We also define the **fudge factor at infinity**

$$G_{F,\infty}(s) = \left(\pi^{-s/2} \Gamma(s/2)\right)^{r_1} \left((2\pi)^{-s} \Gamma(s)\right)^{r_2}$$

$$G_{F,\infty}(2s) = \prod_v (N_v \pi)^{-N_v s} \Gamma(N_v s).$$

The full **fudge factor** is

$$G_F(s) = \mathbf{D}(\mathfrak{o})^{s/2} G_{F,\infty}(s).$$

with the discriminant appearing at the finite places. Thus

$$G_F(2s) = \mathbf{D}^s \left(\pi^{-s} \Gamma(s)\right)^{r_1} \left((2\pi)^{-2s} \Gamma(2s)\right)^{r_2}.$$

This factor will be the relevant one for the Eisenstein series.

We recall the functional equation for $\zeta(s, \mathfrak{a})$. Let $\mathfrak{a}'$ be the complementary fractional ideal, that is $\mathfrak{a}' = \mathfrak{d}^{-1}\mathfrak{a}^{-1}$, where $\mathfrak{d}$ is the different. Then

$$G_F(s)\zeta(s, \mathfrak{a}) = G_F(1-s)\zeta(1-s, \mathfrak{a}').$$

For the next theorem, we define the set of equivalence classes:

$$[\text{Equ}](\mathfrak{a}, \mathfrak{a}') = \text{equivalence classes of pairs } [\mu, \nu'] \text{ with}$$
$$\mu \in \mathfrak{a}, \nu' \in \mathfrak{a}', \text{ and } \mu\nu' \neq 0.$$

We let S be the trace of $F/\mathbf{Q}$ as usual.

THEOREM 3.1. *The Eisenstein series have an expression:*

$$E(z,s,\mathfrak{a}) = \mathbf{N}y(z)^s \zeta(2s,\mathfrak{a}) + \mathbf{N}y(z)^{1-s}$$
$$+ \mathbf{D}^{-1/2} G_{F,\infty}(2s)^{-1} G_{F,\infty}(2s-1)\zeta(2s-1,\mathfrak{a})$$
$$+ \mathbf{D}^{-1/2} G_{F,\infty}(2s)^{-1} \mathbf{N}\mathfrak{a}^{2s-1} \sum_{[\mu,\nu']} e^{2\pi i S(\mu\nu' x)} \prod_v B_v(\mu,\nu',y,s-\frac{1}{2}),$$

where the sum is taken for $[\mu,\nu'] \in [\mathrm{Equ}](\mathfrak{a},\mathfrak{a}')$. *The first two terms have the meromorphic continuation coming from the Dedekind zeta function, and the sum over* $[\mu,\nu']$ *is entire in* s.

Proof. We decompose the sum in (1) as follows:

$$\sum_{\{\mu,\nu\}} = \sum_{\substack{\nu \in R(\mathfrak{a}) \\ \mu=0}} + \sum_{\mu \in R(\mathfrak{a})} \sum_{\nu \in \mathfrak{a}}.$$

Then for $\mathrm{Re}(s) > 1$ we get:

$$E(z,s,\mathfrak{a}) = \sum_{\nu \in R(\mathfrak{a})} \mathbf{N}y(0,\nu;z)^s \mathbf{N}\mathfrak{a}^{2s} + \sum_{\mu \in R(\mathfrak{a})} \sum_{\nu \in \mathfrak{a}} \mathbf{N}y(\mu,\nu;z)^s \mathbf{N}\mathfrak{a}^{2s}$$
$$= \mathbf{N}y(z)^s \zeta(2s,\mathfrak{a}) +$$

$$\sum_{\mu \in R(\mathfrak{a})} \sum_{\nu \in \mathfrak{a}} \prod_v (N_v \pi)^{N_v s} \Gamma(N_v s)^{-1} \int_0^\infty \exp(-\pi N_v t_v y(\mu_v,\nu_v;z_v)^{-1}) t_v^{N_v s} \frac{dt_v}{t_v} \mathbf{N}\mathfrak{a}^{2s}.$$

We now let

$$T = \prod_v T_v \quad \text{with} \quad T_v = \mathbf{R}^+, \quad \text{and} \quad d^* t_v = \frac{dt_v}{t_v}.$$

Thus $t = (\ldots,t_v,\ldots)$ is the variable in T. With this notation, we have

$$E(z,s,\mathfrak{a}) = \mathbf{N}y(z)^s \zeta(2s,\mathfrak{a}) +$$

$$G_{F,\infty}(2s)^{-1} \sum_{\mu \in R(\mathfrak{a})} \int_T \exp\left(-\pi \mathrm{Tr}(ty|\mu|^2)\right) \Theta\left(y^{-1}t,\mathfrak{a}+\mu x\right) \mathbf{N}t^s d^* t \mathbf{N}\mathfrak{a}^{2s},$$

where for $x = (\ldots,x_v,\ldots)$ we define the Hecke theta series

$$\Theta(c,\mathfrak{a}+x) = \sum_{\alpha \in \mathfrak{a}} \exp\left(-\pi \sum_v N_v c_v |\alpha_v + x_v|^2\right).$$

By the Poisson summation formula-Hecke inversion (Cf. [La 70/94], Chapter XIII, §2) we get

$$\Theta(y^{-1}t, \mathfrak{a} + \mu x)$$
$$= \mathbf{D}(\mathfrak{a})^{-1/2}\mathbf{N}y(z)^{1/2}\mathbf{N}t^{-1/2} \sum_{\nu' \in \mathfrak{a}'} e^{2\pi i S(\mu \nu' x)} \exp\left(-\pi \operatorname{Tr}(yt^{-1}|\nu'|)\right).$$

We have $\mathbf{D}(\mathfrak{a})^{-1/2} = \mathbf{D}^{-1/2}\mathbf{N}\mathfrak{a}^{-1}$. We can write the double sum over μ, ν' as

$$\sum_{\mu \in R(\mathfrak{a})} \sum_{\nu' \in \mathfrak{a}} = \sum_{\substack{\mu \in R(\mathfrak{a}) \\ \nu'=0}} + \sum_{[\mu, \nu']}.$$

The sum over $\mu \in R(\mathfrak{a})$ on the right, with $\nu' = 0$, is again a Mellin transform. The factors involving $\mathbf{N}y(z)^{1/2}$ and $\mathbf{N}t^{-1/2}$ have the effect of translating s by $-1/2$. The sum over $[\mu, \nu']$ is a Bessel sum. Both these sums are the ones stated in the theorem, which is proved.

If we multiply both sides of Theorem 3.1 with $\mathbf{D}^s G_{F,\infty}(2s) = G_F(2s)$, we obtain a more symmetric expression as in the next theorem. In addition, it is convenient to introduce the usual abbreviations, namely we let

$$\xi_E(z, s, \mathfrak{a}) = G_F(2s) E(z, s, \mathfrak{a}) \quad \text{and} \quad \xi_F(s, \mathfrak{a}) = G_F(s) \zeta_F(s, \mathfrak{a}).$$

THEOREM 3.2. *We have the expression:*

$$\xi_E(z, s, \mathfrak{a})$$
$$= \mathbf{N}y(z)^s \xi_F(2s, \mathfrak{a}) + \mathbf{N}y(z)^{1-s} \xi_F(2s - 1, \mathfrak{a}) +$$
$$\mathbf{D}(\mathfrak{a})^{s-1/2} \sum_{[\mu, \nu']} \left(\frac{\mathbf{N}(\nu')}{\mathbf{N}(\mu)}\right)^{s-1/2} e^{2\pi i S(\mu \nu' x)} \prod_v y_v^{N_v/2} K_{N_v(s-1/2)}(N_v \pi y_v |\mu_v \nu'_v|).$$

Furthermore, we have the functional equation

$$\xi_E(z, s, \mathfrak{a}) = \xi_E(z, 1 - s, \mathfrak{a}').$$

Proof. Let us replace s by $1-s$ on the right side. Then by the functional equation

$$G(w)\zeta(w, \mathfrak{a}) = G(1-w)\zeta(1-w, \mathfrak{a}'),$$

with $w = 2 - 2s$, we find the term with $2s - 1$ on the right and $\mathfrak{a}$ replaced by $\mathfrak{a}'$. A similar process starting with this term yields the first term with

$2s$ and also with $\mathfrak{a}$ replaced with $\mathfrak{a}'$. Thus sending $s \mapsto 1 - s$ interchanges the first two terms on the right, while replacing $\mathfrak{a}$ by $\mathfrak{a}'$.

As to the Bessel sum, we use the functional equation of the Bessel function
$$K_{-w} = K_w,$$
with $w = s - 1/2$, changed into $-w$ by $s \mapsto 1 - s$. The products $\mu \nu'$ are unchanged when we interchange $\mathfrak{a}$ and $\mathfrak{a}'$. The factors
$$\mathbf{D}(\mathfrak{a})^{s-1/2} \quad \text{and} \quad (\mathbf{N}(\nu')/\mathbf{N}(\mu))^{s-1/2}$$
go to their inverses. Since $\mathbf{D}(\mathfrak{a})^{-1} = \mathbf{D}(\mathfrak{a}')$ and the pairs $[\mu, \nu']$ go to $[\nu', \mu]$, while $\mathfrak{a}'' = \mathfrak{a}$, the Bessel sum also gets transformed to the corresponding Bessel sum with $\mathfrak{a}$ replaced by $\mathfrak{a}'$. This concludes the proof.

From Theorem 3.2, we may read the residue and constant term for the Laurent expression at $s = 1$ We have:
$$\mathbf{N} y^{1-s} = 1 - (\log \mathbf{N} y)(s - 1) + O(|s - 1|^2)$$
$$\xi_F(2s - 1, \mathfrak{a}) = \frac{1}{2}\text{res}_{s=1}\xi_F(s, \mathfrak{a})\frac{1}{s - 1} + \text{CT}_{s=1}\xi_F(s, \mathfrak{a}) + O(|s - 1|).$$

where $\text{CT}_{s=1}$ denotes the constant term in a Laurent expansion about $s = 1$. Therefore at $s = 1$, the function $\xi_E(z, s, \mathfrak{a})$ has only a simple pole and the first terms of the expansion are as in the next theorem. We introduce the following functions.

The **Bessel series** $\text{Bess}(z, \mathfrak{a})$, defined by
$$\text{Bess}(z, \mathfrak{a}) = \mathbf{D}(\mathfrak{a})^{\frac{1}{2}} \sum_{[\mu, \nu']} \left(\frac{\mathbf{N}(\nu')}{\mathbf{N}(\mu)}\right)^{1/2} e^{2\pi i S(\mu \nu' x)} \prod_v y_v^{N_v/2} K_{N_v/2}(N_v \pi y_v |\mu_v \nu'_v|).$$

where the sum is taken over $[\mu, \nu'] \in [\text{Equ}](\mathfrak{a}, \mathfrak{a}')$.

The **Asai function** $h_F(z, \mathfrak{a})$, defined by
$$\operatorname*{res}_{s=1} \xi_F(s, \mathfrak{a}) h_F(z, \mathfrak{a}) = \xi_F(2, \mathfrak{a}) \mathbf{N} y + \text{Bess}(z, \mathfrak{a}).$$

THEOREM 3.3. *At $s = 1$, the expansion of the Eisenstein series is given by:*
$$\xi_E(z, s, \mathfrak{a}) = \operatorname*{res}_{s=1} \xi_F(s, \mathfrak{a}) \left[\frac{1/2}{s - 1} - \frac{1}{2}\log \mathbf{N} y + h_F(z, \mathfrak{a})\right] + O(|s - 1|).$$

Note. Our normalization of the Asai function is $1/2$ that of Asai's normalization in [As 70].

§4. Modularity of $h_F(z, \mathfrak{a})$

Before dealing with $h_F(z, \mathfrak{a})$, we mention some properties of the Eisenstein series which will now become relevant. First an invariance property. We let:

$$\mathbf{G}_v = \mathrm{SL}_2(\mathbf{R}) \text{ if } v \text{ is real, and } \mathrm{SL}_2(\mathbf{C}) \text{ if } v \text{ is complex};$$

$$\mathbf{G}_{F,\infty} = \prod_{v \in S_\infty} \mathbf{G}_v.$$

We let $\mathfrak{o} = \mathfrak{o}_F$ be the ring of algebraic integers of F. Then $\mathrm{SL}_2(\mathfrak{o})$ gets imbedded into $\mathbf{G}_v$ for each v by our fixed imbedding of $\mathfrak{o}$ into $\mathbf{R}$ or $\mathbf{C}$ corresponding to v, so we have an imbedding

$$\mathrm{SL}_2(\mathfrak{o}) \hookrightarrow \mathbf{G}_{F,\infty}$$

on the diagonal. An element $\sigma \in \mathrm{SL}_2(\mathfrak{o})$ will be identified with its image, so we can write σ as a vector

$$\sigma = (\ldots, \sigma_v, \ldots),$$

where σ_v is the image of σ in $\mathbf{G}_v$. It is immediate that $\mathrm{SL}_2(\mathfrak{o})$ is a discrete subgroup of $G_{F,\infty}$, which we also denote by Γ.

The group $\mathbf{G}_F$ operates on $\mathbf{h}_F$ via diagonal action and then operates as a discrete subgroup on $\mathbf{h}_F$.

Let $\mathfrak{a}$ be a fractional ideal $\neq (0)$. Let

$\mathbf{S}(F, \mathfrak{a})$ = the set of all matrices

$$\sigma = \begin{pmatrix} \xi & \eta \\ \mu & \nu \end{pmatrix} \in \mathrm{SL}_2(F) \quad \text{such that } (\mu, \nu) = \mathfrak{a},$$

$\mathbf{S}_\infty(F, \mathfrak{o})$ = subgroup of $\mathrm{SL}_2(F)$ consisting of matrices of the form

$$\begin{pmatrix} \varepsilon^{-1} & \lambda \\ 0 & \varepsilon \end{pmatrix} \quad \text{with } \lambda \in F \text{ and a unit } \varepsilon.$$

Then we have a bijection

$$\mathbf{S}_\infty(F, \mathfrak{o}) \setminus \mathbf{S}(F, \mathfrak{a}) \longrightarrow \mathrm{Equ}^*(\mathfrak{a})$$

which to each matrix in $\mathbf{S}(F, \mathfrak{a})$ as above associates the pair (μ, ν).

Note that $\mathrm{SL}_2(\mathfrak{o})$ operates on the right of $\mathrm{Equ}^*(\mathfrak{a})$, in a manner corresponding to matrix multiplication on the right of $\mathbf{S}(F, \mathfrak{a})$.

With the above notation, we may then express the **primitive Eisenstein series** as a sum

$$E^*(z, s, \mathfrak{a}) = \sum_{\sigma \in \mathbf{S}_\infty \setminus \mathbf{S}(\mathfrak{a})} \mathbf{N} y(\sigma\langle z\rangle)^s \mathbf{N}\mathfrak{a}^{2s}, \tag{1}$$

where we abbreviated $\mathbf{S}_\infty(F, \mathfrak{o}) = \mathbf{S}_\infty$ and $\mathbf{S}(F, \mathfrak{a}) = \mathbf{S}(\mathfrak{a})$.

Although strictly speaking, we shall not need the following remarks leading to Proposition 4.1, and they are essentially well-known, we include them here for the convenience of the reader. They have to do with the operation of $\mathrm{SL}_2(\mathfrak{o})$ on the projective line $\mathbf{P}^1(F)$, and have to do with the compactification of $\mathrm{SL}_2(\mathfrak{o}) \setminus \mathfrak{h}_F$ by the cusps, which correspond to the ideal classes, but we do not go into these considerations here.

The group $\mathrm{SL}_2(F)$ operates on the projective line $\mathbf{P}^1(F)$ as usual. If $z \in F$ and

$$M = \begin{pmatrix} \xi & \eta \\ \mu & \nu \end{pmatrix} \in \mathrm{SL}_2(F),$$

then the operation is given by $z \mapsto (\xi x + \eta)/(\mu z + \nu)$, with the possible value ∞. We may also represent this operation on vectors

$$\begin{pmatrix} \alpha_1 \\ \alpha_2 \end{pmatrix} \longmapsto M \begin{pmatrix} \alpha_1 \\ \alpha_2 \end{pmatrix}$$

where $\begin{pmatrix} \alpha_1 \\ \alpha_2 \end{pmatrix}$ represents the element $\alpha = \alpha_1/\alpha_2 \in F \cup \{\infty\}$.

To each $\alpha \in F$, $\alpha \neq 0$ we associate an ideal class c_α as follows. We write $\alpha = \alpha_1/\alpha_2$ with $\alpha_1, \alpha_2 \in \mathfrak{o}$, $\alpha_1\alpha_2 \neq 0$, and we let $\mathfrak{a} = (\alpha_1, \alpha_2)$ be the ideal generated by α_1 and α_2. If we write $\alpha = \beta_1/\beta_2$ with $\beta_1, \beta_2 \in \mathfrak{o}$, then there exists $\lambda \in F^*$ such that

$$\mathfrak{b} = \lambda\mathfrak{a}, \quad \text{where} \quad \mathfrak{b} = (\beta_1, \beta_2),$$

namely $\lambda = \beta_2\alpha_2^{-1}$, so the ideal class c_α of $\mathfrak{a}$ is well defined.

PROPOSITION 4.1. *The association $\alpha \mapsto c_\alpha$ induces a bijection between orbits of $\mathrm{SL}_2(\mathfrak{o})$ in F^* and the set of ideal classes of F.*

The proof is routine and well-known. Cf. for instance [Si 61/80], Chapter III, §1, Proposition 20, and [Ge 80], Chapter I, Proposition 1.1. But the matter is older, cf. Siegel's comment p. 207: "It was Blumenthal who first gave a method of constructing a fundamental domain for Γ in $\mathfrak{H}_n$, but his proof contained an error since he obtained a fundamental domain with just one cups and not h cusps. This was set right by Maass."

PROPOSITION 4.2. *The Eisenstein series are invariant under the action of* $\mathrm{SL}_2(\mathfrak{o})$, *that is for* $\sigma \in \mathrm{SL}_2(\mathfrak{o})$ *and any non-zero fractional ideal* $\mathfrak{a}$. *we have*

$$E^*(\sigma\langle z\rangle, s, \mathfrak{a}) = E^*(z, s, \mathfrak{a}),$$

and similarly when E^* *is replaced by* E.

Proof. It is clear that the result for E^* implies the result for E. As for E^*, the result comes from the fact that multiplication by σ on the right permutes the elements of $\mathbf{S}_\infty \setminus \mathbf{S}(\mathfrak{a})$.

THEOREM 4.3. *For* $\sigma \in \mathrm{SL}_2(\mathfrak{o})$, $\sigma = \begin{pmatrix} * & * \\ \gamma & \delta \end{pmatrix}$, *let*

$$J_v(\sigma, z) = |\gamma_v x_v + \delta_v|^2 + |\gamma_v|^2 y_v^2,$$

and

$$\mathbf{N}J(\sigma, z) = \prod_v J_v(\sigma, z)^{N_v}.$$

Then $h_F(z, \mathfrak{a})$ satisfies the modular relation

$$h_F(\sigma\langle z\rangle, \mathfrak{a}) + \frac{1}{2}\log \mathbf{N}J(\sigma, z) = h_F(z, \mathfrak{a}).$$

Proof. Let a_{-1} and $a_0(z)$ denote the residue and constant term respectively for the Eisenstein series $E(z, s, \mathfrak{a})$, so

$$E(z, s, \mathfrak{a}) = \frac{a_{-1}}{s-1} + a_0(z) + O(|s-1|).$$

From Proposition 4.2, we conclude that for $\sigma \in \mathrm{SL}_2(\mathfrak{o})$,

(2) $$a_0(\sigma\langle z\rangle) = a_0(z).$$

Up to a constant factor C, by Theorem 3.3, we know that

$$Ca_0(z) = -\frac{1}{2}\log \mathbf{N}y + h_F(z, \mathfrak{a}).$$

Applying (2) proves the theorem.

Finally, we want to apply the transformation theory of §1, §2 to the present case, as does Asai, so we restrict $h_F(z, \mathfrak{a})$ to $z = y$ so then $x = 0$, which we write as $h_F(y, \mathfrak{a})$. We let

$$c_{-1}(\mathfrak{a}) = \operatorname*{res}_{s=1} \xi_F(s, \mathfrak{a})$$

$$h_0(y, \mathfrak{a}) = c_{-1}(\mathfrak{a})^{-1} \mathbf{D}(\mathfrak{a})^{1/2} \sum_{[\mu, \nu']} \left(\frac{\mathbf{N}(\nu')}{\mathbf{N}(\mu)}\right)^{1/2} \mathbf{N}y^{1/2} \prod K_{N_v/2}(N_v \pi y_v |\mu_v \nu'_v|),$$

$$= c_{-1}(\mathfrak{a})^{-1} \operatorname{Bess}(y, \mathfrak{a}),$$

where the sum is taken over $[\mu, \nu'] \in [\operatorname{Equ}](\mathfrak{a}, \mathfrak{a}')$. Let

$$c_0 = c_0(\mathfrak{a}) = \xi_F(2, \mathfrak{a})/c_{-1}(\mathfrak{a}).$$

THEOREM 4.4. *The function $h_0(y, \mathfrak{a})$ has exponential decay whenever $\mathbf{N}y \to \infty$, and we have*

$$h_F(y, \mathfrak{a}) = h_0(y, \mathfrak{a}) + c_0 \mathbf{N}y.$$

Furthermore, we have the transformation rule

$$h_F(y^{-1}, \mathfrak{a}) + \log \mathbf{N}y = h_F(y, \mathfrak{a}).$$

Proof. This comes from the definitions and Theorem 3.3, as well as applying Theorem 4.3 with

$$\sigma = \begin{pmatrix} 0 & -1 \\ 1 & 0 \end{pmatrix}.$$

§5. Harmonicity of $h_F(z, \mathfrak{a})$

So far we have dealt only with the algebraic properties of h_F. Now we deal with differential geometric properties and we consider the behavior of $E(z, s, \mathfrak{a})$ vis a vis the Laplace operators corresponding to each factor.

The metric form on $\mathbf{h}_3$ is represented by

$$\frac{1}{y^2} \left(dx_1^2 + dx_2^2 + dy^2 \right).$$

and the corresponding volume form by $dx_1 dx_2 dy/y^2$. The formulas for $\mathbf{h}_2$ are even better known. The positive Laplace operator on $\mathbf{h}_2$ is given by

$$\Delta_2 = -y^2 \left(\frac{\partial^2}{\partial x^2} + \frac{\partial^2}{\partial y^2} \right).$$

The positive Laplacian $\boldsymbol{\Delta}_3$ on $\mathbf{h}_3$ is given by

$$\boldsymbol{\Delta}_3 = -y^2\left(4\frac{\partial^2}{\partial x \partial \overline{x}} + \frac{\partial^2}{\partial y^2}\right) + y\frac{\partial}{\partial y}$$

$$= -y^2\left(\frac{\partial^2}{\partial x_1^2} + \frac{\partial^2}{\partial x_2^2} + \frac{\partial^2}{\partial y^2}\right) + y\frac{\partial}{\partial y}.$$

It is the unique (up to constant factor) $SL_2(\mathbf{C})$-invariant differential operator on $\mathbf{h}_3$. For each absolute value v, we let $\boldsymbol{\Delta}_v$ be the corresponding Laplacian.

We can exhibit eigenfunctions of the Laplace operator as follows. Let $\alpha \neq 0$ be a complex number and let

$$e_s(\alpha, z) = yK_{2s-1}(|\alpha|y)e^{2i\operatorname{Re}(\alpha x)}.$$

Note that if we denote by $e_s^B(\alpha, z)$ the corresponding function in Asai, then his normalization is related to ours by $2e_s^B(\alpha, z) = e_s(\alpha, z)$.

PROPOSITION 5.1. *We have the eigenfunctions*

$$\boldsymbol{\Delta}_v y_v^{N_v s} = N_v^2 s(1-s)y_v^{N_v s} \quad \text{and} \quad \boldsymbol{\Delta}_v e_s(\alpha, z) = N_v s(1-s)e_s(\alpha, z).$$

Proof. The first formula follows from a direct computation. The second one is a consequence of the differential equation satisfied by the K-Bessel function $K_s(y)$, which is standard, and is derived directly from our definition of the K-Bessel function in [JoL 96], Lemma 3.1.

From (1) in §4, applying the above eigenfunction relation to each term in the sum, and to all fractional ideals $\mathfrak{ab}$ with $\mathfrak{b}$ integral $\neq 0$, we find:

PROPOSITION 5.2. *For all v and all Laplacians $\boldsymbol{\Delta}_v$ we have*

$$\boldsymbol{\Delta}_v E(z, s, \mathfrak{a}) = N_v^2 s(1-s)E(z, s, \mathfrak{a}),$$

and similarly for E^* instead of E.

We then obtain the corresponding result for the function $h_F(z, \mathfrak{a})$.

PROPOSITION 5.3. *This function is harmonic for each $\boldsymbol{\Delta}_v$, that is*

$$\boldsymbol{\Delta}_v h_F(z, \mathfrak{a}) = 0.$$

Proof. Let a_{-1} and $a_0(z)$ be as in the proof of Theorem 4.3. From Proposition 5.2 we find

$$\Delta_v E(z, s, \mathfrak{a}) = -N_v^2 a_{-1} + O(|s-1|),$$

since the right side of the first formula in Proposition 5.1 contains a factor $-(s-1)$. But we also have

$$\Delta_v E(z, s, \mathfrak{a}) = \Delta_v a_0(z) + O(|s-1|).$$

Therefore

$$\Delta_v a_0(z) = -N_v^2 a_{-1}.$$

Furthermore, by direct computation,

$$\Delta_v \log \mathbf{N} y(z) = N_v^2.$$

Hence

$$\Delta_v(a_0(z) + a_{-1} \log \mathbf{N} y(z)) = 0.$$

But from Theorem 3.3, one sees that

$$a_0(z) + a_{-1} \log \mathbf{N} y(z)$$

up to a constant factor is equal to

$$-\frac{1}{2}\log \mathbf{N} y + h_F(z, \mathfrak{a}) + \frac{1}{2}\log \mathbf{N} y.$$

This proves that $\Delta_v h_F(z, \mathfrak{a}) = 0$, and concludes the proof of the proposition.

§6. Regularized products

We show here that the Eisenstein series give examples of regularized products, as defined in [JoL 93a] and [JoL 93b]. See also the definition of regularized product type in [JoL 94], Chapter I, §6. To save space, we do not reproduce the relevant definitions here.

THEOREM 6.1. *For each z and $\mathfrak{a}$, the Eisenstein functions $E^*(z, s, \mathfrak{a})$, $E(z, s, \mathfrak{a})$ are of regularized product type. So are $E_F^*(z, s)$ and $E(z, s)$. Furthermore, $E(z, s, \mathfrak{a})$ is of order 1 and has polynomial growth in vertical strips.*

Proof. We first prove the property of regularized product type for $E(z, s, \mathfrak{a})$. This property is actually a corollary of our general Cramér theorem from [JoL 93b]. It suffices to verify the conditions of this theorem. By combining equation (1) in §4 and (6) in §3, we see that $E(z, s, \mathfrak{a})$ has a Dirichlet series representation in a right half plane (see also (1) in §3). By Theorem 3.2, the function

$$\xi_E(z, s, \mathfrak{a}) = G_F(2s) E(z, s, \mathfrak{a})$$

satisfies a functional equation, actually is invariant under $s \mapsto 1 - s$. Since G_F is of regularized product type, it remains to show that $E(z, s, \mathfrak{a})$ is of finite order, actually order 1, and has polynomial growth in vertical strips. This is done by classical routine arguments as follows.

Since $E(z, s, \mathfrak{a})$ has a Dirichlet series representation in a right half plane, it is bounded in a slightly smaller right half plane. By the functional equation, we have

$$E(z, 1-s, \mathfrak{a}) = G_F(2-2s)^{-1} G(2s) E(z, s, \mathfrak{a}).$$

Since $G_F(2s)/G_F(2-2s)$ has a polynomial growth in vertical strips and is of order 1, the function $E(z, s, \mathfrak{a})$ has polynomial growth in vertical strips in some left half plane and is of order 1 in this left half plane, i.e. $O(e^{|s|^{1+\varepsilon}})$ for $|s| \to \infty$, s in that left half plane. Observe that the proof of Theorem 3.1, or Theorem 3.2, expresses E as a sum of terms involving the Dedekind zeta function and gamma function, and a Bessel series times $G_{F,\infty}(2s)^{-1}$. For $\sigma = \text{Re}(s)$ in a finite interval (so s itself is in a vertical strip), the Bessel function satisfies an exponential estimate

$$|K_s(c)| \leq C e^{-2c},$$

uniformly for σ in the interval. The number of elements in a fractional ideal with absolute values bounded by B (with $B \to \infty$) is polynomial in B, and the absolute values of elements in such a $\mathbf{Z}$-lattice are bounded from below. The elements can be split up into annuli according to the maximum absolute value, and one can estimate the Bessel series by an integral

$$\int_\delta^\infty \int_\delta^\infty x^{\pm M} y^{\pm M'} e^{-cxy} dx dy, \quad \text{with some } c, M, M' > 0.$$

and with some fixed $\delta > 0$. In the application, M, M' are bounded when σ lies in a finite interval. Hence, the Bessel series is bounded in the strip.

The factor $G_F(2s)^{-1}$ has order 1 in vertical strips. Theorem 3.2 shows that the function ξ_E (as a function of s) is entire, except for trivial poles of ξ_F. Hence, the Eisenstein function has only a finite number of poles in a given vertical strip. Thus, we have proved that the Eisenstein function is of order 1. Finally, the functional equation

$$E(1-s) = \Phi(s)E(s)$$

with a fudge factor $\Phi(s)$ which is a quotient of gamma factors shows that E has polynomial growth on vertical lines far to the right and to the left. Since it has order 1, we can apply the Phragmen-Lindelöf theorem to see that it has polynomial growth in vertical strips. This concludes the proof.

Note that the fudge factors in the functional equation of the Eisenstein functions involve both gamma factors and zeta factors with the Riemann zeta function. Thus we are already on the third rung of the ladder of functions which are regularized products.

Note also that the functional equations are multiplicative and we are not dealing with the logarithmic derivatives, which are of regularized harmonic series type.

The Eisenstein functions satisfy not only the ladder theorem, but they satisfy conditions which allow us to get an asymptotic expansion for the associated theta series. More precisely, combining (1) from §4 and the Dirichlet series representation of $\zeta_F(s, \mathfrak{a})$ we obtain a Dirichlet series representation for $E(z, s, \mathfrak{a})$, which we write in the form

$$E(z, s, \mathfrak{a}) = \zeta_F(s, \mathfrak{a}) E^*(z, s, \mathfrak{a})$$

$$= \mathbf{N}\mathfrak{a}^s \sum_{(\mu)} \mathbf{N}(\mu)^{-s} \sum_{\mathbf{S}_\infty \backslash \mathbf{S}(\mathfrak{a})} \mathbf{N} y(\sigma \langle z \rangle)^s$$

$$= \sum_k a_k \lambda_k^{-s}.$$

Thus we have the **associated theta series** defined by

$$\Theta(z, t, \mathfrak{a}) = \Theta(t) = \sum a_k e^{-\lambda_k t}.$$

In [JoL 93a], [JoL 93b] and [JoL 94], we have systematically used the three axioms **AS 1, AS 2, AS 3**, of which the main axiom is **AS 2**, asserting the existence of an asymptotic expansion for $\Theta(t)$ at the origin, in terms of generalized polynomials. In [JoL 93a], we gave a criterion to get such an asymptotic expansion. We can now use this criterion to prove:

THEOREM 6.2. *The theta series* $\Theta(z, t, \mathfrak{a})$ *satisfies the asymptotic conditions* **AS 1**, **AS 2** *and* **AS 3**.

Proof. The result comes from a direct application of [JoL 93a], §7. By the polynomial growth asserted in Theorem 6.1, the Eisenstein series $E(z, s, \mathfrak{a})$ is in the domain of the inverse Mellin transform. Hence, Theorem 7.4 and Theorem 7.5 of [JoL 93a] apply to conclude the proof of Theorem 6.2.

The above two results show how easily our axioms can be applied in certain concrete specific situations.

One can further apply the results of [JoL 94] to obtain an explicit formula and theta inversion formula associated to the Dirichlet series expansion for $E(z, s, \mathfrak{a})$. Thus one can do for the Eisenstein series themselves what we pointed out for the scattering determinant in another special case, see [JoL 96], §7.

§7. Heat kernel, spectral theoretic applications and associated zeta functions

We describe how one can introduce the heat kernel to produce theta relations via a spectral decomposition and then we can apply the Gauss transform to get zeta functions, as we pointed out already in [JoL 94] and [JoL 96].

Let $\mathbf{L}_F$ be the heat operator on $\mathbf{h}_F$, acting on functions $u(t, x)$ of a real variable t and variable $x \in \mathbf{h}_F$. Specifically, $\mathbf{L}_F$ acts via an operator $\mathbf{L}_v$ at each place v, and

$$\mathbf{L}_v = \mathbf{\Delta}_v + N_v^2 \frac{\partial}{\partial t},$$

so

$$\mathbf{L}_v = \mathbf{\Delta}_v + \frac{\partial}{\partial t} \quad \text{if } v \text{ is real,}$$

$$\mathbf{L}_v = \mathbf{\Delta}_v + 4\frac{\partial}{\partial t} \quad \text{if } v \text{ is complex.}$$

Let $\mathbf{K}_{\mathbf{h}_F}$ be the heat kernel on $\mathbf{h}_F$, associated to the Laplacian which acts on $C_c^\infty(\mathbf{h}_F)$ (smooth compactly supported functions). It is immediate that

$$\mathbf{K}_{\mathbf{h}_F} = \prod_{v \in S_\infty} \mathbf{K}_{\mathbf{h}_v},$$

where $\mathbf{K}_{\mathbf{h}_v}$ is the heat kernel on the hyperbolic half plane $\mathbf{h}_2$ if v is real and the heat kernel on hyperbolic $\mathbf{h}_3$ if v is complex. We recall that

$$e^{t/4}\mathbf{K}_{\mathbf{h}_2}(t,\rho) = \frac{1}{4\pi t} e^{-\rho^2/4t} g(t,\rho),$$

where

$$g(t,\rho) = t^{-1/2} \int_\rho^\infty \frac{u e^{(\rho^2-u^2)/4t}}{(\cosh u - \cosh \rho)^{1/2}} \frac{du}{\sqrt{2\pi}}$$

and

$$e^{t/4}\mathbf{K}_{\mathbf{h}_3}(t,\rho) = \frac{1}{(\pi t)^{3/2}} e^{-\rho^2/t} \frac{\rho}{\sinh \rho}.$$

Warning. The factor N_v in the definition of the norm on $\mathbf{h}_F$ was natural (and follows Asai), since, for example, all Eisenstein series then have a functional equation which amounts to invariance under the map $s \mapsto 1-s$. Other normalizations occur in the literature. For instance, if we let E_{EGM} denote the Eisenstein series as defined in [EGM 85], then

$$E_{\text{EGM}}(z,s,\mathfrak{a}) = E(z,s/2,\mathfrak{a}).$$

Our normalization entails a scaling of the heat kernel when v is complex. Note that the usual $\rho^2/4t$ (present for $\mathbf{h}_2$) becomes ρ^2/t for $\mathbf{h}_3$ (the complex case) and $(4\pi t)^{3/2}$ becomes $(\pi t)^{3/2}$. If one denotes by $\mathbf{K}_{\mathbf{R}}$ the real heat kernel, found elsewhere in the literature (e.g. [JoL 94], [JoL 96]), then

$$\mathbf{K}_{\mathbf{C},\mathbf{h}_3}(t,\rho) = \mathbf{K}_{\mathbf{h}_3}(t,\rho) = \mathbf{K}_{\mathbf{R},\mathbf{h}_3}(\rho, t/4).$$

More appropriately, we may then use the notation $\mathbf{K}_v$ for the heat kernel at the place v, where $\mathbf{K}_v$ is $\mathbf{K}_{\mathbf{h}_2}$ or $\mathbf{K}_{\mathbf{C},\mathbf{h}_3}$ according as v is real or complex.

Given a Hecke character $\chi = [\beta]$ (notation of §1, see the **OR** conditions), we define the corresponding **primitive Hecke-Eisenstein** series as in §4, (1) by

(1) $$E^*(z,s,\mathfrak{K},\chi) = \sum_{\sigma \in S_\infty \backslash S(\mathfrak{a})} \chi(y(\sigma\langle z\rangle)) \mathbf{N} y(\sigma\langle z\rangle)^s \mathbf{N}\mathfrak{a}^{2s}$$

where, as in §2, $\chi(y) = \prod y_v^{N_v \beta_v} = \mathbf{N}(y^\beta)$. By relations **OR 1** and **OR 2** of §1, the series (1) is, at least formally, invariant under the action of $SL_2(\mathfrak{o})$. The formalism of §1 applies directly and similar estimates as before extend the results of §3 through §6 to the Eisenstein series with Hecke characters to yield meromorphic continuations, functional equations and Fourier expansions. These were actually carried out in the Hilbert modular case (when F is totally real) in [Ef 87] and [Zo 82].

Let

$$X_F = SL_2(\mathfrak{o}) \setminus \mathbf{h}_F.$$

In the Hilbert modular case, [Ef 87] and [Zo 82] also deal with the eigenfunction expansion in $L^2(X_F)$. See especially [Ef 87] p. 41, Definition 1.7, p. 83, Theorem 9.8 for one cusp and p. 100 in the general case. Similar results are valid in the general case with arbitrary number fields. The proofs are similar and we merely summarize the situation.

The space $L^2(X_F)$ has an orthogonal decomposition consisting of a discrete part and a continuous part, as follows. The discrete part is the direct sum orthogonal decomposition of the subspace of $L^2(X_F)$ generated by the eigenfunctions. The continuous part is determined by the family

$$E^*(z, \frac{1}{2} + ir, \mathfrak{K}, \chi)$$

parametrized by the ideal classes $\mathfrak{K}$ all $r \in \mathbf{R}$ and all Hecke characters χ. As recalled in §1, the group of Hecke characters is isomorphic to $\mathbf{Z}^{r_1+r_2-1}$, and thus it constitutes a discrete family, but the variable r parametrizes a continuous family. Note that by using r to parametrize the continuous family, we give priority to the notation of the spectral gang. (Too many r's!)

In particular, we may periodize the heat kernel $\mathbf{K}_{\mathbf{h}_F}$ with respect to $SL_2(\mathfrak{o})$ to get the heat kernel $\mathbf{K}_{X_F}$ on X_F. The following result is obtained in the Hilbert modular case in [Ef 87], III, 4, and can be proved similarly in the general case. We let $\{\varphi_k\}$ be total orthonormal family of eigenfunctions in $L^2(X_F)$. We let $\{\lambda_k\}$ be the eigenvalues and we let

$$\mu_k = \lambda_k - 1/4,$$

following our standard normalization.

THEOREM 7.1. *Let* $\mathbf{K}_{X_F}(t,z,w)$ *be the heat kernel on* X_F. *Define*

$$E^*_{\text{tot}}(z,w,s,\chi) = \sum_{\mathfrak{K}} E^*(z,s,\mathfrak{K},\chi)\overline{E^*(w,s,\mathfrak{K},\chi)}.$$

Let $r_\chi = r - i\beta_\chi$ *where* β_χ *is the sum of the exponents of the character associated to* χ. *Then there is a constant* c_F *such that one has the expansion:*

$$e^{t/4}\mathbf{K}_{X_F}(t,z,w) = \sum_{\gamma \in \text{SL}_2(\mathfrak{o})/\{\pm 1\}} e^{t/4}\mathbf{K}_{\mathfrak{h}_F}(t,\gamma z,\tilde{w})$$

$$= \sum_k \varphi_k(z)\overline{\varphi_k(w)}\, e^{-\mu_k t}$$

$$+ c_F \int_{-\infty}^{\infty} \sum_\chi E^*_{\text{tot}}\left(z,w,\frac{1}{2}+ir,\chi\right) e^{-r_\chi^2 t} dr.$$

The convergence of the series (i.e., the sum over all Hecke characters χ) is addressed in [Ef 87] and [Zo 82]. It is similar to the convergence of a Fourier series. No new convergence problems arise by using the complex places of the number field.

Note that the equality of the two expansions for $\mathbf{K}_{X_F}$ is a theta inversion formula. For a totally imaginary number field, it fits exactly our previous formalism considered in [JoL 94] and [JoL 96]. The integral involving the Eisenstein series in Theorem 7.1 above corresponds precisely to the term $E_\Phi(t)$ in [JoL 94], Chapter IV, Theorem 1.2; and to the term $E_A R_{\text{left}}$ or $E_{-a} R_{\text{right}}$ in [JoL 96], see the "Basic Assumptions" of §5. The notation in these references was not accidental. We had in mind the Eisenstein series as an eventual manifestation of such terms, which do not occur for compact quotients.

For arbitrary F, because of the real places, the theta inversion formula is generalized because of the integral expression for the heat kernel on the hyperbolic upper half plane. Hence our formalism has to be extended to the more general situation involving such integral representations. We have already mentioned this possibility explicitly in the above cited references. See for instance [JoL 94], Chapter V, §4, Remark 1 and [JoL 96], §8 Remark 1.

Having the theta inversion, we may proceed as in [JoL 94] and [JoL 96]. We take the Gauss transform. Let

$$\Theta(t,z,w) = e^{t/4}\mathbf{K}_{X_F}(t,z,w).$$

Then the **Gauss transform** is formally defined by

$$\text{Gauss}(\Theta)(s) = 2s \int_0^\infty e^{-s^2 t} \Theta(t, z, w)\, dt.$$

It is explained in [JoL 94] and [JoL 96] how to regularize such an integral. Then $\text{Gauss}(\Theta)(s)$ is a generalized zeta-type function which admits an additive functional equation under $s \mapsto -s$ and has a generalized Bessel series representation in a right half plane. For $z \neq w$, the additive fudge term in the functional equation is expressible in terms of $E^*_{\text{spec}}(z, w, s)$, which is the Gauss transform of

$$c_F \int_{-\infty}^\infty \sum_\chi E^*_{\text{tot}}(z, w, \tfrac{1}{2} + ir, \chi) e^{-r^2 t} dr.$$

If the number field F is totally complex, then the heat kernel is split, in the terminology of [JoL 96] and hence the extra spectral integrals for $\mathbf{h}_2$ do not occur. In this totally complex case, the Gauss transform is equal to a Bessel series in the sense of [JoL 96], in a right half plane and hence the results of [JoL 96] apply directly.

The theta inversion formula of Theorem 7.1 comes from the heat kernel itself, before taking the trace of the heat kernel (integrating the diagonal over the manifold). As a result, we avoid various difficulties which occur because of the presence of elliptic elements in the group $\text{SL}_2(\mathfrak{o})$ (see remarks of Gangolli [Ga 68], pp. 153 and 190). If we take $z \neq w$, then all elements in the group side of the theta inversion formula contribute to a Bessel series, or more generally to an integral of a Bessel series expansion. If we take $z = w$, then we must consider separately the case when z is or is not a fixed point of some elliptic element. If z is not a fixed point, then one separates out the identity element alone. If z is a fixed point, then one must consider as well all elliptic elements which fix z. The totality of these elements, together with the integral involving E^*_{spec}, form the E_Φ or $E_A R$ terms, in the notation of [JoL 94] and [JoL 96]. Further theta inversion formulas which occur from the regularized trace of the heat kernel thus do cause additional complications caused by elliptic elements.

A detailed exposition for the above results and further applications belongs to subsequent publications.

§8. Heat Eisenstein series

To conclude this paper, let us reformulate the spectral expansion of Theorem 7.1 in terms of a new Eisenstein series which we call the heat Eisenstein series. By doing so, we obtain a spectral expansion involving only a finite number of terms representing the continuous spectrum. In fact, the number of terms is equal to the number of points at infinity, namely the class number. We are developing this approach currently in the setting of other symmetric spaces, most notably the symmetric space associated to $SL_n(\mathbf{R})$.

In the notation of §7, let χ be any Hecke character. The space of all Hecke characters is parameterized by $\mathbf{Z}^{r_1+r_2-1}$. In other words, there is a torus $\mathbf{T}_{\mathfrak{K}}$ isomorphic to $(\mathbf{R}/\mathbf{Z})^{r_1+r_2-1}$ such that any Hecke character is an eigenfunction of the Euclidean Laplacian on $\mathbf{T}_{\mathfrak{K}}$. Let $\mathbf{K}_{\mathbf{T}_{\mathfrak{K}}}$ denote the associated heat kernel on $\mathbf{T}_{\mathfrak{K}}$, and let $p_{\mathfrak{K}}$ denote the projection which maps points in X_F to points onto the torus $\mathbf{T}_{\mathfrak{K}}$. The map $p_{\mathfrak{K}}$ amounts to first decomposing the space $\mathbf{h}_F$ in terms of its Iwasawa coordinates and then projecting any point $z \in \mathbf{h}_F$ onto the associated quotient of a unipotent subgroup. Consider the (formal) series

(1) $E_{\mathfrak{K}}(t, z, w, s)$

$$= \mathbf{N}\mathfrak{a}^{2s} \overline{\mathbf{N}\mathfrak{a}^{2s}} \sum_{\sigma, \sigma' \in \mathbf{S}_\infty \backslash \mathbf{S}(\mathfrak{a})} \mathbf{K}_{\mathbf{T}_{\mathfrak{K}}}(t, p_{\mathfrak{K}}(z), p_{\mathfrak{K}}(w)) \mathbf{N}y(\sigma\langle z \rangle)^s \overline{\mathbf{N}y(\sigma'\langle w \rangle)^s},$$

which we take to be defined for $\operatorname{Re}(s)$ sufficiently large. By taking $t \to 0$, we obtain the formal equality

(2) $$\lim_{t \to 0} E_{\mathfrak{K}}(t, z, w, s) = \sum_{\chi} E^*(z, s, \mathfrak{K}, \chi) \overline{E^*(w, s, \mathfrak{K}, \chi)}.$$

This suggests that one can use the **heat Eisenstein series** defined in (1) to reformulate the spectral expansion obtained in Theorem 7.1. Indeed, there is considerable work which needs to be completed in the study of the heat Eisenstein series. At this point, we assert the following property of (1). The series (1) admits a meromorphic continuation to all $s \in \mathbf{C}$ and for any smooth, bounded function ψ on X_F, we have, in the notation of Theorem 7.1, the spectral expansion

(3) $$\psi(z) = \sum_{\mathfrak{K}} \langle \psi, \varphi_k \rangle \varphi_k(z) + c_F \sum_{\mathfrak{K}} \lim_{t \to 0} \int_{-\infty}^{\infty} \langle \psi * E_{\mathfrak{K}} \rangle (t, z, \tfrac{1}{2} + ir) dr,$$

where
$$\langle w * E_{\mathfrak{R}}\rangle(t,z,s) = \int_{X_F} \psi(w)E_{\mathfrak{R}}(t,z,w,s)d\mu(w)$$

denotes the usual convolution (symmetric L^2 inner product) on X_F. Various questions which naturally arise in the study of Eisenstein series, namely results analogous to those obtained in this paper, can be asked about our heat Eisenstein series and are currently under consideration as well as the verification of (3).

References

[As 70] T. Asai. *On a certain function analogous to* $\log \eta(z)$. Nagoya Math. J., **40** (1970), 193–211.

[Ef 87] I. Efrat. *The Selberg trace formula for* $PSL_2(\mathbf{R})^n$. Memoir AMS, **359** (1987).

[EfS 85] I. Efrat and P. Sarnak. *The determinant of the Eisenstein matrix and Hilbert class fields*. Trans. AMS, **290** (1985), 815–824.

[EGM 85] J. Elstrodt, E. Grunewald and J. Mennicke, *Eisenstein series on three dimensional hyperbolic spaces and imaginary quadratic fields*. J. reine angew. Math., **360** (1985), 160–213.

[EGM 87] J. Elstrodt, E. Grunewald and J. Mennicke, *Zeta functions of binary hermitian forms and special values of Eisenstein series on three-dimensional hyperbolic space*. Math. Ann., **277** (1987), 655–708.

[Ga 68] R. Gangolli. *On the length spectra of some compact manifolds of negative curvature*. Acta Math., **121** (1986), 151–192.

[Ge 89] G. Van der Geer. Hilbert modular surfaces, Springer Verlag, 1980.

[JoL 93a] J. Jorgenson and S. Lang. Basic analysis of regularized series and products. Springer Lecture Notes **1564**, 1993.

[JoL 93b] J. Jorgenson and S. Lang. *On Cramér's theorem for general Euler products with functional equations*. Math. Ann., **297** (1994), 383–416.

[JoL 94] J. Jorgenson and S. Lang. Explicit formulas for regularized products and series. Springer Lecture Notes **1593**, 1994.

[JoL 96] J. Jorgenson and S. Lang. *Extension of analytic number theory and the theory of regularized harmonic series from Dirichlet series to Bessel series*. Math. Ann., **306** (1996), 75–124.

[Ku 68] T. Kubota. *Über diskontinuierlicher Gruppen Picardschen Typus und zugehörige Eisensteinsche Reihen*. Nagoya Math. J., **32** (1968), 259–271.

[La 73/87] S. Lang. Elliptic Functions, Addison Wesley, 1973, (second edition Springer Verlag, 1987).

[La 70/94] S. Lang. Algebraic Number Theory, Addison Wesley, 1970, (second edition Springer Verlag, 1994).

[Sa 83] P. Sarnak. *The arithmetic and geometry of some hyperbolic three manifolds*. Acta Math., **151** (1983), 253–295.

[Si 61/80] C. L. Siegel, Advanced Analytic Number Theory, Lecture Notes Tata Institute, 1961, (reprinted in book form, 1980).

[Sz 83] J. Szmidt, *The Selberg trace formula for the Picard group $SL(2, Z[i])$*, Acta Arith., **42** (1983), 291–424.

[Zo 82] P. Zograf, *Selberg trace formula for the Hilbert modular group of a real quadratic number field*, J. Soviet Math., **19** (1982), 1637–1652.

Jay Jorgenson
Department of Mathematics
Oklahoma State University
Stillwater, OK 74078
U.S.A.

Serge Lang
Department of Mathematics
Yale University Box 20-8283
New Haven, CT 06520-8283
U.S.A.

Permissions

Springer-Verlag is grateful to the AMS, *Annals of Mathematics*, and the *American Journal of Math* for granting permission to reprint the following articles:

American Mathematical Society:

[1952b] Hilbert's nullstellensatz in infinite dimensional space, *Proc. AMS* **3**, No. 3 (1952) pp. 407–410.
[1952c] On Chevalley's proof of Luroth's theorem (with John Tate), *Proc. AMS* **3**, No. 4 (1952) pp. 621–624.
[1960a] Existence of invariant bases (with E. Kolchin), *Proc. AMS* **11**, No. 1 (1960) pp. 140–148.
[1960c] Some theorems and conjectures in diophantine equations, *Bull. AMS* **66**, No. 4 (1960) pp. 240–249.
[1961] Review: Éléments de géometrie algébrique by A Grothendieck, *Bull. AMS* **67** No. 3 (1961) pp. 239-246.
[1970b] Review of Diophantine Equations by L.J. Mordell, *Bull. AMS* **76** (1970) pp. 1230–1234.
[1971a] Transcendental numbers and diophantine approximations, *Bull. AMS* **77** No. 5 (1971) pp. 635-677.
[1974a] Higher dimensional diophantine problems, *Bull. AMS* **80** No. 5 (1974) pp. 779–787.
[1977b] Primitive points on elliptic curves (with H. Trotter), *Bull. AMS* **83**, No. 2 (1977) pp. 289–292.
[1982b] Units and class groups in number theory and algebraic theory, *Bull. AMS* **6** No. 3 (1982) pp. 253–316.
[1986a] Hyperbolic and diophantine analysis, *Bull. AMS* **14**, No. 2 (1986) pp. 159–205.
[1987a] Diophantine problems in complex hyperbolic analysis, *Contemporary Mathematics AMS* **67** (1987) pp. 229–246.
[1990a] The error term in Nevanlinna theory II, *Bull. AMS* **22**, No. 1 (1990) pp. 115–125.
[1990b] Old and new conjectured diophantine inequalities, *Bull. AMS* **23**, No. 1 (1980) pp. 37–75.

[1995a] Mordell's Review, Siegel's letter to Mordell, diophantine geometry, and 20th century mathematics, *Notices AMS* (1995) pp. 339–350.
[1995b] Some History of the Shimura-Taniyama Conjecture, *Notices AMS* (1995) pp. 1301–1307.
[1996b] Comments on Chow's Works, *Notices AMS* **43**, No. 10 (1996) pp. 1117–1124.

Annals of Mathematics:

[1952a] On quasi-algebraic closure, *Ann. of Math.* **55**, No. 2 (1952) pp. 373–390.
[1953] The theory of real places, *Ann. of Math.* **57**, No. 2 (1953) pp. 378–391.
[1956a] Unramified class field theory over function fields in several variables, *Ann. of Math.* **64**, No. 2 (1956) pp. 285–325.
[1956b] On the Lefschetz principle, *Ann. of Math.* **64**, No. 2 (1956) pp. 326–327.

American Journal of Math:

[1954a] Some applications of the local uniformization theorem, *Am. J. Math.* **76**, No. 2 (1954) pp. 362–374.
[1954b] Number of points of varieties in finite fields (with A. Weil), *Am. J. Math.* **76**, No. 4 (1954) pp. 819–827.
[1956e] Algebraic groups over finite fields, *Am. J. Math.* **78** (1956) pp. 555–563.
[1957a] Sur les revêtements non ramifiés des variétés algébriques (with J.-P. Serre), *Am. J. Math.* **79**, No. 2 (1957) pp. 319–330.
[1957b] On the birational equivalence of curves under specialization (with W.L. Chow), *Am. J. Math.* **79**, No. 3 (1957) pp. 649–652.
[1957c] Divisors and endomorphisms on abelian varieties, *Am. J. Math.* **79**, No. 4 (1957) pp. 761–777.
[1958a] Reciprocity and correspondences, *Am. J. Math.* **80**, No. 2 (1957) pp. 431–440.
[1958b] Principal homogeneous spaces over abelian varieties (with J. Tate), *Am. J. Math.* **80**, No. 3 (1958) pp. 659–684.
[1958c] Algebraic groups and the Galois theory of differential fields (with E. Kolchin), *Am. J. Math.* **80**, No. 1 (1958) pp. 103–110.
[1959a] Rational points of abelian varieties over function fields (with A. Néron), *Am. J. Math.* **81**, No. 1 (1959) pp. 95–118.
[1964a] Diophantine approximation on toruses, *Am. J. Math.* **86**, No. 3 (1964) pp. 521–533.
[1965d] Asymptotic approximation to quadratic irrationalities I, *Am. J. Math.* **87**, No. 2 (1965) pp. 481–487.
[1965e] Asymptotic approximation to quadratic irrationalities II, *Am. J. Math.* **87**, No. 2 (1965) pp. 488–496.

Permissions

[1972a] Isogenous generic elliptic curves, *Am. J. Math.* **94** (1972) pp. 661–674.
[1973a] Frobenius automorphisms of modular function fields, *Am. J. Math.* **95** (1973) pp. 165–173.
[1975b] Division points of elliptic curves and abelian functions over number fields, *Am. J. Math.* **97**, No. 1 (1972) pp. 124–132.

Springer-Verlag would also like to extend thanks to various other publishers for granting permission to reprint the following articles:

[1955] Abelian varieties over finite fields, *Proc. NAS* **41**, No. 3 (1955) pp. 174–176 and [1956c] L-series of a covering, *Proc. NAS*, **42**, No. 7 (1956) pp. 422–424 are reprinted with permission from the National Academy of Sciences.

[1956d] Sur les séries L d'une variété algébrique, *Bull. Soc. Math. France* **84** (1956) pp. 385–407 and [1965a] Report on diophantine approximations, *Bull. Soc. Math. France* **93** (1965) pp. 177–192 and [1966b] Asymptotic diophantine approximations, *Proc. NAS* **55**, No. 1 (1966) pp. 31–34 and [1979c] Modular units inside cyclotomic units (with D. Kubert), *Bull. Soc. Math. France* **107** (1979) pp. 161–178 and [1996a] La conjecture de Bateman-Horn, *Gazette des mathématiciens* No. 67 (1996) pp. 82–84 are reprinted with permission from Société Mathématique de France.

[1960b] Integral points on curves, *Pub. IHES* No. 6 (1960) pp. 27–43 is reprinted with permission from Imprimerie de Presses Universitaires de France.

[1960d] On a theorem of Mahler, *Mathematika* **7** (1960) pp. 130–140 and [1962a] A transcendence measure for E-functions, *Mathematika* **9** (1962) pp. 157–161 are reprinted with permission from *Mathematika*, University of London.

[1962b] Transcendental points on group varieties, *Topology* **1** (1962) pp. 313–318 and [1965c] Algebraic values of meromorphic functions, *Topology* **3** (1965) pp. 183–191 and [1966a] Algebraic values of meromorphic functions II, *Topology* **5** (1966) pp. 363–370 are reprinted with permission from Elsevier Science.

[1965b] Division points on curves, *Annali Mat. pura ed applicata*, Serie IV **70** (1965) pp. 229–234 is reprinted with permission from *Annali di Matematica*.

[1965f] Some computations in diophantine approximation (with W. Adams), *J. reine angew. Math.* **220** Heft 3/4 (1965) pp. 163–173 and [1972b] Continued fractions for some algebraic numbers (with H. Trotter), *J. reine angew. Math.* **255** (1972) pp. 112–134 and [1974b] Addendum to "Continued fractions of some algebraic numbers" (with H. Trotter), *J. reine angew. Math.* **267** (1974) pp. 219–220 are reprinted with permission from Walter de Gruyter GmbH & Co.

[1975a] Diophantine approximations on abelian varieties with complex multiplication, *Advances in Math.* **17** (1975) pp. 281–336 is reprinted with permission from Academic Press.

[1981a] Finiteness theorems in geometric class field theory (with N. Katz), *Enseignement mathématique* **27** (3–4) (1981) pp. 285-314 is reprinted with permission from Universite de Geneve.

[1982a] Représentations localement algébriques dans les corps cyclotomiques, Seminaire de Théorie des Nombres 1982, Birkhäuser, pp. 125–136 and [1983a] Conjectured diophantine estimates on elliptic curves, in Volum I of *Arithmetic and Geometry*, dedicated to Shafarevich, Birkhäuser (1983) pp. 155–171 and [1984a] Vojta's conjecture, *Arbeitstagung Bonn 1984*, Springer Lecture Notes **1111** (1985) pp. 407–419 and [1984b] Variétés hyperboliques et analyse diophantienne, *Séminaire de théorie des nombres*, 1984/85, pp. 177–186 are reprinted with permission from Birkhäuser Boston.

[1988a] The error term in Nevanlinna theory, *Duke Math. J.* **56**, No. 1 (1988) pp. 193–218 is reprinted with permission from Duke University Press.

ISBN 0-387-95030-3